Comparative Studies on Pandemic Control Policies and the Resilience of Society

AF800519

Simon X. B. Zhao · Kam Tong Chan ·
Selçuk Çolakoğlu · Qiaoyun Zhang · Bo Yan
Editors

Comparative Studies on Pandemic Control Policies and the Resilience of Society

Springer

Editors
Simon X. B. Zhao
Faculty of Humanities and Social Sciences
Beijing Normal University-Hong Kong Baptist University United International College
Zhuhai, Guangdong, China

Kam Tong Chan
Faculty of Humanities and Social Sciences
Beijing Normal University-Hong Kong Baptist University United International College
Zhuhai, Guangdong, China

Selçuk Çolakoğlu
Faculty of Humanities and Social Sciences
Beijing Normal University-Hong Kong Baptist University United International College
Zhuhai, Guangdong, China

Qiaoyun Zhang
Faculty of Humanities and Social Sciences
Beijing Normal University-Hong Kong Baptist University United International College
Zhuhai, Guangdong, China

Bo Yan
Faculty of Humanities and Social Sciences
Beijing Normal University-Hong Kong Baptist University United International College
Zhuhai, Guangdong, China

ISBN 978-981-19-9995-6 ISBN 978-981-19-9993-2 (eBook)
https://doi.org/10.1007/978-981-19-9993-2

© The Editor(s) (if applicable) and The Author(s), under exclusive license to Springer Nature Singapore Pte Ltd. 2023
This work is subject to copyright. All rights are solely and exclusively licensed by the Publisher, whether the whole or part of the material is concerned, specifically the rights of reprinting, reuse of illustrations, recitation, broadcasting, reproduction on microfilms or in any other physical way, and transmission or information storage and retrieval, electronic adaptation, computer software, or by similar or dissimilar methodology now known or hereafter developed.
The use of general descriptive names, registered names, trademarks, service marks, etc. in this publication does not imply, even in the absence of a specific statement, that such names are exempt from the relevant protective laws and regulations and therefore free for general use.
The publisher, the authors, and the editors are safe to assume that the advice and information in this book are believed to be true and accurate at the date of publication. Neither the publisher nor the authors or the editors give a warranty, expressed or implied, with respect to the material contained herein or for any errors or omissions that may have been made. The publisher remains neutral with regard to jurisdictional claims in published maps and institutional affiliations.

This Springer imprint is published by the registered company Springer Nature Singapore Pte Ltd.
The registered company address is: 152 Beach Road, #21-01/04 Gateway East, Singapore 189721, Singapore

Foreword by Adrian Bailey

The global COVID-19 pandemic exposes divergence in global affairs. While there is agreement among epidemiologists and clinicians that the Delta and Omicron variants, dominant since late 2021, are associated with high infection rates and relatively low death rates, and public health officials can point to the effectiveness of mass vaccinations in reducing the severity of COVID, two diametrically opposed public health strategies have emerged. That is, China stands as the last remaining major economy to adopt and retain a "Zero COVID" strategy, while other nations continue to open up and "live with COVID." This book—the second to be published on global COVID by researchers and students at Beijing Normal University-Hong Kong Baptist University United International College (BNU-HKBU UIC)—addresses this dramatic global divergence.

The volume makes three important contributions to knowledge. First, it examines COVID-19 using a lens of comparative globalization. It highlights the latest development of the pandemic from a comparative perspective. The comparisons unfold in areas of public health, international relations, and social media responses. The general discourse on COVID-19 has changed since the first weeks and months of the crisis, the sudden appearance of the virus, the initial lack of effective vaccines, and the swift progress of the disease spreading around the world could be perceived as posing a terrible but perhaps temporary threat to the processes of globalization. Reflections on the "late" or "liquid" modernity and globalization occupied the central mentality of researchers. Now that effective vaccines are available, the threat is not perceived as being so fatal; however, the endless mutations and variants of the virus mean that the world is now learning to live with COVID-19, although some are still insisting on a zero COVID strategy. The virus is not going away, and the changes that it has wrought on the governmental, economic educational, commercial, cultural, and social spheres are profound and enduring. The divergence caused by the two separate paths navigating the pandemic triggers thinking over the future and fate of globalization.

Second, the volume brings together scholars and practitioners with social science, media studies, and humanities expertise. Its perspective on global COVID is unique in blending social science and humanities scholarship. The book's contributors draw on their disciplinary strengths and specialities to offer an objective account of the comparative and contemporary characteristics of COVID-19 and to trace the overlapping and conflicting values that are associated with diverging public health strategies. The editorial team has done a deft job of blending these diverse disciplinary insights and the overall volume can achieve a contribution greater than the sum of the individual parts.

Third, as the second in the global COVID-19 series, the book continues to demonstrate the powerful insights to be gained from the nexus of research and teaching. Every chapter of the book has involved, at some level, students collaborating with instructors to deliver ground-breaking research. The American philosopher John Dewey famously said that "Education is not preparation for life, education is life itself." By actively engaging students in the life of research, professors at BNU-HKBU UIC are giving students the best possible preparation for an uncertain future. It is also the case that the theories and values which lend coherence to this account of global COVID-19 are themselves discussed and debated in the classrooms of the researchers and students, with these discussions enriching the integrity of the volume and, in return, the findings from the book enriching the classroom.

With these distinct contributions to knowledge on global COVID-19, I am confident that—like its predecessor—this book will not go unnoticed. The Division as a community should feel very proud of what it has achieved, not least because the book has been prepared under the very strictures (closed borders, zero mobility, frequent lockdowns, etc.) of which it writes. I thus congratulate the vision, insight, and boldness of Professor Simon Zhao Xiaobin, and his dedicated team, who poured their devotion, energy, and perseverance into this book project. I also thank all the contributors to this volume whose works closely follow the latest happenings during the "biggest global challenge the world is facing in modern times" (United Nations) and impacts brought upon by diverging control policies in a timely and up-to-date manner.

This book will be of particular interest to government and public-sector policymakers; entrepreneurs and commentators interested in the recovery of markets, institutions, and networks after systemic crisis; globalization scholars charting the particularities of global processes including public health discourses, understandings of risk, and the changing relationship between state and society. Selected chapters on the cultural impact of the pandemic will be of particular interest to the academic community and the wider public.

Adrian Bailey
Chair Professor, Associated Vice President
Beijing Normal University-Hong Kong Baptist University United International College
Zhuhai, China

Foreword by John Corbett

The global pandemic of COVID-19 has entered new stages when the Delta and Omicron variants became dominant since late 2021. The high infection rates, relatively low death rates caused by the two variants, and the availability of vaccines have brought the effectiveness of the two major pandemic control policies, namely, the zero COVID strategy and "live with COVID," to the fore. In the context that the two divergent pandemic control approaches are shaping or shaped by public health, international relations, and social media responses, keen researchers at Beijing Normal University-Hong Kong Baptist University United International College (BNU-HKBU UIC) have captured this academic and practical meaning of the divergence.

This volume is the second collection of research studies undertaken by scholars at BNU-HKBU United International College to address the global impact of the ongoing COVID-19. While the first volume offered a general overview of the impact as well as the responses to the pandemic on a changing world, the present volume highlights on the latest development of the pandemic from a comparative perspective. The comparisons unfold in areas of public health, international relations, and social media responses. The general discourse on COVID-19 has changed ever since the publication of the first volume in 2021. In the first weeks and months of the crisis, the sudden appearance of the virus, the initial lack of effective vaccines, and the swift progress of the disease spreading around the world could be perceived as posing a terrible but perhaps temporary threat to the processes of globalization. Reflections on the "late" or "liquid" modernity and globalization occupied the central mentality of researchers. Now that effective vaccines are available, the threat is not perceived as being so fatal; however, the endless mutations and variants of the virus mean that the world is now learning to live with COVID-19, although some are still insisting on a zero COVID strategy. The virus is not going away, and the changes that it has wrought on the governmental, economic educational, commercial, cultural, and social spheres are profound and enduring. The divergence caused by the two separate paths navigating the pandemic triggers thinking over the future and fate of globalization.

This book marks a continuing attempt from the Division of Humanities and Social Sciences at BNU-HKBU United International College to chart the ongoing effects of the pandemic on the world. Into their community of inquiry, they have brought colleagues from other institutions, and graduate and undergraduate students. Every chapter of the book has involved, at some level, students collaborating with instructors to deliver ground-breaking research. The American philosopher John Dewey famously said that "Education is not preparation for life, education is life itself." By actively engaging students in the life of research, professors at BUN-HKBU UIC are giving students the best possible preparation for the uncertain future. Therefore, I congratulate the vision, insight, and boldness of Professor Zhao Xiaobin and his editorial team who poured their devotion, energy, and perseverance into this book project. I also thank all the contributors to this volume whose works closely follow the latest happenings during the "biggest global challenge the world is facing in modern times" (United Nations) and impacts brought upon by diverging control policies in a timely and up-to-date manner.

This book will be of particular interest to government and public-sector policymakers; some chapters on the cultural impact of the pandemic will again be of wider interest to the academic community and indeed the wider public.

John Corbett
Professor, Associate Dean for Teaching and Learning
Faculty of Humanities and Social Sciences
Beijing Normal University-Hong Kong Baptist University United International College
Zhuhai, China

Acknowledgments

We are extremely grateful for the support from the Springer Nature team, with who we have collaborated very well in our studies and research of the global pandemic of COVID-19. We are also thankful for all the reviewers who took the time in reading, reviewing, and providing precious feedback to the component chapters. We offer our gratitude to the leadership at the Beijing Normal University-Hong Kong Baptist University United International College (BNU-HKBU UIC) whose academic and administrative support were instrumental in driving forward this book project. Last but not the least, we thank all contributors of the book project for pouring their efforts, wisdom, time, and thoughts into expanding and deepening our knowledge of pandemic control policies.

Editorial Team

Introduction

While wrapping this book project up in the September of 2022, the whole world is experiencing the third year of ravage brought by the COVID-19 pandemic. However, some positive signals look like lights at the end of the tunnel.

Since late 2021, a growing number of countries around the world have significantly eased COVID restrictions. Most European countries, such as Britain, France, Denmark, Norway, Ireland, and the Netherlands, have ended or eased their respective COVID restrictions.[1] Singapore furthered easing COVID restrictions in August 2022, ending almost all COVID restrictions. Japan raised the daily entry cap for visitors on 7 September 2022 and removed pre-departure testing mandates for travelers who have received three COVID-19 vaccine doses. Australia, once imposing the longest COVID lockdowns in the world, lowered the quarantine time for anyone who contacts COVID-19 but shows no symptoms from 9 September onwards. The mask mandate for domestic flights in Australia was also eliminated.[2]

The ease of COVID-related restrictions came following the emergences of the Delta and Omicron variants, as well as the availability and rollout of COVID vaccination. The higher infection rates of the variants had rendered suppression and elimination of all infected cases more and more costly. Rates of severe and death cases were lowered by the mass rollout of vaccination programmes. Many countries have pivoted to "live with COVID." They tolerate the accumulation of new cases, from thousands to tens of thousands, while trying to drag the severe and death cases down to prevent medical resources from being overwhelmed. Meanwhile, these countries reboot the economic activities and re-open with the rest of the world in a continued effort to sustain globalization.

[1] Jamey Keaten, "More Countries in Europe, Recently the Pandemic's Epicenter, Ease COVID Restrictions," *Los Angeles Times*, February 2, 2022, https://www.latimes.com/world-nation/story/2022-02-02/more-countries-easing-covid-restrictions.

[2] Elise Schoening and Lizzie Wilcox, "The Latest Updates on International Gathering and Travel Restrictions," North Star Meeting Group, September 13, 2022, https://www.northstarmeetingsgroup.com/coronavirus-countries-cities-reopening-COVID-19-new-cases.

In stark contrast, some other countries, notably China, are still pursuing a zero-COVID strategy, resorting to strict restriction measures when confronting the coronavirus. The goal of the zero COVID strategy is to quickly identify and eliminate positive cases so that as few people as possible in the community get infected: zero infection, less medical resources being overwhelmed, and free from any potential sequelae of COVID infection. Worries about the public health crisis imposed by the possible transmission of coronavirus oftentimes outweigh the urgency for economic resumption.

Subsequently, the September 2022 wave of COVID-19 pandemic in China witnessed a new wave of lockdowns when less than two thousand confirmed cases were reported across a country of 1.4 billion. Thirty-three cities, including seven provincial capitals and one municipality under the direct rule of the State Council, joining the overall 103 cities across 26 provinces, entered semi- or full-scale city-wide lockdowns. More than 65 million[3] residents, from Dalian in the northeast to Sanya on the southernmost tip, from Shenzhen on the mouth of Southern China to Xining and Chengdu far into the west, were affected. After Shanghai (February to June 2022), Chengdu became the second Chinese city with a population of over twenty million being forced to lock down.

The dichotomy of the two major pandemic control policies, the "zero COVID" strategy and "live with COVID" strategy, has not only indicated two different approaches to pandemic control, but resulted in two separate worlds, two mentalities, and two sets of pandemic narratives. The dichotomy of policies reflects a broad array of differences in cultural settings, social backgrounds, political arrangements, economic affordances, value system, ideologies, as well as psychological factors. Why is this policy? Why is this policy at this stage? Why not another policy? Why policies between countries are so divided, and even conflicting? What are the implications of the divided policies on international relations? What are the cultural and psychological responses to the policies? All the questions warrant an international, comparative study of pandemic control policies as well as social resilience.

The crystallization of the contemplations is *Comparative Studies on Pandemic Control Policies and the Resilience of Society*. The volume contains three sections: "Public Health and Social Economic Studies out of the Pandemic," "International Relations in the Post-Pandemic World: Globalization, Global Governance and Leadership," and "Cultural and Social Media Responses: Perceptions, Emotions, and Narratives." Through the three sections, this volume aims to provide facts-based, data-supporting documentation of pandemic-control policies across the globe and a comparative analysis of the deep-seated backgrounds and factors in determining

---

[3] Moting Jiang and Yunhong Bao, "(Yiqing Boji Fanwei Chixu Kuoda Quan/ 33 Cheng Bufen Huo Quanyu Guankong) 疫情波及范围持续扩大 全国 33 城部分或全域管控 [Pandemic Continues to Explode, 33 Cities across Country Enter Semi-/Full-Scale Lockdowns]," Caixin, September 3, 2022, https://www.caixin.com/2022-09-03/101935462.html.

the selection of a particular policy. This volume also concerns with social resilience associated with a policy. This series of articles will produce a timely and informative edition of comparative studies of pandemic control policies and social resilience, which will shape public discussions and debates in many years to come.

In 2021, researchers at the Beijing Normal University-Hong Kong Baptist University United International College (BNU-HKBU UIC) blazed the trail of pandemic studies with the edited volume *COVID-19 Pandemic, Crisis Responses and the Changing World: Perspective in Humanities and Social Sciences.* This volume marked the latest effort and endeavour by UIC researchers to astutely capture the ethos of the comparative nature in the pandemic studies for an up-to-date, insightful, and state-of-the-art research in the course of the pandemic.

BNU-HKBU United International College is nothing short of an ideal institution to produce pandemic studies from comparisons of control policies and social resilience. Located in Zhuhai, China, one of China's frontier cities in its economic reform and opening-up, this ground-breaking, English-medium, and internationally oriented liberal arts college sits right where China and the world meet each other. Particularly so is during the global pandemic of COVID-19. While locating inside the world's most notable zero COVID-pursuing country and dissecting the concrete conditions undergirding such a policy, many of the UIC contributors of the book are also living the varied experiences brought about by the distinctive pandemic control policies in other countries in their cross-border travels. For example, some scholars witnessed first-hand Singapore's "circuit breaker" period in response to the initial shock and panic regarding the COVID-19 pandemic and the subsequent phase of ease of restrictions. Some scholars, who were once trapped in Hong Kong during the fifth wave of the local pandemic outbreak, had exposure to the seemingly impossibility of a reconciliation of zero COVID strategy and "live with COVID" endeavours. Some other scholars poured their European experiences into their studies of the pandemic. In a sense, contributors from UIC are on the front line of producing a personal, professional, detail-rich, readable, and academically valuable account regarding the dichotomy of zero COVID strategy and "live with COVID" strategy.

In the first section "**Public Health and Social Economic Studies out of the Pandemic**," contributors probe the public health policy and economic factors underneath the selections of zero COVID strategy and "live with COVID" strategy from a comparative studies perspective. **Kam Tong Chan, Xiaoman Yang,** and **Xinyu Yi** adopt an institutionalism perspective as their basic framework in a comparative examination of the control policies at the early stage of the pandemic. The authors divide the following six case regions in the chapter, namely, Mainland China, Hong Kong, Japan, Singapore, the United Kingdom, and Denmark into three categories: dynamic zero COVID strategy, "live with COVID" strategy, and mitigation policy. The authors have invoked the perspectives of welfare ideologies, regimes, economic considerations, and governance in examining selections of differing pandemic control policies.

Following and extending Chan, Yang, and Yi's study, **Simon X. B. Zhao, Bo Yan,** and **Yutong Chen** observe the up-to-date development of the pandemic in 2022. In *Between Zero COVID and "Live with COVID:" Comparative Studies in Pandemic Control Policies of China, Singapore, and Hong Kong at the Stage of Omicron*, the authors zoom in on three cases of the most significant representativeness in comparative studies on pandemic control policies: Mainland China, Hong Kong, and Singapore. China is the world's most prominent and the last holdout of zero COVID nation, distinguishing itself from the rest of the world. Singapore is among the first countries to switch from pursuing zero COVID to live with COVID, whose pandemic control path represents and sheds lights upon many nations ending zero COVID to bring economy and social lives back to normal. Hong Kong establishes as an ideal observatory sample of how the once effective zero COVID strategy, embodied by mass testing and strict lockdowns, had difficulty in being applied outside China and in coping with the pandemic in the age of Omicron. By investigating Hong Kong's struggle in the fifth wave of the pandemic of February 2022, the chapter meticulously documents the social conditions behind the unique effectiveness of zero COVID strategy in China and the scientific preparedness behind Singapore's swift actions to transition to live with COVID.

Economic factors, while not given a prominent consideration in the early days of the pandemic when saving lives was prioritized, are emphasized in considerations over pandemic control policies in *Economic Rationale of Different Pandemic Control Policies and their Impact on Economy.* In the early days of the pandemic, China, Australia, and Singapore chose a zero COVID strategy not only because such an elimination strategy saved lives but also because the countries' economies fared much better than their "live with COVID" counterparts. However, in the era of the Delta and Omicron variants, the heavy costs of eliminating the highly infectious variants, the lowered death rates dragged down by the rollout of vaccination, and the urgency to resume economic vitality had driven many of the previous zero COVID nations to pivot to "live with COVID." China, while maintaining the status as the world's last holdout of zero COVID strategy, faced mounting economic pressure brought by the fallout of frequent lockdowns in the nation's various economic powerhouses, particularly, Shanghai, Shenzhen, and Xi'an in 2022. The authors conducted an original set of research by firstly dividing the pandemic from 2020 to 2022 into two phases in its examination of pandemic control navigations in China, Australia, Singapore, and the United States; and secondly following closely the Shanghai lockdown in 2022, a focal point in pandemic control approaches comparison and a timely and representative city struggling between public health and economic development. By referring to various data and putting Shanghai in comparison in terms of lives lost and economic downturn. The authors thus question the very existence of a "best" or "one-size-fits-all" control strategy, or one that could be effective at all stages of a pandemic. Authors in *Social Resilience and COVID-19 Recovery Outcome: A Comparative Analysis of Resilience Attributes in 107 Countries Using Random Forest Algorithm* conclude that a combination of multiple factors, in their studies, the demographic character, accesses to medical services and resources, and government responses are important factors in pandemic recovery and resilience. In *"Surprisingly Real": Regressive*

Structural Transformation in the Wake of the Corona Crisis, **Markus Heidingsfelder** and **Lihua Chen** examine the two functional systems, namely, the health system and the political system during the COVID-19 pandemic. They present that individual efforts of "renationalization" come at high political and economic costs, and a "deglobalization" could not be implemented.

The second part of the book, titled "**International Relations in the Post-Pandemic World: Globalization, Global Governance and Leadership**," navigates through the COVID-19 pandemic from the perspective of political science and international relations.

Employing the concept of "Kindleberger Trap," **Selçuk Çolakoğlu, Chaofan Chen,** and **Chen Jiawei** approach the global leadership competition between China and the United States from the perspective of global public goods provisions, denying the emergence of a "Kindleberger Trap." Following chapter, authors in *Political Considerations and Rationales for Different Control Policies in COVID-19 Pandemic: An Analysis from the State Theory Perspective* trace the rationales behind countries' differing pandemic control policies, notably zero COVID strategy and "live with COVID" strategy. The authors suggest that the zero COVID strategy is embedded in the Neo-Weberian approach of the state theory; while the liberalist approach of state theory, rooted in the idea of social contract, has resulted in a limited capacity of state, which is straightjacketed by citizens' (dis)satisfaction of some restriction measures and the state's capability to mobilize the masses to implement a zero COVID strategy. Again, this chapter confirms that each pandemic control is rooted in very concrete social, political, and cultural conditions.

Chan Siu-han in *COVID-19 Pandemic as the Cosmopolitan Moment of World Risk Society* invokes Ulrich Beck's "world risk theory." Chan disagrees with the idea that the COVID-19 pandemic brought an end to the globalization. Instead, the COVID-19 pandemic, standing as "the first truly global illness and risk that expresses the profound level of globalization our world has already attained," compels us to confront the reality of our independence as well as our common destiny in the global world more than ever. Calling the COVID-19 pandemic "the cosmopolitan moment of world risk society," Chan further suggests that COVID-19 brings forth an ever more globalized world. Authors in *Politicization of Anti-Pandemic Measures in Europe: Cleavage Politics and Divided Publics* also touched upon the relations between the pandemic and globalization. Using the cleavage theory, the authors demonstrate that within the European context, people's anti-pandemic measures and unwillingness to receive vaccination were positively correlated with their negative views of the European Union and globalization. Sweden caused a controversy among Chinese after a nationalistic Chinese newspaper reported its distinctive pandemic control policies almost to the opposite of China's. The controversy was captured by one set of authors in the book. By inquiring into images of Sweden constructed by Chinese in China and Sweden, the authors provide Sweden as another perspective to global pandemic narratives.

Bo Yan, Simon X. B. Zhao, Yizhou Dong, and **Shihao Sun** approach the pandemic narrative from the perspective of conspiracy theories. In *Sino-US Great Game Out of the Pandemic: A Conspiracy Theories Perspective*, the authors analyze and debunk three COVID-related conspiracy theories in the relationship between China and the United States. The pandemic was first seen by many observers in the West as "China's Chernobyl," which was instantly debunked by China's immediate suppression of virus spread and quick recovery in economic production. The leftists in the United States mobilized the pandemic narrative as "God's gift to the US left" in ousting President Donald Trump. Yet, the pandemic as well as socio-economic conditions in the United States continue to worsen under the new president. Faced with a declining US in the quagmire of pandemic and inflation, hypernationalists in China interpreted China's nationwide power crunch in September 2021 as China's shrewd move to "corner overtake" the United States. This conspiracy theory was debunked by the official media of China. Wrapping up Part II of the book, this conspiracy chapter stands as a powerful narrative that the pandemic requires global efforts to tackle, and any force trying to mobilize the pandemic narrative for "I win, you lose" game will only be backfired.

Part III of the book, titled "Cultural and Social Media Responses: Perceptions, Emotions, and Narratives," continues observations of the pandemic through the comparative lens. **Qiaoyun Zhang** and **Yushan Wang** discuss the cultural incentives of risk perception and response in the perspectives of individuality and collectivism. The different viewpoints reflect a varied understanding of individual body politics and of social organization and groupness. **Ye Tao**, **Chuanli Xia, Hong Zhao,** and **Ning Liu** investigate the mediating effects of social media in promoting willingness and reducing negative emotions of Chinese netizens regarding getting COVID vaccination. In *Media Discourses and China's Social Mobilization at the Early Crisis Stage of the COVID-19 Pandemic*, the authors turn to China's state-sanctioned media. The state media produced three major discourses of militarism, altruism, and authority, and helped the Chinese state effectively mobilize the society and rapidly controlled the pandemic. In the meantime, Chinese netizens did not stop to make their voices heard via social media. **Qingyuan Zhao, Liang Nie,** and **Xiaying Xu** compare three of the Mainland China's provincial capitals and investigate whether the demands and sentiments expressed in online posts by netizens in the cities were associated with local economic development. In the final chapter of the book, **Lu Ziran** and **Li Li** turn to feature films dramatizing China's responses to the COVID-19 pandemic. Surveying recurrent aspects of the characterization of key agents in the epidemic narrative as it appears in feature films, the chapter considers the necessary simplifications that the Chinese epidemic narrative demands and identifies some of the complexities absent in such narratives to date.

The book then ends with a conclusion by **Selçuk Çolakoğlu** and **Kam Tong Chan.**

Simon X. B. Zhao
Beijing Normal University-Hong Kong
Baptist University United International
College
Zhuhai, China

Bo Yan
Beijing Normal University-Hong Kong
Baptist University United International
College
Zhuhai, China

Contents

Editors and Contributors

About the Editors

Simon X. B. Zhao is Professor and Associate Dean Division of Humanities and Social Sciences BNU-HKBU United International College (UIC), used to teach at HKU as a founding Director of the International Center for China Development Studies and specializing in urban, regional, and global studies and publishing in the broad areas of urban regional studies, geopolitical and economic development, spatial and global transformation, geography of international trade and finance, and development of international financial centers (IFCs). He is an editorial board member for four international journals and published more than 150 papers in international refereed journals and ISI (SSCI) Citation Journals, book chapters, consultative reports, etc. Simon Zhao was a Member of the Central Policy Unit, HKSAR Government (Pan-PRD Panel) and chaired and co-led many substantial academic research projects, providing wide consultancy services to China national, provincial, and local governments as well as private sector. His latest edited volume *COVID-19 Pandemic, Crises Response and the Changing World: Perspectives in Humanities and Social Sciences* (by Springer) is well received globally.

Kam Tong Chan is Associate Professor in Social Work and Social Administration, BNU-HKBU United International College. He is a Registered Social Worker and was awarded Top 10 Outstanding Social Worker Professional in 2012 by the National Ministry of Civil Affairs of the People's Republic of China. Dr. Chan's academic interests focused on a non-profit organization, social service administration, social work supervision, and social enterprises and social innovation. He oversees many research projects funded by Hong Kong SAR Government, Mainland Chinese Government, and Korean institutions. Dr. Chan is also the author of many research publications on comparative studies of social enterprises, social management, and social work.

Selçuk Çolakoğlu is Associate Professor at Faculty of Humanities and Social Sciences, BNU-HKBU United International College. Çolakoğlu received his B.A. from the Department of International Relations at Ankara University, obtained his M.A. from the Department of International Relations at Middle East Technical University (METU) in Ankara, and completed his Ph.D. in the Department of International Relations at Ankara University in 2003. His research interests include political and security issues in international relations, particularly in East Asia, Central Eurasia, and Turkish Foreign Policy. He has published monographs, 30 articles, and book chapters. His books are *Northeast Asia in International Relations* (in Turkish, Ankara: USAK Press, 2009); *Korean Society, Culture, Politics* (in Turkish, Ankara: Orion Press, 2008); *China in World Order and Turkey-China Relations*, (Ankara: USAK Press 2012). He has held positions in universities in South Korea, Turkey, the United States, and China.

Qiaoyun Zhang received her Ph.D. in Anthropology from Tulane University (USA) in 2016 and was a research fellow at the International Institute for Asian Studies (2017–2018, Leiden) and associate researcher at the School of Anthropology and Museum Ethnography of the University of Oxford. She has published more than 10 articles in leading SSCI, A&HCI, and CSSCI journals including *Journal of Contemporary China*, *Asian Ethnology*, and *China Information*. Currently, she is conducting research on social development in post-pandemic era. Zhang's research focuses on cultural recovery and preservation after disasters and engages with topics of risk and culture, state-ethnic relations in China, sustainable development, and intangible cultural heritage preservation.

Bo Yan received Ph.D. in History from Nanyang Technological University, Singapore. He is a Research Fellow in the Faculty of Humanities and Social Sciences. Yan's main research interests include modern Chinese history, postwar United States, modern Southeast Asian history, and the history of the global 1960s. He has published in the *Journal of Asian Studies*, *Twentieth Century China*, and others. Yan's Ph.D. dissertation was listed as the International Convention of Asia Scholars (ICAS) Book Prize 2021-Best Dissertation in Humanities. Bo Yan is currently conducting studies and research on the global pandemic of COVID-19.

Contributors

Kam Tong Chan Beijing Normal University-Hong Kong Baptist University, United International College, Zhuhai, China

Siu Han Chan Beijing Normal University-Hong Kong Baptist University, United International College, Zhuhai, China

Chaofan Chen Beijing Normal University-Hong Kong Baptist University, United International College, Zhuhai, China

Jiawei Chen Beijing Normal University-Hong Kong Baptist University, United International College, Zhuhai, China

Lihua Chen Beijing Normal University-Hong Kong Baptist University, United International College, Zhuhai, China

Yutong Chen Beijing Normal University-Hong Kong Baptist University, United International College, Zhuhai, China

Selçuk Çolakoğlu Beijing Normal University-Hong Kong Baptist University, United International College, Zhuhai, China

Yizhou Dong Beijing Normal University-Hong Kong Baptist University, United International College, Zhuhai, China

Yangtian Feng Beijing Normal University, Beijing, China

Ruijie Guo Beijing Normal University-Hong Kong Baptist University, United International College, Zhuhai, China

Markus Heidingsfelder Beijing Normal University-Hong Kong Baptist University, United International College, Zhuhai, China

Xinyu Huang Beijing Normal University-Hong Kong Baptist University, United International College, Zhuhai, China

Li Li Macao Polytechnic University, Macao SAR, China

Tao Li Beijing Normal University, Beijing, China

Zhiyi Liao Beijing Normal University-Hong Kong Baptist University, United International College, Zhuhai, China

Chally Chai Lee Lim Beijing Normal University-Hong Kong Baptist University, United International College, Zhuhai, China

Liming Liu Beijing Normal University-Hong Kong Baptist University, United International College, Zhuhai, China

Jiaqi Liu Beijing Normal University-Hong Kong Baptist University, United International College, Zhuhai, Guangdong, China

Ning Liu Beijing Normal University-Hong Kong Baptist University, United International College, Zhuhai, China

Yu Liu Beijing Normal University-Hong Kong Baptist University, United International College, Zhuhai, Guangdong, China

Ziran Lu Macao Polytechnic University, Macao SAR, China

Liang Nie United International College, Beijing Normal University-Hong Kong Baptist University, Zhuhai, China

Siyao Qin Beijing Normal University-Hong Kong Baptist University United International College, Zhuhai, China

Shihao Sun Beijing Normal University-Hong Kong Baptist University, United International College, Zhuhai, China

Ye Tao Beijing Normal University-Hong Kong Baptist University, United International College, Zhuhai, China

Michal Vavřík Beijing Normal University-Hong Kong Baptist University United International College, Zhuhai, China

Yushan Wang Beijing Normal University-Hong Kong Baptist University, United International College, Zhuhai, China

Chuanli Xia Beijing Normal University-Hong Kong Baptist University, United International College, Zhuhai, China

Xiaying Xu United International College, Beijing Normal University-Hong Kong Baptist University, Zhuhai, China

Bo Yan Beijing Normal University-Hong Kong Baptist University, United International College, Zhuhai, China

Xiaoman Yang Beijing Normal University-Hong Kong Baptist University, United International College, Zhuhai, China

Chengxi Yao Beijing Normal University-Hong Kong Baptist University, United International College, Zhuhai, China

Fangxin Yi Beijing Normal University, Beijing, China

Xinyu Yi Beijing Normal University-Hong Kong Baptist University, United International College, Zhuhai, China

Qiaoyun Zhang Beijing Normal University-Hong Kong Baptist University, United International College, Zhuhai, China

Hong Zhao Beijing Normal University-Hong Kong Baptist University, United International College, Zhuhai, China

Qingyuan Zhao United International College, Beijing Normal University-Hong Kong Baptist University, Zhuhai, China

Simon X. B. Zhao Beijing Normal University-Hong Kong Baptist University, United International College, Zhuhai, China

Public Health and Social Economic Studies out of the Pandemic

Comparison of COVID-19 Control Policies and Public Health Regimes: The Institutionalism Perspective

Kam Tong Chan, Xinyu Yi, and Xiaoman Yang

1 Introduction

In December 2019, the first COVID case was discovered in Wuhan, China. With the disease's high infectivity, the virus has now spread to most of the regions in the world and plagued millions of people. Confronted with the pandemic, many countries have come about different COVID control policies. Apart from the dichotomies of "zero COVID infection" and "living with COVID virus" approaches emerged in the beginning, now many states have changed their COVID Control Strategies in reaction to the virus variants' continuous mutations. This article will compare and contrast the COVID Monitoring Policies among six different countries and regions: Mainland China, Hong Kong SAR of China, Japan, Singapore, the United Kingdom, and Denmark. The institutionalism perspective is adopted as the basic framework in examining the various COVID 19 Control policies adopted by some countries or regions. There are two basic concepts in institutionalism, namely the institutional theory which focuses on the formal institutions and structures of the government. The second is the neo-institutionalism which focuses on the interaction and the effects of those institutions on society. By probing into the divergent policies and the values and assumptions underlying them, we can derive new inspirations and insight for increasing the social resiliency facing the COVID challenge. The main reasons of adopting such two perspectives are because the social origin and the institutional framework are important components in shaping the policy initiatives.

According to Scott (2013), institutions incorporate the regulative, normative, and cultural-cognitive elements. These elements work jointly with associated activities

K. T. Chan (✉) · X. Yi · X. Yang
Beijing Normal University-Hong Kong Baptist University,
United International College, Zhuhai, China
e-mail: kamtongchan@uic.edu.cn

© The Author(s), under exclusive license to Springer Nature Singapore Pte Ltd. 2023

S. X. B. Zhao et al. (eds.), *Comparative Studies on Pandemic Control Policies and the Resilience of Society*, https://doi.org/10.1007/978-981-19-9993-2_1

and resources, providing stability and meaning to social life. The concept of legitimacy in institutional theory is perceived as the foundation of survival for any institution. Consequently, institutions are continuously endeavouring to influence society's perceptions about them, namely, to make the public see them as "legitimate measures in society." To quest legitimacy, the institutions alike may undergo some isomorphic change. The terminology of isomorphism encapsulates the ways and reasons institutions seek to attain legitimacy by becoming similar to other units in the population that are confronted with the same environmental factors. There are three subbranches of isomorphism, namely coercive, mimetic, and normative:

1. Coercive isomorphism is operated under the regulative elements in institutions and propels them to abide by expedience, rule, laws, or sanctions. For instance, a new COVID prevention law is initiated, and companies have to follow it.
2. Normative isomorphism is driven by the normative elements within an institution and pushes it to be compliant with social obligation, certifications, or accreditations. For example, the suppliers of COVID-19 testing kits are expected to act professionally and follow the guidelines specific to their industry.
3. Mimetic isomorphism is influenced by the cognitive elements, which drives the institutions to imitate other more successful institutions, especially when the environment is uncertain, or the goals are ambiguous. The myriad imitations of the successful COVID prevention policies provide an apt illustration.

Apart from the legitimacy perspective, this article will also consider how institutional theory interplays with the government's policymaking. As theorized by neo-institutionalism, there are three ways institutions may affect public policies. They respectively belonged to the concepts of historical institutionalism, rational choice institutionalism (RCI), and sociological institutionalism (Mahmud, 2017). Historical institutionalism states that the functioning of a state's polity, politics, and policies depends largely on its historical context. According to its proponents, the formal and informal rules, conventions, and norms developed within a state or government agency over the years will affect how it initiates and implements its policies. For instance, the Scandinavian countries have a historical preference for the social democratic model. In contrast, rational choice institutionalism argues that policy making is the sensible choice based on the best available resources to maximize the utilities or achieve the institutional goals. For example, in the face of COVID, many governments seek help from suitable experts to gain technical assistance, like think tanks and universities, to formulate highly efficient countermeasures. Lastly, sociological institutionalism holds that the aims of policy formulation are not only to improve efficiency in resolving problems but also to improve social legitimacy to accommodate behavioural patterns. Considering the moral factors and social appropriateness, sociological institutionalism may manifest in the ways different neighbourhoods deal with quarantine policies. In some areas where individual freedom is highly valued, forced quarantine is morally unacceptable and be discarded as a policy alternative. Considering the features of sociological institutionalism, it may give birth to institutional isomorphism.

According to Dimaggio and Powell (1983), the more uncertainties, the more imitations, and the fewer the number of existing models, the quicker the isomorphism. Facing the unprecedented challenge of COVID-19, there is a rising incidence of isomorphism across the globe and within different countries. Additionally, different countries and regions adopt various policies with the influences of historical, rational choice, and sociological institutionalism. Consequently, it's necessary and rewarding to look into multiple COVID policies through the lens of institutional theories.

This chapter and the subsequent chapter are closely linked together. In fact, this chapter will provide the theoretical framework and the dynamic of six countries or cities at the early stages of COVID, while the latter chapter will focus more on the three countries or cities in the recent two years, particularly with the changes of fundamental shift of paradigm in policy making.

2 The Major and Distinct Measures Adopted by Major Countries and Region

2.1 Mainland China

China adopts the socialist system. Since the outbreak of the epidemic in early 2020, the Chinese government had always insisted on putting people's safety and interests as the priority, so it was very resolute in eliminating the epidemic in the early days. In the prevention and control of the epidemic, the Chinese government adopted an active policy led by the government and assisted by residents' autonomous organizations. With the continuous control of the epidemic, China's anti-epidemic policy has gradually changed from a "Zero COVID Infection" policy to a "Dynamic zero-clearing" policy.

The "zero COVID" policy was proposed by the Chinese government when the outbreak broke out in early 2020. The Chinese government have adhered to the principle that the safety of the people's lives always comes first, controls the spread of the epidemic at all costs, and resolutely eliminates the epidemic everywhere. Various Provincial Governments across China have adopted the "four early measures[1]" which include early detection, early reporting, early isolation, and early treatment (State Council Information Office of the People's Republic of China, 2020). The Chinese government has imposed blockade measures on cities or regions where the epidemic has developed severely, mobilized community residents to assist in the regional blockade of the community, and required residents in the region to work from home, conduct online teaching, and stay at home and not going out. The government have arranged designated hospitals and established "square cabin hospital" (the temporary

[1] The "four early measures" refer to as early as possible detection of infected cases and timely reporting of the outbreak to the government, as well as the isolation of infected persons at the early stage of infection and timely and targeted treatment, as proposed by the Chinese government in the "Chinese Action to Combat the New Coronary Pneumonia Epidemic".

shelter which provide basic medical care for the infected patients) to receive and treat confirmed cases, community residents have taken turns to conduct door-to-door inspections, and established a medical treatment expert group to achieve "receive all the confirmed cases and treat them as they should" (State Council Information Office of the People's Republic of China, 2020). After the development of the new coronavirus pneumonia vaccine, China have provided vaccination to the Chinese people free of charge in batches to achieve the effect of national prevention. China has also further improved the national health certification system, realizing the "three-in-one" health code, itinerary code, and vaccine pass (Wang, 2021).

As starting from August 2021, China's epidemic prevention policy changed from "zero COVID Infection" to "dynamic zero-clearing" of precise prevention and control. The "dynamic clearing" policy refers to the government's rapid response to timely detection of infection sources and cases, taking relevant public health intervention measures, and deploying measures such as community closure management and large-scale nucleic acid testing within 24 h after the outbreak. This policy aims to minimize the occurrence of the epidemic and obtain the greatest prevention and control effect with the least social cost. Through big data, the government can pinpoint the scope of the epidemic more precisely, and the restrictions usually only involve one district, one community, or one building in the city, thus reducing the disruption to people's daily life.

In the face of the epidemic situation in the first half of 2022, Shenzhen and Shanghai have adopted different epidemic prevention policies. Shenzhen has adopted the "strict closure and control" epidemic prevention measures, that is, set up closure and control areas, control areas, and prevention areas according to different epidemic situations to carry out strict personnel access management. And after multiple rounds of nucleic acid testing for all staff, the Shenzhen government quickly adopted a week-long silence period. During the period of silence, the government took strict control measures to ensure basic living, shutting down buses and subways, and not leaving the city unless necessary, thus bringing the epidemic under control quickly and achieving "dynamic zero-clearing" in a short period of time and restoring social order. Compared with Shenzhen, during the epidemic period in the first half of 2022, Shanghai adopted "precise prevention and control" epidemic prevention measures, that is, nucleic acid testing was carried out in some key target groups to minimize the scope of risk areas. However, this kind of prevention and control measures is difficult to effectively find the hidden source of infection and transmission chain in the face of the hidden and highly contagious virus strain of Omicron, so this has also led to the continuous epidemic in Shanghai. It has been nearly 3 months and caused many people to be infected with the COVID-19.

In the next chapter, it will provide some of the recent development and the new measures such as "focused COVID control" methods in the 2022.

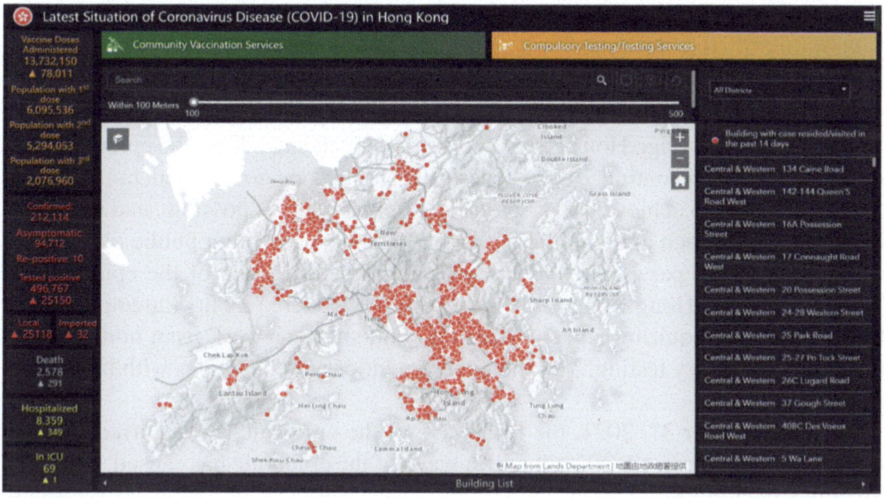

Fig. 1 Latest situation of coronavirus disease (COIVD-19) in Hong Kong (as at March 8, 2022)

2.2 Hong Kong Special Administrative Region of China

Since the first case of novel coronavirus pneumonia was discovered on January 23, 2020, Hong Kong has experienced the fifth wave of the epidemic. In the face of these five waves of epidemics, Hong Kong SAR has introduced many iconic anti-epidemic policies. Influenced by the Chinese mainland government, the Hong Kong SAR government's epidemic prevention policies are like those in mainland China. In the face of the new coronavirus pneumonia, it adopts an anti-epidemic policy led by the government throughout the process.

The Hong Kong SAR government proposed a "social distancing measures[2]" policy when controlling the second wave of the epidemic in Hong Kong in March 2020. The "social distancing measures" refers to the Hong Kong government's efforts to limit the number of people gathering, control dining time, and close various cultural and entertainment venues to reduce gatherings to reduce the risk of virus transmission. The policy will continuously adjust the number of people and time limit according to the development of the epidemic, to reduce the impact of the epidemic on the tertiary industry and the normal life and order of Hong Kong residents. The Hong Kong government established a real-time tracking of confirmed cases in the early stage of the outbreak. The system is updated once a day and provides dynamic information including the places visited by the confirmed person within 14 days, the living place of the confirmed person, and the nucleic acid testing location (Fig. 1).

[2] The "social distancing measures" was first proposed in a press release by the Hong Kong SAR Government. It refers to the government's efforts to reduce the number of people gathering and maintain social distancing through the Prevention and Control of Disease Regulation by limiting the number of people gathering, controlling meal times, and closing some entertainment venues.

In September 2020, the Hong Kong SAR government launched the "Universal Community Testing Program[3]", which aims to popularize nucleic acid testing in the community to achieve territory-wide nucleic acid monitoring (Tu, 2021). The Hong Kong SAR government has also implemented the "COVID-19 Vaccination Program[4]" to encourage Hong Kong people to get vaccinated. As the epidemic situation in Hong Kong continued to be severe, the Hong Kong government implemented a "vaccine pass" to achieve universal vaccination as soon as possible, and stipulated that public place such as shopping malls, barbershops, and other public places need to show a vaccine pass before entering during the fifth wave of the epidemic. In January 2021, the Hong Kong SAR government conducted its first attempt to close the Jordan area, requiring citizens not to go out, study or work from home, and to undergo compulsory testing. Yet, the government provided citizens with basic daily necessities.

Since December 31, 2021, the fifth wave of the epidemic has occurred in Hong Kong. And as on June 19, 2022, the fifth wave of the epidemic has caused 1,214,053 infections in Hong Kong and 9,182 deaths (Latest Situation of Coronavirus Disease [COVID-19] in Hong Kong). Due to the rapid development of the epidemic and the shortage of medical testing in Hong Kong, the Hong Kong SAR government took sewage samples from buildings or communities and tested them. Among them, areas with positive sewage tests would be designated as high-risk infection areas. The government then distributed rapid test kits to citizens in these areas and implement one-day inspections to try to detect confirmed cases as soon as possible (Chen, 2022). In addition, due to the shortage of medical resources in Hong Kong, many confirmed patients could not receive corresponding treatment promptly. Hong Kong SAR government made a request to the Chinese mainland government for medical assistance, including but not limited to Hong Kong's nucleic acid detection capabilities, medical material supply, community isolation, and the construction of treatment facilities.

All in all, Hong Kong's anti-epidemic policy is basically in line with the mainland Chinese government, and the policy strength is on the rise. However, in the later stage of the pandemic, the Hong Kong SAR government changed its policies to a more 'focused' way of isolation to maintain the social stability.

In the following chapter, it will further examine the recent policies to control the fifth wave of outbreak of pandemic.

[3] The "universal Community Testing Program" is an initiative of the Hong Kong Special Administrative Region, with the support of the Chinese government, to provide free testing for the new coronavirus to all citizens of Hong Kong in order to locate the infected and break the chain of transmission. The program is based on the community setting up sampling sites where people can go to their community sampling sites and wait for the test results at home.

[4] The "COVID-19 Vaccination Program" is a Hong Kong Government program covering all Hong Kong citizens. The program provides free vaccination against the new crown to protect public health and enable the community to resume normal operations gradually under the epidemic.

2.3 Japan

Japan is a capitalist country that attaches great importance to human rights and freedoms. Therefore, the Japanese government has adopted a relatively soft epidemic prevention policy in the face of the new coronavirus pneumonia, and its anti-epidemic policy focusing more on alleviating the development of the epidemic. In Japan's epidemic prevention policy network, the government plays a dominant role, while other local public organizations and designated public institutions (such as the Japanese Red Cross) play a supporting role.

To protect the basic human rights and freedoms of the people, most of the anti-epidemic policies introduced by the Japanese government are non-mandatory. The government has focused on adopting measures to prevent infection, such as encouraging people to wash their hands frequently and to wear masks when going out. Under the state of emergency, the government proposed holidays for businesses and schools and called on people to work from home or conduct distance education. In addition, necessary behaviours such as people going out to purchase supplies for life, visiting medical institutions, and maintaining life skills such as appropriate outdoor exercise or walking are allowed. In addition, to alleviate people's anxiety and panic about the epidemic, local public organizations have integrated local resources and set up new coronavirus consultation windows and special lines. People who think they may be suspected of being infected with the new coronavirus can use the consultation for free.

In terms of controlling the spread of the epidemic, the Japanese government restricted different types of restaurants, large shopping malls, supermarkets, museums, cinemas, and other public places to shorten or close their business hours. At the same time, the government also provided the people with certain life assistance and various support materials needed for health management to maintain the normal life of the people. Relevant Japanese laws also required that some people from some countries or regions were prohibited from entering Japan, nucleic acid testing and isolation measures were required for those entering Japan, and comprehensive disinfection measures were also required for entry and exit aircraft. Regarding confirmed cases, the Japanese government encourages hospitalization for severe cases and home treatment for mild cases and discourages asymptomatic people from being tested for the new coronavirus. As the global epidemic continues to worsen, since the development of the new coronavirus pneumonia vaccine, the government has begun to encourage people to be vaccinated and continue to expand the population of vaccines.

According to the plan of the Japanese government, Japan has now fully opened domestic tourism locally. In addition, the Prime Minister of Japan officially announced in his speech on the evening of May 26, 2022, that starting from June 1, the Japanese government will increase the number of people entering the country from 10,000 to 20,000 per day, and greatly relax short-term entry visa application restrictions. From June 10, Japan will reopen tourist entry, but it will be limited to tour groups led by professional guides for the time being.

2.4 United Kingdom

Compared with other countries, the UK's initial responses towards the COVID are relatively slow and ineffective. Haven't been alarmed by COVID's devastating effects on China and Italy, UK's government didn't take the COVID issue seriously in the beginning. Instead, blame games were rife in the council, and the COVID was just viewed as another influenza (Pegg, 2020). Worse still, when the policymakers finally looked at the COVID issue, there was a systematic failure in procuring the personal protective equipment (PPE), which led to a period of paralysis in the UK's healthcare system (Gaskell et al., 2020). Taking a glimpse of the UK's COVID policies, one of the most contentious issues is the "mitigation strategy," namely using the natural rule of "the survival of the fittest" to cut off the population that is too vulnerable to survive and retaining those who are robust enough to withstand the infection. This kind of "herd immunity" notion is a stark contrast to the strict strategies taken by other countries, which aim to eliminate the virus.

Tracing the UK's COVID policies since the pandemic outbreak, it was found some intriguing changes in the UK's attitude. The UK first was rather confident in the face of the virus and deemed that it was just an ordinary epidemic, and the government was well-prepared for its challenges. Nonetheless, with the situation evolving, the British council started to notice the seriousness of the COVID. Confronting the skyrocketing infection and death toll, parliament passed the Coronavirus Act in March 2020, legitimizing the government's emergency power in handling the COVID. Thereafter, sanctions upon public gatherings were introduced, and the school and businesses lockdowns became mandatory. Moreover, wearing masks became legally obligatory in public spaces such as underground, and the COVID free tests were available for the public, along with vaccination. Through the implementation of a series of anti-COVID measures, the UK's COVID infection rate dropped effectively, which led to the third phase of the UK's COVID policy. Considering the socioeconomic damages caused by mass shutdowns and testing, Boris Johnston lifted all the COVID restrictions and announced a "living with COVID" plan. Currently, those who tested positive for the COVID are no longer mandated to self-isolate. Additionally, the free COVID tests are gradually cancelled, and the mandatory mask-wearing rules are eased by changing the word "required" to "strongly recommended (*BBC*, 2022a, 2022b)." In sum, we can see a reverse "U" shape in British COVID policymaking. Passive and reactive at first, then moved to progressive when the situation aggravated, and generally become laissez-faire again when the pandemic is under control.

Undeniably, UK's COVID policy is a classic example of the "living with COVID" strategy. Although the UK government has taken some initiatives during the spike of the pandemic, the overall approaches are still non-interventional. This feature is evident in the UK's judgment in whether they have moved to the "safe zone" of COVID infection numbers. Even though the UK's infection rate is staggeringly higher than China's (around 30,000 per day when Boris proposed the "living with COVID" plan), the country still considered itself to be relatively safe, which is impossible in those areas embracing "zero COVID" policies.

2.5 Singapore

As an island city-state accommodating 5.6 million people, Singapore's limited territory and small population allow its policymakers to devise rigorous and clear-defined policies. Being an archetypal "soft authoritative" regime, Singapore has responded swiftly to the COVID challenges and implemented a comprehensive policy suite. Drawing experience from combating SARS, the Singaporean government urged people to close schools and public venues, scanning people's body temperature when they enter public spaces, reinforcing the mandatory quarantine schedules, and so forth. Similar to other countries like South Korea and China, Singapore utilizes technological tools to track the spread of COVID. Developed by the Government Technology Agency of Singapore (Govtech) and the Health Ministry, the TraceTogether app helps monitor individuals' movement using Bluetooth signals (Woo, 2020). Besides, citizens are required to scan the QR code when using the public spaces, facilitating the process of contact tracing of infected masses.

Instead of the effective contact tracing measures, Singapore is best known for its low fatality rates of COVID with the robust support of its national healthcare system. The high resilience and efficiency of the Singaporean health system are contributed by the three-pronged strategies taken by the government. That is, reducing the imported cases, acting pre-emptively by detecting and isolating the infected individuals, and underscoring the social responsibility (Chua et al., 2020). In sum, with its natural endowments of a relatively small scale of land and population and its policy endeavour, Singapore scores high in curbing the spread of the virus at first and is still among the list of countries that have the least COVID fatalities.

However, even if Singapore successfully implemented the "zero-COVID" plan initially, it was defeated by the reality of surging cases of COVID soon, which propelled it to modify its policies. The very reason for this failure may lie in the negligence of the migrant workers at first, which is a typical case of a "black elephant" incident, namely the emergence of an unexpected event due to some previously cognizant problems unwilling to address (Ho, 2008). Unlike the well sanitary conditions and health welfare enjoyed by the citizens, the living conditions and healthcare were abysmal for the foreign workers in Singapore. The cramped and sordid slums inhabited by migrant workers are breeding grounds for COVID, and that's where the government overlooked initially. Consequently, the COVID cases soared in the foreign workers' dormitories, making Singapore one of the countries severely affected by the COVID infection. In closing, the analytical failure in omitting the potential risks in migrant workers' dormitories causes Singapore to have a high COVID infection rate despite its initial claim of maintaining zero infection.

On October 9, 2021, Lee Hsien Loong, Prime Minister of Singapore, formally announced that the Singapore government would stop its zero-infection policies rather than be paralyzed by the fear of COVID. When classifying Singapore regarding COVID policies, its transition from the "zero COVID" measure to "living with COVID" facing the harsh reality should be noticed. With the spread of the highly infectious Omicron variant and the wide coverage of vaccination, which accounts for 86.6% of Singapore's entire population (Our World in Data, 2022), the country is now considering a transfer from a zero-COVID nation to a COVID resilient nation (Rising & Liang, 2021). The endemic mindset embraced by the Singaporean government is proof that Singapore has accepted the truth that the virus can't be eliminated in the short term, and the measures taken to combat COVID must be sustainable. In summary, Singapore's COVID policy is marked by a transition in attitude facing COVID, from complete elimination to long-term coexistence. Yet the prime minister of Singapore announced the Singapore government would change the policy. This would be further examined in the next chapter.

2.6 Denmark

Belonging to the social democratic welfare regime, Denmark has outperformed its counterparts in handling the COVID. Even its official health system Danish Health Authority (DHA) mis-predicted that the COVID threat was minimal to Denmark in January 2020, the Danish government set up its containment strategies after noticing the severity of COVID in February 2020. Nonetheless, as the containment strategies fell flat due to the skyrocketing infection rates, the government swiftly replaced its containment strategies with repression strategies, imposing rigorous restrictions on social activities (Laage-Thomsen & Frandsen, 2022). Spearheaded the decisive acts towards COVID in Europe, Denmark unhesitatingly shut its borders and announced a national lockdown. To curb the spread of the virus, Denmark prohibited mass public assemblies, shut all needless public venues down, and exhorted citizens not to use public transportation and abstain from traveling, except in cases of emergency (Marinov, 2020). With the effective implementation of the strategies, Denmark has successfully flattened the curve of infections and reduced the hospitalization rate. This triumph is a stark contrast to other Nordic states like Sweden, which failed to act promptly to the COVID challenge and the public's trust in government became the casualty (Nielsen & Lindvall, 2021).

Due to the efficacious earlier radical interventions and a high vaccination rate of around 81.3% (Our World in Data, 2022), Denmark now lifts all the COVID restrictions and declares that COVID is no longer a socially critical disease. The national COVID pass app becomes not compulsory to show when entering public spaces, and the businesses and schools generally resumed in the country. According to Danish officials, as long as the new variant Omicron doesn't burden the nation's hospital capacity, they can live peacefully with the virus (Olsen, 2020).

Despite its early radical strategies to contain the spread of the virus, Denmark now embraces the "living with COVID" strategy because of some socioeconomic considerations. The critical concern of the country now is not the COVID infection rates but the number of people in ICU being infected by COVID, which takes a toll on the nation's hospitalization resources.

3 Comparison of Measures of COVID-19 Control Policies Among the Six Selected Countries and Region

By comparing the anti-epidemic policies of mainland China, Hong Kong, Japan, Singapore, the United Kingdom, and Denmark. We can find that both China and Singapore[5] adopted a "zero COVID" policy at the beginning of the outbreak. With the continuous control of the epidemic, China gradually changed to a "dynamic zero-clearing" policy, while Singapore has gradually transformed into a policy of "living with COVID" due to the intensification of the epidemic due to re-judgment. Although Hong Kong SAR's response to the epidemic prevention and control seems to be relatively slow, the main theme of its epidemic prevention policy is still following the Chinese government's "dynamic zero-clearing" policy. Both the United Kingdom and Denmark adopted a series of mandatory containment policies at the beginning of the epidemic, but due to the variation of socio-economic factors, they eventually turned into a "living with COVID" policy. Japan has always adopted a soft policy on COVID-19, that is, it focuses more on mitigating the development of the epidemic rather than actively eliminating it. To sum up, we can divide the anti-epidemic policies of the above six countries into three categories: "dynamic zero-clearing", "living with COVID" and mitigation policies for comparative discussion (see Table 1).

3.1 "Dynamic Zero-Clearing" Policy

The "dynamic zero-clearing" policy is an active and assertive policy to control the development of the epidemic. The policy is mostly mandatory. When the government finds out about the epidemic, it will quickly block the epidemic area and close or limit the business hours of public places. At the same time, the government will also require companies and schools to move from offline to online, and people are isolated at home and cannot go out. The government will also use large-scale free nucleic acid combined with big data technologies such as itinerary codes, health codes, and vaccine passes to detect confirmed cases in a timely manner, and quickly send them to designated hospitals or makeshift hospitals for treatment by medical treatment experts according to different conditions. In terms of preventive measures, the government will require people to be vaccinated, wear masks to travel, and take nucleic acid testing and isolation measures for those entering the country.

[5] It is found that Singapore have changed the policy of "Zero COVID" to "Living with COVID" as announced by their prime minister in in March 2022 in a nationally televised address titled "Into a New Phase", such changes will be examined in the following chapter.

Table 1 Comparison of three types of anti-COVID-19 control policies

Type of anti-COVID-19 policies	"Dynamic zero-clearing"	"Living with COVID"	Mitigation policies (soft policy)
People gathering management	Block off the epidemic area, control the dining time of restaurants, close all kinds of public places, work from home, online teaching (mandatory)	Open public places, businesses and schools resume normal offline operations, it is recommended to wear masks in public places (non-mandatory)	Shorten business hours in public places, encourage people to work from home, and teach online (non-mandatory)
Treatment of diagnosed patients	Quickly send them to designated hospitals or makeshift hospitals according to the condition, set up a medical treatment expert group	Many confirmed cases do not receive timely medical treatment	Encourage severe cases to be hospitalized and mild cases to be treated at home
Nucleic acid testing and vaccination	Organize large-scale free nucleic acid testing and require people to be vaccinated	No mandatory requirements for nucleic acid testing and vaccination	Asymptomatic people are not encouraged to undergo nucleic acid testing, and the public is encouraged to be vaccinated
Immigration management	Nucleic acid testing and isolation measures	/	Prohibit entry of people from some countries, and implement nucleic acid testing and isolation measures
Data monitoring and tracking	Itinerary code, health code, vaccine pass	/	/

3.2 *"Living with COVID" Policy*

The "living with COVID" policy is a policy that the government does not impose mandatory requirements on epidemic prevention and control based on the premise that the symptoms of virus infection are not too severe, and the people are generally immune. In this policy, the government opens public places and allows businesses and schools to conduct normal offline activities. For the public, it is encouraged but not mandatory to wear masks in public places. As for the treatment of confirmed cases, due to many infected people in these countries, many confirmed cases cannot receive timely medical treatment. And to protect the rights and freedoms of the people, the government will not force the people to undergo nucleic acid testing and vaccination.

3.3 Mitigation Policies

The main prevention and control purpose of mitigation policies is to control the spread of the epidemic, delay its spread, and clear the overall harm. The policy focuses on relatively mild measures, such as the government will shorten the operating hours of public places, and non-mandatory encouragement of people to work from home and online teaching to reduce gatherings. Different from "dynamic zero-clearing" and "living with COVID", the policy encourages hospitalization for severe cases in the treatment of confirmed cases, while home treatment for mild cases. The government encourages people to get vaccinated but discourages asymptomatic people from nucleic acid testing.

4 Discussion and Considerations in the Choice of COVID Control Policies

4.1 Will the Public Health and Welfare Ideologies and Regime Affect the Choice of COVID Control Policies?

There are various underlying reasons for different COVID strategies taken by the afore-mentioned countries. First, we consider their respective public health and welfare ideologies and regime contribute to their divergent reactions to COVID control strategies. According to Esping Anderson (1990), UK and Denmark belong respectively to Liberal and Social Democratic welfare regimes, that's to say, UK's passive COVID policy and its strong emphasis on individual responsibility are evident manifestations of the residual welfare approach that the country has taken for a long time. The notion of "herd immunity" and the unhesitant embrace of "living with COVID" policies both eminently show the UK's entrenched belief in minimal government intervention and the self-regulating power of society. In contrast, Denmark's vigilant response to COVID and the following resolute anti-COVID measures implemented in the state and community levels reflect the country's adherence to the Social Democratic welfare ideology. Assuming the power of government intervention and a universal citizen right, the Danish government decidedly carried out a series of lockdown strategies at first, then immunized its citizens through mass vaccination. Although the country has now accepted the "living with COVID" strategy, it's based on the premise of universal immunization and the less severe symptoms of Omicron. In sum, Denmark's strategies show the characteristics of the Nordic welfare state.

For the remaining Asian countries, we recognize that they belong to the conservative welfare regime according to Aspalter's (2006) classification. The emphasis on the redistribution concerning the immaterial resources, which involve public health issues, leads those countries to react relatively proactively to the COVID challenge.

Nonetheless, there are still nuanced differences among those countries. As a country traditionally underlines state control, China confronts COVID under the direction of the central communist government. The public's obedience and trust towards the government facilitate the anti-COVID process to some extent. On the other hand, Hong Kong SAR, as a special administrative region in China, doesn't have a high level of public trust in government. Consequently, the implementation of COVID policies in Hong Kong SAR doesn't go as smoothly as their counterparts in the mainland. Singapore, being renowned as governed by a "soft authoritarian" government, was the exemplar in handling COVID in the beginning. However, it has also been trapped by the pitfall of neglecting the foreign workers' dormitories due to its administrative limitations. Japan, standing out of other Asian countries with its vague COVID attitude, tend to conceal the COVID cases.

In summary, the welfare regime and ideologies can offer some explanations of the divergence among different countries' COVID policies.

4.2 Economic Considerations and Governance

With the situation unfolding, many countries that initially uphold the "zero COVID" approach or adopt stringent COVID policies are transforming to embrace "dynamic zero-clearing" or "living with COVID" strategies. This transition can be accounted for by the socioeconomic burden posed by the pandemic and the governance challenges resulting from the sky-high infectivity of the virus. Pressured by the continuous spikes in COVID cases and the rise in asymptomatic infections, China and Hong Kong now accept a "dynamic clearing" measure in handling COVID. Nevertheless, the huge costs induced by massive COVID free tests and lockdowns still pose tremendous challenges in these two areas. Aware of the fallouts draconian COVID strategies resulting in the economy and national treasury, some countries start to treat COVID as an endemic disease and scrap their strict COVID policies. Among those states that announce the "living with COVID" plan, Denmark and Singapore have already achieved a high vaccination rate, allowing them to loosen their COVID regulations while not burdening their hospital capacity. Nonetheless, some liberal countries like the UK choose to live with COVID regardless of its still surging infection cases. In stark contrast to other countries, Japan holds a quite ambiguous attitude towards COVID, which may be resulted from the government's will in maintaining social order.

4.3 *Reflection of the COVID Control Policies from the Perspective of the Institutional Theories*

4.3.1 Governance

Through the comparison above, we can find from the macro level that different countries have different foci when adopting epidemic prevention policies. Among them, some countries put the economy first in the face of epidemic prevention and control, and all epidemic prevention and control policies must be implemented based on protecting and maintaining economic development and stability. From the perspective of historical institutionalism, this is due to the specific historical background of these countries. The United Kingdom, the United States, and other Western liberal countries advocate the concept of "freedom" and the "democratic" institution, so they pay more attention to the economy and emphasize the capital of laissez-faire development. And after the outbreak of the epidemic, the economies of the United Kingdom and European countries have shrunk significantly, leading them to choose to put economic development and stability first under enormous economic pressure. On the contrary, China has prioritized social stability and community governance in the face of epidemic prevention and control. From the perspective of historical institutionalism, China belongs to a socialist system. The government adheres to the principle that the safety of the people always comes first, controls the spread of the epidemic at all costs, and resolutely eliminates the epidemic everywhere. This also prompted China to choose the "dynamic zero-clearing" epidemic prevention policy. Of course, the price of China's choice to put social stability in the first place is that the long-term strict closed management has also caused a certain impact on economic development. But, through strict closed management, China quickly controlled the development trend of the epidemic in a relatively short period and at a low cost, while maintaining social stability, enabling China to develop based on sound social and economic development. It can quickly recover from the epidemic and develop gradually and steadily.

At the micro level, we can find that the epidemic prevention and control policies adopted by governments is different between intra-cities and inter-cities. In the event of an outbreak, there will be many tests and boundaries between many cities to limit people's movement between cities, and there will be similar tensions and strict boundaries even between different districts within the same city. Regarding China's epidemic prevention and control policy, there is a voice that China's epidemic prevention policy is a bit excessive, and in many cases, there is no need to restrict people's activities or even lock down the city. But, according to rational choice institutionalism, the Chinese government attempts to seek appropriate methods to control the development of the epidemic from professional experts and scholars, to formulate efficient countermeasures. China's expert think tank said that the way to control the development of the epidemic quickly and effectively is to quickly find the source of infection, find out and cut off the chain of transmission. Therefore, the Chinese government will detect the epidemic in a timely manner through strict testing measures during the daily epidemic monitoring stage. When an epidemic occurs, by setting boundaries between districts and cities in the city to limit the scope of people's activities, the scope of the spread of the epidemic can be quickly reduced and controlled, thereby reducing the negative impact of the epidemic on social stability. At the same time, to minimize the impact of the epidemic on people's daily lives, people have gradually raised their awareness of self-protection and always remain nervous about outsiders.

4.3.2 Citizen Participation and Engagement

In epidemic prevention and control policies, we not only see the impact of the role of the government on policy formulation, but we can also use sociological institutionalism to analyze its impact on policy formulation from the perspective of citizen participation.

Although Hong Kong has now been turned to mainland China for help with epidemic prevention and control and has gradually adopted the strategy of "dynamic zero-clearing" in terms of epidemic prevention and control policies. Yet, many of China's epidemic prevention policies have not been well implemented in Hong Kong SAR. The most important issue is the role of the resident association in epidemic prevention and control. In mainland China, we can find that the residents of almost every community are very enthusiastic and have a strong sense of self-protection. They will all spontaneously participate in the work of epidemic prevention and control as volunteers, and strictly control the entry and exit of the community based on the health code. Therefore, in mainland China, the role of citizens is relatively strong in participating in epidemic prevention and control. On the contrary, Hong Kong SAR has been influenced by the history of British rule and attaches great importance to personal freedom. Therefore, it is morally unacceptable for Hong Kong citizens to enforce quarantine and limit the scope of activities. Therefore, the policy of resident association to participate in epidemic prevention and control work has almost been discarded in Hong Kong. In addition, due to the difference in the structure of the administrative system, the resident associations in mainland China belong to the administrative system and have certain administrative capabilities. And because it is composed of the power of the residents, it has greater power in the community. In Hong Kong, the resident association does not belong to the administrative system of the government, so it also has the administrative capacity entrusted by the government. In addition, the community in Hong Kong lacks a certain power of neighbourhood, so the implementation of many things cannot be realized by the resident associations. Therefore, during the epidemic, we can often find that when Hong Kong closes the management community, it is basically the police, not the resident associations.

5 Conclusion

The COVID-19 epidemic that has lasted for three years is the common enemy of people all over the world. Every country has some successful anti-epidemic experiences that can be used for reference by all countries in the world. In this article, we conducted research on six countries or cities, including mainland China, Hong Kong SAR, Japan, Singapore, the United Kingdom, and Denmark from the aspects of personnel gathering management, treatment of confirmed patients, data monitoring and tracking, entry and exit management, nucleic acid detection and vaccination, etc. Analysis and comparison of epidemic prevention policies. We divide the epidemic prevention policies of these six countries into three categories: "dynamic zero-clearing", "living with COVID" and mitigation policies. We put forward new thinking on epidemic prevention and control policies by analysing different policies and the values behind them, to improve the coping capabilities of countries in the face of the challenges brought by the epidemic, and finally jointly achieve a great victory in the fight against the COVID-19 epidemic.

References

Aspalter, C. (2006). The East Asian welfare model. *International Journal of Social Welfare, 15*(4), 290–301.

BBC News. (2022a). Denmark COVID restrictions lifted despite increase in cases. *BBC News.* https://www.bbc.com/news/world-europe-60215200

BBC News. (2022b). What are the UK's 'Living with COVID' plans? *BBC News.* https://www.bbc.com/news/explainers-52530518

Cairney, P. (2021). The UK government's COVID-19 policy: Assessing evidence-informed policy analysis in real time. *British Politics, 16*(1), 90–116.

Chan Ho-Yin, C. A.-N. (2021). COVID-19, community response, public policy, and travel patterns: A tale of Hong Kong. *Transport Policy*, 173–184. https://doi.org/10.1016/j.tranpol.2021.04.002

Chen, S. W. (2022, February 15). *The new crown epidemic is severe, and Carrie Lam announced a number of new epidemic prevention measures at the press conference.* CCTV. https://news.cctv.com/2022/02/15/ARTIjqGTKV5yWsi0TxNxPIKe220215.shtml

Chua, A. Q., Tan, M. M. J., Verma, M., Han, E. K. L., Hsu, L. Y., Cook, A. R., ... Legido-Quigley, H. (2020). Health system resilience in managing the COVID-19 pandemic: Lessons from Singapore. *BMJ Global Health, 5*(9), e003317.

CNA. (2022, March 24). *Watch: PM Lee announces relaxation of Singapore's COVID-19 measures in national address, ministerial taskforce gives details.* CNA. https://www.channelnewsasia.com/singapore/live-pm-lee-hsien-loong-national-address-COVID-19-march-24-singapore-2581356

DiMaggio, P. J., & Powell, W. W. (1983). The iron cage revisited: Institutional isomorphism and collective rationality in organizational fields. *American Sociological Review*, 147–160.

Esping-Andersen, G. (1990). *The three worlds of welfare capitalism.* Princeton University Press.

Gaskell, J., Stoker, G., Jennings, W., & Devine, D. (2020). COVID-19 and the blunders of our governments: Long-run system failings aggravated by political choices. *The Political Quarterly, 91*(3), 523–533.

Hakim, A. J., & Victory, K. (2021). Mitigation policies, community mobility, and COVID-19 case counts in Australia, Japan, Hong Kong, and Singapore. *Public Health.* https://doi.org/10.1016/j.puhe.2021.02.001

Hall, P. A., & Taylor, R. C. (1996). Political science and the three new institutionalisms. *Political Studies, 44*(5), 936–957.
Ho, P. (2008). *Governing at the leading edge: Black Swans, Wild Cards, and Wicked Problems.* Prime Minister's Office.
Hong Kong Government. (n.d.). New rules to reduce gatherings. *COVID-19 Thematic Website – Together Against the Virus – Home* https://www.coronavirus.gov.hk/chi/reduce-gatherings.html
Jackson, M., & O'Connor, M. (2022). COVID: Face mask rules and COVID passes to end in England. *BBC News.* https://www.bbc.com/news/uk-60047438
Laage-Thomsen, J., & Frandsen, S. L. (2022). Pandemic preparedness systems and diverging COVID-19 responses within similar public health regimes: A comparative study of expert perceptions of pandemic response in Denmark, Norway, and Sweden. *Globalization and Health, 18*(1), 1–18.
Latest Situation of Coronavirus Disease (COVID-19) in Hong Kong. (2022). COVID-19 thematic website. https://chp-dashboard.geodata.gov.hk/COVID-19/zh.html
Liang. W. N. L. M. (2022, January). The "Dynamic Clearing" Strategy for the Prevention and Control of the Novel Coronavirus Pneumonia in my country. *Chinese Medical Journal*, pp. 239–242. https://kns.cnki.net/kcms/detail/detail.aspx?dbcode=CJFD&dbname=CJFDZHYX&filename=ZHYX202204002&uniplatform=NZKPT&v=74lwvkc-Vi0f20-vB4Tt78ogqKPfjWMTY09a0A-Y0XQPImuc9fg41lHbEXuhpOIi
[LIVE] PM Lee, task force on latest COVID-19 updates|Oct 9, 2021. (2021). Youtube.
Mahmud, R. (2017). *Understanding institutional theory in public policy.* University of Bergen, Bergen, Norway.
March, J. G., & Olsen, J. P. (2005). Elaborating the "new institutionalism". In *The Oxford handbook of political institutions.*
March, J. G., & Olsen, J. P. (2010). *Rediscovering institutions.* Simon and Schuster.
Marinov, M. (2020). Denmark and COVID 19. BSR Policy Briefing Series. *Centrum Baltic*, 2–12.
Naoyuki Yoshino, N. H. (2020). The COVID-19 crisis: Policy recommendations for Japan. *DE GRUYTER.* https://doi.org/10.1515/ev-2020-0017
Nielsen, J. H., & Lindvall, J. (2021). Trust in government in Sweden and Denmark during the COVID-19 epidemic. *West European Politics, 44*(5–6), 1180–1204.
Olsen, J. M. (2022). Denmark scraps most COVID-19 restrictions. *Northern Beaches Review.* https://www.northernbeachesreview.com.au/story/7603774/denmark-scraps-most-COVID-19-restrictions/?cs=5
Our World in Data. (2022). *Coronavirus (COVID-19) Vaccinations.* Our World in Data. https://ourworldindata.org/COVID-vaccinations?country=OWID_WRL
Pegg, D. (2020). *COVID-19: Did the UK government prepare for the wrong kind of pandemic?* https://www.theguardian.com/world/2020/may/21/did-the-uk-government-prepare-for-the-wrong-kind-of-pandemic
People's Daily. (2020, September 2). *Hong Kong officially launched the universal community testing program – "Let Hong Kong overcome the epidemic as soon as possible".* Central People's Government of the People's Republic of China. http://www.gov.cn/xinwen/2020-09/02/content_5539233.htm
Rising, D., & Liang, A. (2021). Singapore strategy of living with COVID raises concern, hope. *abc NEWS.* https://abcnews.go.com/Health/wireStory/living-COVID-singapore-strategy-raises-concerns-hope-80316695
Scott, W. R. (2013). *Institutions and organizations: Ideas, interests, and identities.* Sage Publications.
State Council Information Office of the People's Republic of China. (2020). *China's action to fight the new coronary pneumonia epidemic.* http://www.scio.gov.cn/zfbps/32832/Document/1681801/1681801.htm
Terry Lum, C. S. (2020, May). COVID-19 and long-term care policy for older people in Hong Kong. *Journal of Aging & Social Policy*, 373–379. https://doi.org/10.1080/08959420.2020.1773192

The Government of Hong Kong Special Administrative Region. (2022a, September). *"COVID-19 Vaccination Programme"—About the Programme*. Government of Hong Kong Special Administrative Region. https://www.COVIDvaccine.gov.hk/zh-CN/programme

The Government of Hong Kong Special Administrative Region. (2022b, September). *Government tightens social distancing measures in view of changes in epidemic situation*. The Government of Hong Kong Special Administrative Region Press Releases. https://www.info.gov.hk/gia/general/202202/09/P2022020900022.htm?fontSize=1

Tokyo Online. (2022, May 26). *Japanese official announcement! From June, Chinese people entering Japan do not need to be quarantined; from June 10, Japan will restrict the opening of tourist visas*. https://www.163.com/dy/article/H8B597J60517BACO.html

Tu, Y. F. (2021, April). Customs clearance and clearing in the distant future: Analysis of the current situation of the epidemic in Hong Kong. *Compilation of Global Health and Medical Trends*.

Wang, T. D. L. (2021). Strategy adjustment in foreign prevention and control of COVID-19 and the related focus in China. *Shanghai Preventive Medicine, 33*(12), 1121–1126. https://doi.org/10.19428/j.cnki.sjpm.2021.21569

Woo, J. J. (2020). Policy capacity and Singapore's response to the COVID-19 pandemic. *Policy and Society, 39*(3), 345–362.

Xiaohan Wang, L. S. (2021). Policy disparities in fighting COVID-19 among Japan, Italy, Singapore and China. *International Journal for Equity in Health*. https://doi.org/10.1186/s12939-020-01374-2

Xu, Z., H. L. (2021, August). The Chinese Model of Novel Coronavirus Pneumonia Epidemic Prevention and Control: Summary and Reflections. *Chinese Journal of Infectious Diseases, 39*(8), 455–460. https://doi.org/10.3760/cma.j.cn311365-20210208-00049

Kam Tong Chan Associate Professor in Social Work and Social Administration, BNU-HKBU United International College. He is a Registered Social Worker and was awarded Top 10 Outstanding Social Worker Professional in 2012 by the National Ministry of Civil Affairs of the People's Republic of China. Dr. Chan's academic interests focused on non-profit organization, social service administration, social work supervision, and social enterprises and social innovation. He oversees many research projects funded by Hong Kong SAR Government, Mainland Chinese Government, and Korean institutions. Dr. Chan is also author of many research publications on comparative studies of social enterprises, social management, and social work.

Xinyu Yi is from Beijing Normal University - Hong Kong Baptist University United International College. Yi majors in Social Work and Social Administration, and her research interests are in social policy and social work practice.

Xiaoman Yang studies at the Beijing Normal University-Hong Kong Baptist University United International College. Her research focuses on sociology, social policy, social work, and applied social data science. Her recent publication includes *Illegal Gambling Behavior Among Elderlies in a Chinese Village: a Social Work Perspective*.

Between Zero COVID and "Live with COVID:" Comparative Studies in Pandemic Control Policies of China, Singapore, and Hong Kong at the Stage of Omicron

Simon X. B. Zhao, Bo Yan, and Yutong Chen

1 Introduction

This chapter is a delineation of how three places, namely, (Mainland) China, Singapore, and Hong Kong, have navigated the global pandemic of COVID-19 up to the stage when the Omicron variant have become the dominant variant.

Since the pandemic, there gradually developed two major pandemic control strategies: zero COVID strategy and "live with COVID" strategy. China has resorted resolutely to a zero COVID strategy, seeking the elimination of positive COVID-19 cases. When necessary, Chinese cities have quickly conducted mass testing upon millions of residents and imposed city-wide lockdowns. The effectiveness of China's zero COVID strategy in the first two years of the pandemic was, however, called into question following the Shanghai lockdown in 2022 and the subsequent economic fallout. Singapore imitated China in adopting a zero COVID strategy during the initial months of the pandemic. Exemplified by the circuit breaker (April–June 2020), Singapore also imposed various typical zero COVID restrictions, such as closure of border, mass testing, and nation-wide lockdown. In view of the highly infectious, far less lethal Delta variant, and the availability of vaccines, Singapore swiftly switched to live with COVID. Tolerating increases in positive cases, Singapore focused on reducing severe and death cases and resumption of economic and social activities. The government of Hong Kong SAR, with the purpose of re-opening with Mainland China out of inextricable economic connections, followed a China-style zero COVID strategy. After weathering the first four waves of the pandemic, Hong Kong however, found itself struggling in the fifth.

S. X. B. Zhao (✉) · B. Yan · Y. Chen
Beijing Normal University-Hong Kong Baptist University,
United International College, Zhuhai, China
e-mail: sxzhao@uic.edu.cn

© The Author(s), under exclusive license to Springer Nature Singapore Pte Ltd. 2023
S. X. B. Zhao et al. (eds.), *Comparative Studies on Pandemic Control Policies and the Resilience of Society*, https://doi.org/10.1007/978-981-19-9993-2_2

This chapter benefits from a growing body of research articles and media coverage dedicated to evolving situations of COVID-19 for the past two years. In addition, this chapter is enlivened by first-hand experiences by its authors. The authors of the chapter happened to meet head-on with various control/quarantine measures in the said places: China, Singapore, and Hong Kong. One author was in Singapore throughout the entire circuit breaker period from April to June 2020 and the transitional phase after Singapore loosened its restriction measures. Another author was trapped in Hong Kong during the tumultuous fifth wave of the pandemic. He and his family has made strenuous efforts navigating COVID-related red tapes in Omicron-hit Hong Kong. His documentation of his experience is of value of reference in an understanding of how doggedly adherence to zero COVID in Hong Kong had inflicted inconvenience and exhaustion upon cross-border travellers.

This chapter attempts to answer the following questions: *What lies behind the unique effectiveness of China's zero-COVID Policy in its first two years of fighting the pandemic? What has Singapore done right? In contrast, what may explain Hong Kong's fifth wave of the pandemic?* This chapter explains Hong Kong's struggle in the fifth wave of the pandemic: unlike China, Hong Kong is incapable of implementing the zero COVID strategy; compared with Singapore, Hong Kong has lacked a scientific preparedness regarding the pandemic in aspects of vaccination, medical authorization, protection of the elderly and children, and economic normalization. The chapter selects the three cases for their representativeness. China is the world's most prominent and the last holdout of zero COVID nation, distinguishing itself from the rest of the world. Singapore is among the first countries to switch from pursuing zero COVID to live with COVID. Its pandemic control path represents and sheds lights upon many nations ending zero COVID to bring economy and social lives back to normal. Hong Kong establishes as an ideal observatory sample of how the once effective zero COVID strategy, embodied by mass testing and strict lockdowns, had difficulty in being applied outside China and in coping with the pandemic in the age of Omicron. In the context that the Omicron variant has called for rethinking over suitability of pandemic control strategies in different social settings, this chapter is meaningful both academically and practically. It further debunks any concept of "best," "one-size-fits-all," contextless pandemic control policy or policy as "doctrine."

2 Analytical Results

2.1 What's Behind the Effectiveness of China's Zero COVID Strategy in the First Two Years of the Pandemic?

Since the onset of the coronavirus outbreak in Wuhan at the turn of 2020, China has resorted resolutely to a strategy named "zero COVID" to combat the pandemic. This zero COVID strategy, as the title suggests, seeks the elimination of confirmed cases. The strategy included mass testing to find out positive cases, tracing the trajectory of confirmed cases and their close contacts and put them into mandatory quarantine; in rare cases, the strategy also imposed lockdowns or even closure of border.

China's zero COVID strategy turned out to be very effective when it came to the number of confirmed positive cases. Within just three months since the outbreak, China has suppressed further infections of coronavirus and achieved economic growth. Following China's step, many countries around the world also adopted zero COVID strategy to various degrees when they first encountered COVID-19, such as New Zealand, Singapore, Brunei, Tonga, Vietnam, as well as Australia. When measured by the number of confirmed cases, the zero COVID countries indeed outperformed countries planning to "live with COVID." Even among these zero COVID nations, China's implementation of zero COVID has demonstrated an unmatched level of effectiveness.

Why is China's zero COVID strategy able to achieve this indisputable effectiveness in the first two years of its implementation? This section would approach the question from the perspective of social prerequisites that underpinned the effectiveness of China's zero COVID policy. The section would unfold from the following aspects: an ability to undergo mandatory testing on a large scale; a capability of utilizing big data, a society that places no value on privacy, an organization by gridding communities, and a vertical administration that could hold underperformed officials accountable and sack them immediately.

An ability to undergo mandatory testing on a large scale. The elimination of the number of confirmed cases requires actively seeking hidden cases before further infections. Thus, compulsory mass testing is conducted in Chinese cities even when only a handful COVID-19 positive cases are detected. China does have the capability to test millions of citizens within a very short time span. For example, to combat a wave of COVID-19 cases outburst in Qingdao, China, the authorities launched a city-wide screening and tested its 10.9 million within five days since three cases were reported on 11 October 2021.[1] The operation eventually identified 9 more cases. In June 2021, Guangzhou, the southern metropolis of China, instantly underwent a blanket testing on its 18 million residents in three days after just one positive case was discovered.[2]

A privacy-sacrificing society. When a positive case was detected in China, an extensive epidemiological investigation (EI) surrounding the case was instantly conducted by local health authorities. The authorities would then publicize travel histories and personal information of the concerned cases. The degree of detailedness of such disclosure is only possible in a privacy-sacrificing society. For instance, when Shanghai, China has its first COVID-19 cases on 21 January 2020, the health authorities immediately implemented a series of EI to identify the cases' close contacts and their precise geographical locations. Because of such precision, the health authorities in Shanghai were able to test the potentially related cases without the need to conduct blanket testing across China's largest city. Only close contacts and residents in the same or sur- rounding communities and workplaces with infected cases were tested

[1] Xing et al. (2020).

[2] Guo (2021).

for the virus. Eventually, only 27,592 individuals were tested, against a roughly 27 million of the total population in Shanghai.

EI in other countries such as Brunei,[3] Qatar,[4] and Japan,[5] EI concentrated on characteristics related to the cases, such as incubation time, median age, symptoms, and percentages of severe or critical illness. However, the spotlight of EI in China, however, was its stress on the confirmed cases' travel histories, activities, and geographical locations they visited. For example, on 14 March 2022, Meihekou, Jilin Province, a Northeastern Chinese city publicized its seventh governmental announcement regarding local COVID cases. In documenting the travel history of one asymptomatic case, the detail-filled public announcement went:

6 March
13:19–13:24 KFC at *Ouya* shopping mall
16:19–16:25 *Ershazi* fruit grocery at *Gaoli* street
17:27–17:33 Jinming grocery store next to *Shuijunhuating* community
17:40 walked back to block 34 at *Chengshijingdian* community
7 March
07:55 picked up kid at *Milan* kindergarten (no entrance)
08:00 went to work at *Dongbeilvfa* corporation at No. 517 *Kangmei* Avenue
09:58–11:04 central municipal hospital (wearing mask throughout the course)
09:58–10:05 scanned copies of medical records at general service desk, central hospital
10:06–10:07 photocopy service, level one
10:08–10:13 scanned copies of medical records at general service desk, level 1
10:14–10:16 waiting area next to cashier counter[6]

The above was only part of a lengthy and blow-by-blow description of travel histories of one asymptomatic case. After publicizing the travelling histories of the cases and locations they visited, the announcement ordered residents to report to local CDC if their travelling routes overlapped with those of the positive cases at the said time. This detail-filled EI served one purpose: to find any potential infected contacts and quarantine them before virus spread out.

Pervasiveness of Big Data. The success of tracking down trajectories of members in the society rests on the pervasiveness of big data in society. WeChat is a prime example of how "health code" was invasively and mandatorily inserted into people's daily lives in China. Today, for most Chinese, WeChat has been used not just for communications, but overwhelmingly for showing COVID testing results or vaccination status when required. Apart from health codes, a travel code is instituted

[3] Wong, Chaw, et al. (2020, October 7).

[4] Al Kuwari et al. (2020).

[5] "Active Epidemiological Investigation on SARS-CoV-2 Infection Caused by Omicron Variant (Pango Lineage B.1.1.529) in Japan: Preliminary Report on Infectious Period" (Japan: National Institute of Infectious Diseases Disease Control and Prevention Center, National Center for Global Health and Medicine, January 5, 2022), https://www.niid.go.jp/niid/en/2019-ncov-e/10884-covid19-66-en.html.

[6] Weixin (2022, March 14).

nation-wide to monitor travel trajectory of a person in China. A travel code documents a person's travel history of the last 14 days. When a person has a history of visiting or passing a city which contains mid- or high-risk areas, the travelling code will be marked by an asterisk. "A history of visiting or passing" is defined as an over 4 h of stay.[7] When a person's travel code is marked by an asterisk, the person is then subject to various travel and movement restrictions. Such restrictions include an incapability to access to hospitals, schools, theatres, shopping malls, public transportation, and other public facilities.

Gridding communities. If there is no imposed and mandatory quarantine for those suspected, then the above-described detailed EI would mean nothing. Not only confirmed cases are quarantined, the close contacts and secondary contacts are also quarantined to further interrupt the transmission. Chen et al. (2020) have documented the China's governmental efforts in combatting the spread of COVID-19 by means of quarantining close contacts of confirmed cases.[8]

To accommodate, provide necessary assistance to, and monitor the quarantined cases, millions of grid organizers (*wanggeyuan* 网格员) across China have played a major role. In addition, throughout the first two years of fighting the pandemic, China's methods of gridding the residential communities have turned out an essential factor laying the foundation of the effectiveness of China's zero COVID strategy. Originating from a 2005 urban management practice in Dongcheng District, Beijing[9] the grid organization was highlighted in the *Decision of the Central Committee of the Communist Party of China on Some Major Issues Concerning Comprehensively Deepening the Reform* at the Third Plenary Session of the 18th Central Committee of the Communist Party of China on November 12, 2013.[10]

As a prime example as well as means of "participation of all sectors of the society…and social self-management and residents' self-management," grid organisers have played expansive roles during the pandemic. The very first duty of grid organisers was to conduct clearance of personal information. They conducted blanket investigation of household registration information of their "grids," closely monitoring population flow, especially those from mid- or high-risk areas. For example, Chen Xingyi, a grid organiser in Guanghan, Sichuan Province, recounted that local grid organisers issued one pandemic control notice to every household, asking them to report to community if any resident had returned from outside.[11] Huang Guiyun, a grid organiser in Qingdao, Shandong Province, recounted that "since the pandemic, we grid organisers all went out without sleep, and rolled out blanket investigation of household information." Huang had to telephone those absent from home to ascertain

[7] Tencent (2022).

[8] Chen et al. (2020).

[9] Wang (2020).

[10] China.org.cn (2014). The translation of "网格化" is "networked" by China.org.cn, but "grid" by LawinfoChina, a Peking University-based database dedicated to translations of legal documents in China. This chapter uses "grid." Readers, however, should note that "networked" and "grid" are used interchangeably regarding this research topic.

[11] Guanghan.gov.cn (2022).

the exact location of the residents and reported the information to the community. Key information that Huang asked during investigations included: how many people in a family, their contacts, places where community residents came back from, their health status, and so on.[12] The second duty is to publicize to residents within their grids required information on pandemic control. For example, grid organisers in Nanhan community, Taiyuan, Shanxi Province disseminated information on self-protection and latest news by way of home visits and social media online.[13] The third duty of grid organisers is to organize and assist mandatory mass testing. In a grid community in Chengyang District, Qingdao, Gao Shihua, a grid organiser was reported to wake up at 4 am to prepare for mass testing. Gao described that the work included testing facilities, handling supplies, setting up tables and chairs, posting slogans, pulling up caution lines, explaining to residents the process of mass testing, helping residents in presenting required health QR code to pandemic control officers, and aiding the elderly and others with special needs.[14] The last major duty of grid organisers is to take care of, and meanwhile to conduct surveillance upon the quarantined. Previously, grid organisers used paper seal and closed-circuit television to contain the movement of the quarantined. They needed to come to the quarantine room to check if the paper seal had been torn down. Later, they resorted to high-tech means. A report from Wenzhou news website would give a glimpse of how grid organisers conducted surveillance by high-tech means. Since December 2021, grid organisers have affixed a white box-like intelligent sensor to doors of homes quarantining those returned to Wenzhou from other places and particularly those from mid- and high-risk areas. Once the door opened, the system would send alert and the pandemic control monitor system would be notified immediately. Watchman would instantly contact grid organiser, and a grid organiser would then rush to scene to check why the door opened on a 24-h basis.[15]

Vertical administrative structure. China possesses a vertical administrative structure which associated the rise and fall of governmental officials directly to their pandemic control performances. Officials were held accountable if their performance in controlling the pandemic was considered unsatisfactory by authorities. In March 2022 alone, at least 26 government officials were dismissed for poor performances in handling the pandemic across cities in Jilin, Shandong, and Guangdong provinces.[16] Party secretary of the Jilin Agricultural Science and Technology University was instantly sacked for negligence of the pandemic in the university, where a cluster infection made the university an epicentre in the latest wave of pandemic in Jilin Province.[17]

Facing a likely purge attributable to loopholes in pandemic control measures, local officials across China tightened social control and restricted human movement

---

[12] Xiao (2022).

[13] Li and Guo (2022).

[14] Xiao (2022).

[15] Wenzhou xinwen (2021).

[16] Liu (2022).

[17] Linghu (2022).

in an often-excessive manner. In August 2021, Zhengzhou, capital city of Henan Province, ordered passengers to have negative results of nucleic test for COVID-19 within 48 h to take public transport. Bus drivers were requested to refuse those who could not demonstrate this 48-h negative results. After the third round of mass testing, the negative results for most residents were expired, residents must then pay to take a test. The 48-h requirement and the cost of test (RMB 75) had scared off most passengers. Consequently, buses ran empty, or just a couple passenger at most.[18] On 20 January 2022, in view of a human movement peak during the Chinese New Year season, Dong Hong, magistrate of Dancheng County, Zhoukou, Henan Province, attracted waves of online criticism by his leaked words. It is seen from the leaked video that Dong declared: "[anyone] returning from mid- or high-risk areas, regardless of vaccination certificates, regardless of nucleic test within 48 h, as long as you return [to Dancheng], you will be quarantined first and then arrested."[19] In an attempt to clarify after the video went viral, Dong himself responded that his remarks were edited and distorted. He emphasized that what he was saying was: "…if you don't obey the pandemic control regulations at provincial, municipal, and county levels, [if you] insist on malicious homecoming regardless of dissuasion, [you will be] firstly quarantined and then arrested. I say this to ensure safety of the masses."[20] The very expression of "malicious homecoming" instantly ignited an even larger bombardment from the internet. On 2 April 2022, Nanhe District of Xingtai, Hebei Province issued an announcement, imposing a de facto lockdown over the entire area even no confirmed cases, suspected cases, nor asymptomatic cases were found.[21] These measures taken by local authorities seemed harsh and radical though, were indeed effective in discouraging movement and congregation, the two prerequisites for virus infection.

2.2 *What Has Singapore Done Right About Switching from Zero COVID to Live with COVID?*

This section will analyse factors contributing to the effectiveness of pandemic control policy in Singapore, "another story of success."[22] By reviewing Singapore's navigation through the pandemic, this section would suggest that scientific preparedness was the key to Singapore's pandemic control efforts. Singapore made full use of the time bought through adopting zero COVID strategy at the initial stage of the pandemic to procure vaccines and roll out vaccination. When the high infection rates

[18] Paper.cn (2021, August 12).

[19] DW News (2022).

[20] DW News (2022).

[21] Weixin (2022, April 2).

[22] Yan et al. (2020).

of the Delta variant made zero COVID unsustainable and the high rates of vaccination significantly reduced the severe and death cases, Singapore swiftly switched to "live with COVID" and brought economic and social lives back to normal.

Singapore did adopt a zero COVID strategy in the initial months of the pandemic. Its response was exemplified by its nation-wide lockdown named "circuit breaker" from 7 April 2020 to 1 June 2020 to pre-empt escalating COVID-19 infections. During the circuit breaker, all restaurants, hawker centres, attractions, museums, sports and recreation facilities, religious activities were closed or suspended. Workplaces were closed except for those providing essential services like food courts and hospitals. Schools were closed, and education was in full mode of home-based online learning. The circuit breaker period also witnessed a parliament bill restricting social gatherings in both private and public spaces.[23] Stay-at-home, home schooling, and work-from-home became norms during the period. Access to markets and groceries were also limited. For example, 4 popular wet markets in Singapore permitted entry on alternate dates, depending on the last digit of visitors' IDs.[24] Restriction measures were gradually relaxed by three phases exiting the circuit breaker. Dropping from the peak of 1,425 recorded on 20 April 2020, the confirmed cases in Singapore remained double-digit throughout the rest of 2020 and the first half of 2021.[25]

Singapore's pandemic efforts, however, did not terminate at eliminating positive cases, nor making the zero COVID measures such as lockdown, closure of border, as well as the abruption of business activities permanent. Singapore bought time for eventual re-opening with this "window period." Intensive efforts were under way for preparation for normalization. One significant aspect of government's efforts during this period was the active procurement of COVID-19 vaccination which laid the groundwork for re-opening. According to a government news briefing, Singapore government had formed a Therapeutic and Vaccines expert panel (TxVax panel) in April 2020 to source for vaccine candidates around the world. The panel was comprised of 18 scientists and clinicians across hospitals, A*STAR (Agency for Science, Technology and Research) and the private sector, conducting an extensive study of 35 promising candidates to screen them for safety and efficacy.[26] The panel was instrumental in monitoring clinical trials, diversifying vaccine options, assessments, and procurement and timely delivery. The Government of Singapore made advanced purchase agreement with major vaccine providers, such as Pfizer-BioNTech, Moderna, and Sinovac. Singapore's efforts in securing the COVID-related vaccines were so quickly that Singapore was just several days behind the United Kingdom and the United States in administering an approved vaccine. The UK became the world's first country to deploy an approved COVID-19 vaccine on 8 December 2020.[27] The Food and Drug Agency (FDA) of the United States issued the first emergency use authorization (EUA) for use of the Pfizer-BioNTech COVID

[23] Gov.sg (2020, April 11).

[24] Gov.sg (2020, April 21).

[25] Ministry of Health (n.d.-b).

[26] Gov.sg (2020, December 23).

[27] Gov.uk (2021, December 8).

19 vaccine in persons aged 16 years and older for the prevention of COVID-19 on 11 December 2020.[28] Singapore approved the Pfizer-BioNTech COVID-19 vaccine on 14 December 2020. With the arrival of Pfizer-BioNTech COVID-19 vaccine on 21 December 2020, Singapore became the very first country in Asia to receive the vaccine. On 3 February 2021, Singapore became the first nation in Asia to approve the use of Moderna vaccine against COVID-19. A fortnight later, Singapore received the first shipment of Moderna's COVID-19 vaccine.[29]

Once COVID-19 vaccines were acquired, Singapore immediately and resolutely initiated rolling out the vaccination to achieve herd immunity. It is noted that while in term of lockdown measures, China was stringent, but Singapore was less so; when it came to vaccination, it was the other way around. Singapore demonstrated a tighter and more mandatory attitude towards vaccination. Although 87% of the population in China were with a complete initial protocol and 89% were partly vaccinated,[30] China did not adopt a mandatory vaccination plan matching the degree that it imposed lockdowns and other restriction measures. Some local authorities did attempt to force residents to get vaccinated. For example, in Wancheng, Hainan Province, the government issued a statement in March 2021 forcing residents to get vaccinated. The statement warned that those unvaccinated residents would be barred from taking public transport, entering public spaces like markets and hotels, and running service businesses; unvaccinated residents would also be blacklisted in government's preferential policy considerations, and their children's education and employment would be negatively affected.[31] Such a statement immediately attracted a public outcry which resulted in the issuing government to retract the statement and offer a public apology.[32] In April 2021, China urged local authorities to halt compulsory coronavirus vaccination. The National Health Commission of China, at the same time, proposed that vaccination against COVID-19 be voluntary and be based upon informed consent.[33] Comparatively, China's vaccination policy was more incentivizing rather than mandatory.[34]

Singapore, however, adopted a more mandatory approach to vaccinating its residents. Although, Singapore authorities maintained that it was not compulsory to take COVID-19 vaccination, in practice, unvaccinated residents would face many inconveniences like refusal of service, as well as denial of entry to certain places. Within seven months since January 2021 when Singapore government kicked off its vaccination programme, 72% of the population had completed full regimen or received two doses of COVID-19 vaccines and 81% had received at least one dose.[35] To further boost residents to get vaccinated, Singapore authorities have instituted various

[28] U.S. Department of Health & Human Services (2022, August 31).

[29] ChannelNewsAsia (2021, February 17).

[30] Our world in data.

[31] NBD (2021).

[32] Paper.cn (2021, March 31).

[33] Huang and Feng (2021).

[34] For example, see 163 (2021).

[35] Poa (2022).

restriction measures. For example, in early October 2021, the authority announced that from 13 October 2021, unvaccinated individuals would not be allowed to enter shopping malls or dine-in at hawker centres and coffee shops, although take-aways foods were allowed.[36] University authorities also gradually tightened vaccination requirements upon university staff and visitors. For instance, on 27 December 2021, the National University of Singapore issued its 40th circular regarding COVID-19 in the year, requiring all its staff and students to declare their vaccination status. Unvaccinated staff who needed to return to the workplace would require a valid negative Pre-Event Testing result from Ministry of Health-approved COVID-19 test provider at their own cost.[37] A month later, the university issued another circular. This time, the university barred all unvaccinated or partially vaccinated staff and visitors from entering the campus.[38] Similar arrangements were also found in the Singapore Management University: the SMU required all the staff and alumni who wanted to access SMU's facilities and services be fully vaccinated; the unvaccinated would be barred from January 2022.[39] Expanding to other social aspects, the Singapore Ministry of Manpower issued the updated "Advisory Guidelines on COVID-19 Vaccination at the Workplace" on 27 December 2021, imposing a vaccine mandate on all employees who wished to return to the workplace from 15 January 2022.[40] As a migration hub for expatriates and workers, Singapore mandated vaccination for all those applying for work passes, long-term passes, and permanent residence in Singapore from 1 February 2022 onwards.[41]

The high vaccination coverage has laid the foundation in making Singapore one of the world's first countries to pivot from zero COVID to live with COVID. Gan Kim Yong, Lawrence Wong, and Ong Ye Kung, heads of Singapore's Multi-ministry Task Force in tackling the coronavirus pandemic, published an article in the *Straits Times* on 24 June 2021. The article, titled "Living normally with COVID-19" became a prelude to Singapore's shift to living with COVID. With vaccination rollout, easier testing and treatment and social responsibility, the ministers predicted that Singapore's response to COVID-19 could be very different.[42] On 9 October 2021, in his speech titled "Protect the Vulnerable, Secure the Future," Prime Minister Lee Hsien Loong announced that a "zero COVID" strategy was no longer feasible, and Singapore had changed the strategy into "living with COVID."[43] The speech marked the official ending of Singapore's zero COVID strategy.

The swift switch was also made possible by scientific knowledge of the Delta variant of the coronavirus. Prime Minister Lee Hsien Loong explained that the emergence of the Delta variant had put Singapore's pandemic control efforts in a changed

---

[36] Co (2021).

[37] National University of Singapore (2021).

[38] National University of Singapore (2022).

[39] Singapore Management University (2022).

[40] Ministry of Manpower (2022).

[41] NDTV (2021).

[42] Gan et al. (2021).

[43] Prime Minister's Office Singapore (2021, October 9).

situation. Further in his explanation of such a change, Lee listed several factors associated with the scientific knowledge of the evolving virus. The first was the emergence of the highly infections Delta variant had made stamping out positive cases through lockdowns and restriction measures impossible. The second was that with vaccination, COVID-19 was no longer a dangerous disease for most Singapore residents. Lee cited the data which showed that more than 98% of local cases had mild or no symptoms, only 2% or less developed more serious illness, and of these, 0.2% died or needed ICU treatment. Regarding the transition from zero COVID to live with COVID, Lee stated the need to update the mindsets, encouraged residents to go about daily lives as normally as possible, and urged the elderly to get vaccinated. "Home Recovery,"[44] stated PM Lee, would be the norm for COVID-19 cases. Again, the speech by PM Lee explicitly revealed that essence of "living with COVID." With the knowledge that the virus caused mainly mild or no symptoms and the availability of vaccination, the COVID-19 had become a "manageable disease," residents were encouraged to not fear the disease, to resume normal activities, the medical system would be relieved of the risk of being overwhelmed, and the government would burden itself by spending on mass testing and cabin hospitals. The annual growth of Singapore's economy was 7.6% for 2021, recovering from a 4.1% contraction the previous year.[45]

With the appearance of the Omicron variant since November 2021, Singapore again resorted to scientific preparedness and adjusted regarding the epidemiological traits of the variant. In view that the Omicron variant came with an even higher infection rate than the Delta variant but caused milder symptoms, PM Lee warned that Singapore might be forced to take a few steps back.[46] Indeed, witnessing a surge in COVID-19 cases occupying 89% of the 1650 isolation beds and 67% of intensive care unit beds in early November 2021,[47] Singapore extended its COVID-19 restrictions to late November and accelerated booster vaccination. Citing the international data showing that protection against the Omicron variant from a primary vaccination series became weaker compared to that against the Delta variant,[48] Singapore has placed booster shots of vaccine at the centre of the strategy to fight Omicron.[49] Singapore authority has not just made vaccination mandatory for most employees since January 2022 (as discussed previously), but also started booster vaccination programme simultaneously.

---

[44] Prime Minister's Office Singapore (2021, October 9).

[45] Chen (2022).

[46] Baharudin (2021).

[47] Teo (2021).

[48] Gov.sg (2022, January 21).

[49] The Pioneer (2021).

As of May 2022, 96% of eligible population (92% of total population) in Singapore have completed full regimen in vaccination, 93% of the total population received at least one dose, and 75% of the total population received booster shots.[50] Ranked as one of the world's most vaccinated countries, Singapore in April 2022 eased most of the pandemic related restrictions. There was no longer a group size limit. Safe distancing was no longer required. All workers might return to workplaces regardless of their vaccination status. Mask wearing was only required in indoor settings. The government no longer issued notices to close contacts or used tracing applications.[51] Singapore has entered "COVID-19 Resilience."[52]

2.3 What Does Hong Kong Do Wrong About COVID?

After almost two years of rather satisfactory management of the COVID-19 pandemic, the Hong Kong SAR was hit hard by the Omicron variant of the coronavirus in February and March of 2022. What became known was a COVID-19 debacle. After the daily confirmed cases stayed single- or double-digit for most time since the outbreak of the coronavirus pandemic, the number of confirmed cases in Hong Kong has skyrocketed since 5 February 2022, reaching a peak at 76,341 on 2 March 2022. More severely, the steep upsurge in confirmed cases led to soaring death cases, resulting in the highest death rate in the world in early March 2022.[53]

The Hong Kong Special Administrative Region government has followed closely Mainland China's zero COVID strategy in weathering the first four waves of the pandemic. The fifth wave (February to March 2022), when the Omicron variant became the dominant variant of the coronavirus, soon cautioned the world the disaster of pursuing zero COVID when social, cultural, and political prerequisites were missing in Hong Kong's context. Compared with Singapore, Hong Kong did not make use of the two-year's window period to prepare for normalizing the society, such as rolling out vaccination like Singapore did, which resulted in the world's highest COVID-related death rates and many avoidable deaths by its elderly citizens.

[50] Ministry of Health (n.d.-a).

[51] Ministry of Health (2022).

[52] Prime Minister's Office Singapore (2022, March 24).

[53] Cheng and Cheung (2022).

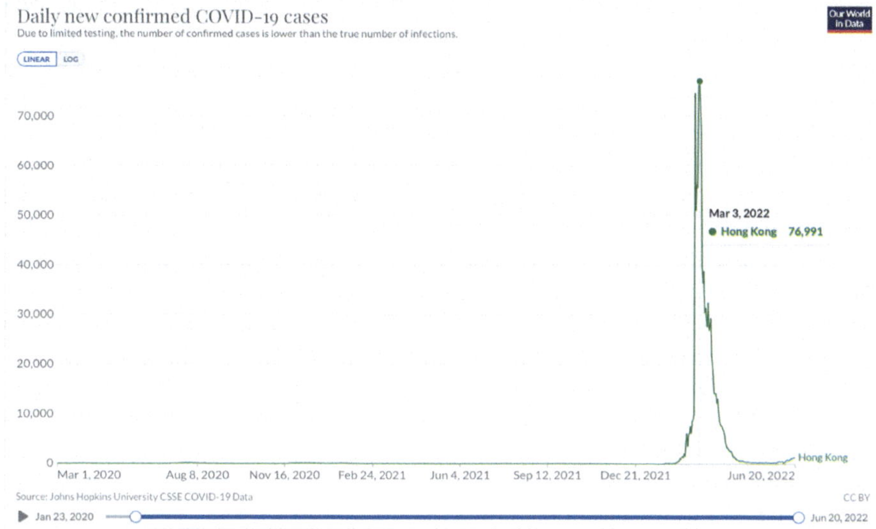

As previously discussed, the effectiveness of China's zero COVID strategy rested upon several factors: China's capability of quick mass testing, characteristic of Chinese society as privacy-sacrificing, pervasiveness of big data, mobilization at grassroots level by grid organizers, as well as a vertical administrative structure. Hong Kong, however, lacked none of these prerequisites to achieve the same degree of effectiveness when trying to copy the China-style zero COVID strategy, especially against the highly infectious Omicron variant.

When it came to testing abilities, Hong Kong did not possess a testing capability nor capture a right timing for mass testing. According to an official report published on the website of the State Council of the People's Republic of China, China currently has 13,100 qualified nucleic acid testing institutions nationwide and nearly 150,000 people working in this regard, capable of processing 51.65 million samples per day.[54] The test results can generally come out within six hours.[55] Speaking of testing capability at city levels, Guangzhou, the southern metropolis of Mainland China completed mandatory mass testing upon its 18 million residents within three days. Comparatively, Hong Kong could only test between 200 to 300 thousand people a day, admitted by Carrie Lam Cheng Yuet-ngor, Chief Executive of the SAR.[56] Hong Kong's testing capability was not uncommendable when compared internationally. It was on par with that of the United Kingdom.[57] However, such a testing capability was still inadequate to underpin a China-style zero COVID policy.

In addition, even if Hong Kong insisted on mass testing, the authority has already missed the critical early timing when mass testing could be the most effective. In

[54] The State Council of the People's Republic of China (2022).

[55] The State Council of the People's Republic of China (2022).

[56] Hassan (2022).

[57] Wain et al. (2022).

response to cases that skyrocketed from late January, the SAR government of Hong Kong announced in February 2022 that Hong Kong would roll out a city-wide mandatory testing in the middle of March. Compared with Mainland Chinese cities where mass testing could be immediately launched after only a few cases emerged, the HKSAR government planned to wait for another three weeks to initiate its mass testing. CE Carrie Lam admitted that it was because Hong Kong's ability (of mass testing) was "not up to it."[58] Even though the government vowed to improve the capability of daily testing to one million residents,[59] the decision, however, attracted criticism from health experts. Yuen Kwok-yung, a veteran microbiologist and one of the top coronavirus experts in Hong Kong, criticized that mass testing could help break transmission chains when there were just a few dozen or a few hundred cases a day. "If we are recording over 50,000 new cases every day, I don't think (mass testing) will be very helpful," Yuen told reporters, as France 24 reported. However, by the day when the HKSAR government announced the decision, February 22, Hong Kong had already witnessed thousands of cases everyday.[60] It only took 10 days that the infected cases skyrocketed to 76,991. The heavy caseload only called the government's mass testing plan into further question.

Amid criticism from local health experts and declining of cases following the 3 March peak, the Hong Kong SAR government cancelled its mass testing plan even before the plan got started. Nevertheless, it could be anticipated that even if the mass testing was launched, the effectiveness would still be limited. A mass testing could help break transmission only if any confirmed or suspected positive cases could be isolated and quarantined in time. China's experience was to quarantine large number of infected cases in expeditiously established cabin hospitals. Hong Kong decided to follow China's practice. With the assistance from Beijing, Hong Kong started to build four mobile cabin hospitals, namely, temporary community isolation and treatment facilities, in Tsing Yi, San Tin, Yuen Long, and Hung Shui Kiu from late February.[61] The cabin hospital in Tsing Yi became the first one to put into use by early March 2022, providing 3,900 bed units.[62] However, one week after the Tsing Yi cabin had been put into use, only 272 people were admitted into the hospital, as one *Wen Wei Po* article reported, making up only 6.97% of the total occupancy.[63]

Hundreds of thousands of confirmed cases were left within communities. For those quarantined at home, Hong Kong also lacked the grassroots level mobilization to serve (and put on surveillance on) the home-quarantined residents in a way that China has done. The same *Wen Wei Po* article invoked Dr. Leung Chi-chiu who explained that reason might be that the confirmed cases in communities communicated with health authorities via government hotlines which ran at a very low efficiency: on one hand, the home-quarantined cases could not secure timely assistance from the

[58] Hassan (2022).

[59] Schnirring and Soucheray (2022).

[60] Our World in Data.

[61] Xinhua (2022).

[62] Wenweipo (2022).

[63] Wenweipo (2022).

HKSAR government, on the other hand, the HKSAR government was unable to obtain precise data so that they could arrange confirmed cases into established cabin hospitals.[64]

Pursuing the zero COVID strategy by the Hong Kong authority has resulted in medical resources being overwhelmed, which led to the world's highest rate of death related to COVID-19, with a death rate above 25 per 100,000.[65] The HK authority ordered every COVID-19 positive case to quarantine in hospitals, being treated with medical resources, and ordered all close contacts to isolate in designated facilities.[66] The limited testing capability resulted in many specimens being delayed; Hong Kong SAR government's practice was to treat preliminary positive results which were not confirmed as confirmed positive cases and quarantine them in hospitals and isolation.[67] The medical resources in Hong Kong were soon overwhelmed. Hospitals, isolation centers, and even morgues were overflowing. By mid-February 2022, the isolation beds had already reached 90% capacity.[68] Various articles have documented the dire situations that the surging COVID-19 cases have overwhelmed the medical resources. David Chan Kwok-shing from the hospital Authority Employees Alliance told reporters of the South China Morning Post that dead bodies had to be stored on gurneys in hallways or beds inside accident and emergency departments at public hospitals and in many cases, a dead body remained in an A&E unit for a day before it could be transferred to the hospital's mortuary, which had already been full.[69] One *Washington Post* article has described the Caritas Medical Center in Hong Kong that one patient who tested positive was isolated in a women's bathroom with an air purifier as no other rooms were available.[70] A 17 March 2022 article has cited a funeral industry representative in Hong Kong that the rising death toll had seen a crunch in Hong Kong's coffins supply with only 300 remaining, forcing the Hong Kong to arrange transportation of coffins from Mainland.[71]

The deaths by the elderly residents of Hong Kong contributed to the largest portion of all fatalities. On 21 February 2022, 32 deaths were reported, 27 of them were over 60 years of age.[72] As reported by Al Jazeera on that on 9 March 2022, the territory had reported 2,365 COVID-19 deaths, 87% were aged 60 or above.[73] Why were there so many elderly deaths? Various reports have pointed to Hong Kong's low rates of vaccination, particularly, the even lower rate of vaccination among the elderly population.[74] The Centre for Health Protection of Hong Kong analyzed the first 102

64 Wenweipo (2022).

65 Taylor (2022).

66 Taylor (2022).

67 Davidson (2022).

68 Davidson (2022).

69 Yeo and Ng (2022).

70 Mahtani and Yu (2022).

71 Hong Kong Free Press (2022, March 17).

72 Hong Kong Free Press (2022, February 22).

73 Cheung (2022).

74 Khan (2022).

deaths since the fifth wave and found that only seven had received two doses of vaccines.[75] The same Al Jazeera article reported that among the elderly deaths in the 2,365 COVID-19 deaths, about 90% were not fully vaccinated. As of 3 March 2022, only 45% of Hong Kong's seniors were vaccinated, and just 15% of elderly residents in Hong Kong's nursing institutions.[76] By February 2022, the vaccination coverage in Hong Kong was only 64%, far lower than that of Singapore and New Zealand.[77] A consensus has reached that vaccination could significantly lower the number of severe and death cases. Scholars have attributed the very low death rate of COVID-19 in New Zealand to its very high vaccination rate, particularly the almost 100% vaccination rate for New Zealand's elderly residents.[78]

Then, why was the vaccination rate in Hong Kong this low? This section would offer two explanations for Hong Kong's low vaccination rate. The first was the rather satisfactory performance of the first two years' pandemic. The low number of infected cases accumulated from the first four waves of the pandemic let Hong Kong residents feel little need to vaccinate themselves. This is what scholars have said of the vaccine complacency[79] or vaccine hesitancy.[80] The second explanation lay in the residents' distrust in the Hong Kong SAR government. The COVID-19 pandemic hit Hong Kong when the territory had barely got out of the series of legal, social, as well as political chaos in the previous year. A vaccination programme, coordinated and pushed forward by the government, was met with cold reaction in Hong Kong due to residents' distrust in the Hong Kong SAR government.[81] The distrust in the Hong Kong government led further to residents' distrust in government's rolled-out vaccinations. One Bloomberg article documented that the convenience of access to vaccination in Hong Kong (evidenced by easy booking, walkable distance, and abundant supply of both China-made Sinovac and Pfizer-BioNTech) did not encourage residents to get vaccinated. A Chau-surnamed interviewee stated that "I won't take the vaccine, because my friends and I just don't want to follow any instructions or recommendations from the government;" still others would "wait and see" as "a lot of news about the adverse reactions spreading on social media and WhatsApp groups."[82] The success of the first two years of pandemic control performance of Hong Kong could not be attributed solely to Hong Kong's attempted zero COVID strategy. Some scholars have attributed Hong Kong's early success in controlling the pandemic to "community capability,"[83] "civil society,"[84] and distrust of the SAR government that

[75] Das (2022).

[76] Einhorn (2022).

[77] Our World in Data.

[78] "News at a Glance: Hong Kong's High COVID-19 Death Rate, Biden's 'Test and Treat,' and a ResearchGate Lawsuit," March 10, 2022, https://doi.org/10.1126/science.adb1942.

[79] Silver (2022).

[80] Tsang (2022).

[81] McLaughlin (2021), Marlow and Tam (2021).

[82] Marlow and Tam (2021).

[83] Hartley and Jarvis (2020).

[84] Wong and Wu (2021), Chan (2021).

"caused the public to take early self-protective measures and initiated societal-wide self-help campaigns."[85] Yet, the Hong Kong SAR government could hardly shake off its duty in encouraging vaccination. Carrie Lam admitted that not enough had been done to vaccinate elderly residents in Hong Kong against COVID-19.[86] The Hong Kong SAR government did not mandate the vaccination among its residents like Singapore did, when the HKSAR government already lagged Singapore in rolling out the vaccination programme.[87] Some hotels in Hong Kong offered cash incentives to staff who took vaccines, the HKSAR government however, rejected that government should also offer such incentive and hoped that private sectors would do more.[88]

In view of that the timely vaccination coverage reduced severe and death cases and greatly helped in making the pandemic a treatable and mild disease, the vaccination in Hong Kong was accelerated from February 2022. Organizations like the University of Hong Kong, mirroring their counterparts in Singapore, required "Vaccine Pass" for all employees and visitors.[89] Repercussions of Hong Kong's lesson, the low vaccination rates and human tragedies, were also felt in Mainland China, the world's last zero COVID holdout. Since April 2022, China has shifted the focus of its vaccination campaign from essential workers to the elderly population.[90]

3 Further Discussions

The previous analysis has revealed that the selection of pandemic control policies is deeply rooted and closely related to various political, cultural, and social prerequisites of a society. The perceived effectiveness of a selected policy was also subject to epidemiological traits of variants of the virus. China's zero COVID strategy has achieved indisputable effectiveness in the first two years of the pandemic. The effectiveness of zero COVID strategy in China rested on a combination of characteristics of Chinese society. China had the capability of conducting quick and mass testing on millions of residents to pre-empt spread of coronavirus when only a handful cases were found. Big data in China's privacy-sacrificing society was then instituted to identify locations and travel histories of positive cases and close contacts. China had both state power and grassroots-level mobilization to quarantine the cases.

These social, cultural, and political prerequisites did not exist in other countries. In terms of restriction measures, experienced by one of the chapter authors who was in Singapore during the circuit breaker, residents in Singapore could still go out to buy

[85] Chan (2021).

[86] Au (2022).

[87] The Singapore government launched the mass vaccination programme in late 2020, while Hong Kong only rolled out it vaccination programme in late February 2021. Davidson (2021).

[88] Davidson (2021).

[89] The University of Hong Kong (2022).

[90] Xie (2022).

food, grocery, to exercise alone or with friends.[91] When it came to the uses of health codes which attached civil rights to code holders' health statuses, it was observed that citizens in many other countries, notably in France,[92] Italy,[93] Bulgaria,[94] and Britain,[95] who protested for weeks against health pass or vaccination passport, people in China overwhelmingly accepted similar arrangement "due to the collective objective of preventing COVID-19 transmission."[96] With regards to mandatory quarantine, Chinese authorities could mobilize power to quarantine any suspected cases without any legal, ethical, and moral challenges. Citizens in China are also overwhelmingly supportive of such mandatory quarantine measures. Comparatively, take the United States as an example, not only self-quarantine for interstate travellers faced legal challenge, but a governor's-imposed quarantine and lockdown measures within the state were often challenged. For example, in March 2020, Governor Gina Raimondo of the state of Rhode Island ordered the state's national guard to go door to door to inform any New Yorker who came to the state that they must self-quarantine for 14 days or fines and jail time if otherwise.[97] This order immediately attracted legal challenge from New York Governor Andrew Cuomo, who called the order "reactionary and unconstitutional," and vowed to sue the state of Rhode Island.[98] Cuomo himself, however, ordered restrictions requiring individuals travelling to New York from areas of high infections to self-quarantine. His order was challenged by an attorney, who filed a lawsuit against Governor Cuomo.[99] It can be said that all these restriction measures that China-style zero COVID could not be applied onto other countries, especially countries that placed high values on personal freedom and civil rights.

The effectiveness of China's zero COVID strategy was also temporal, rather than eternal. In the stage of Omicron, China's zero COVID has met growing challenges posed both by the new variant of the virus and by mental fatigue, social disruption, as well as economic downturn. Once the Omicron variant has become the dominant variant of the pandemic, the zero COVID strategy then faced growing pressure. The Omicron variant demonstrated an infection rate 70 times faster than the Delta variant,[100] rendering locating, quarantining, and hospitalizing cases a mounting task. In March 2022, when China resorted to zero COVID strategy and imposed a lockdown confining practically every resident at home in Shanghai, the restrictions measures generated public discontent.[101] China's economy as well as supply chain,

---

[91] Interview with authors.

[92] France 24 (2021, August 29).

[93] France 24 (2021, September 10).

[94] Reuters (2021).

[95] Pannett (2021).

[96] Li et al. (2022).

[97] New York Post (2020).

[98] "Rhode Island Begins Door to Door Checks for New Yorkers Fleeing Coronavirus."

[99] Singman (2020).

[100] Chung (2022).

[101] Shepherd (2022).

shipping, high-tech manufacturing had severely impacted as a result. According to a Bloomberg article citing the research by economist Zheng Michael Song, China's COVID lockdowns were likely to cost the country at least $46 billion a month, or 3.1% of GDP, with a lockdown in Shanghai could possibly cost China's GDP by 4%.[102]

Therefore, Singapore's timely switch from zero COVID to live with COVID when the Delta variant hit was commendable. Singapore also adopted a zero COVID strategy at the beginning of the pandemic, responding to the outbreak by quickly quarantining and treating the confirmed positive cases, tracing their travel trajectories, and shutting down Singapore's border. The mentality at this stage was to quickly eliminate positive cases. In Lee Hsien Loong's words, "I see a case I will stamp on it."[103]

The reasons accounting for Singapore's zero COVID strategy were manyfold. The first was the inexperience in dealing with the novel virus. In the words of Abdullah and Kim, Singapore's main approach was "learning."[104] Lee Hsien Loong admitted that Singapore had to make a judgment call at the start of the pandemic when little was known about the virus.[105] Singapore's measures, such as mandatory quarantine, shutdown of national border, travel restrictions, epidemiological investigations, were also measures taken by other countries. The second reason was the unavailability of the later widely used vaccinations against COVID-19. Meanwhile, the fatality rate was relatively high. For example, the case fatality rate of COVID-19 was recorded between 3–4% in its initial days, climbing to the peak of 7.34%.[106] The third reason was that the outburst of cases in densely arranged and ill-furnished migrant workers' dormitories.[107] In the context of limited knowledge on the virus, its high fatality rate, and the unavailability of vaccination, Singapore's zero COVID measures seemed to be reasonable to avoid burdens on medical resources which have been overwhelmed in many other countries. Restriction measures were gradually relaxed. Dropping from the peak of 1,425 recorded on 20 April 2020, the confirmed cases in Singapore remained double-digit throughout the rest of 2020 and the first half of 2021.[108]

In view of the availability of vaccines and the evolving characteristics of the coronavirus variants, Singapore has swiftly adjusted its pandemic control policies. Singapore's timely adjustment can be illustrated by Lee Hsien Loong's speech at the Bloomberg New Economy Forum on 17 November 2021:

> when things change, you will have to change along with it, because if you cannot change along with it you will be out of step, and that is very hard. We started off with COVID-19 zero practically, meaning I see a case I will stamp on it, and if I need to trace 200 or 500

[102] Hancock (2022).

[103] Prime Minister's Office Singapore (2021, November 17).

[104] Abdullah and Kim (2020).

[105] Business Standard (2022).

[106] Our World in Data.

[107] Gorny et al. (2021).

[108] "COVID-19 Statistics."

> people to do that, I will try to do that. Not quite like in China where one case means one city locked down, but quite stringently. Now I am not doing that. Now I am saying let the cases go but let us keep the vulnerable people safe.[109]

With a hindsight, Singapore did not stay on zero COVID, namely, or pursue the elimination of positive COVID-19 cases as the ultimate result. Instead, Singapore made use of zero COVID measures to buy time to lay the ground for eventual re-opening, which was in the words of PM LHL, "living with COVID."[110] Singapore's adjustment also reflected a changed mentality regarding the COVID-19. While China insisted on elimination of positive cases, Singapore focused on lowering the number of severe and death cases with the interference of well-timed vaccination. As the Delta variant caused mostly mild or asymptomatic cases, it was economically, socially, and psychologically costly to implement zero COVID measures. Like Lee Hsien Loong explained:

> Each time we tighten up. Businesses are further disrupted. Workers lose jobs. Children are deprived of a proper childhood and school life. Families are separated for even longer. Especially families with loved ones overseas and extended families who have not been able to come together. All these cause psychological and emotional strain, and mental fatigue.[111]

Singapore followed the evolvement of coronavirus variants closely and took steps back and forth in easing and tightening restrictions. The trend, however, was to gradually lift restrictions and bring economic life back to normalcy. Starting from 22 April 2022, except indoor mask-wearing, most COVID-19 related restrictions, such as workplace requirements, safe distancing, capability limits for events, vaccination record checks, were lifted.[112] It was a "decisive step" Singapore has taken to "live with COVID" amidst one of the world's highest vaccination rate and declining COVID cases and hospitalisation.[113] Except that indoor mask-wearing requirement remained in place, business and social lives in Singapore have largely been normalized.

If not for the fifth wave of the pandemic, Hong Kong can be arguably an even bigger success story than Singapore. Hong Kong has outperformed Singapore when measured by Hong Kong's far lower number of confirmed cases than Singapore. The first four waves of the pandemic in Hong Kong only reported 12,649 cases, 1/22 of those in Singapore.[114] Hong Kong's performance during the first four waves of pandemic even led scholars to advocate other countries to learn from "Hong Kong's response to the COVID-19 pandemic."[115] The Hong Kong SAR government's public measures, such as border control, requirements on social distancing high-volume testing, contact racing, and aggressive quarantine were essential in keeping low the number of confirmed positive cases. Hong Kong SAR government's attempted

[109] "PM Lee Hsien Loong at the Bloomberg New Economy Forum 2021."

[110] "Update on the COVID-19 Situation in Singapore by PM Lee Hsien Loong on 9 October 2021."

[111] "Update on the COVID-19 Situation in Singapore by PM Lee Hsien Loong on 9 October 2021."

[112] "Further Easing of Community and Border Measures."

[113] "Further Easing of Community and Border Measures."

[114] Our World in Data.

[115] Wong, Kwok et al. (2020, May 11).

approach to the pandemic was in line with Mainland China, namely, the zero COVID policy.

Behind the Hong Kong SAR government's determination of following China's zero COVID strategy was Hong Kong government's aim to re-open with Mainland China. Striving for a re-opening with Mainland has been a key task in Hong Kong's pandemic control efforts. Keeping a low number of positive cases and well-functioned inspection-control measures were seen as Mainland China's requirements for re-opening with China. There were times that Mainland Chinese health experts visiting and inspecting Hong Kong to examine the readiness of such a re-opening.[116] By November 2021, Hong Kong had met the basic requirements to reopen the border with Mainland China, according to Hong Kong SAR officials.[117]

No matter how determined Hong Kong wanted to re-open with China by adopting the zero COVID strategy, its restriction measures failed to serve such purpose. This chapter will recount stories of one of its authors who experienced first-hand Hong Kong's chaos during the fifth wave of the pandemic. He wanted to bring his wife and their new-born child back to Mainland China from Hong Kong. By February 2022, it was already arduous to book a place of a quarantine hotel in Shenzhen. He was fortunate that after days of snapping up tickets, he eventually managed to secure three for the family. He was however, stopped at the border as his new-born child did not have a negative testing result. Yet, Hong Kong's mass testing scheme did not include children under three months old. He then had to resort to private clinics to have test done on the baby at the expense of more than one thousand Hong Kong dollars.

The personal experiences of the chapter author were vivid example of how Hong Kong's caught-in-between pandemic approach rendered many border-crossers suffering. When Hong Kong had to confront the Omicron variant, the government's zero COVID pursuance and subsequent restriction measures have become a liability instead of assets. As previously analysed zero COVID required instant mass testing ahead of large scale of outburst of positive cases, however, Hong Kong lacked such capability. Upon identifying positive cases and their close contacts, zero COVID strategy demanded quarantine be imposed. Yet, Hong Kong failed to timely locate and transfer the cases to quarantine facilities. Hospitalizing any positive cases only overwhelmed Hong Kong's medical resources, which was exactly what Singapore had tried to avoid. The flip-flop attitude of the Hong Kong government on mass testing had only deteriorated the already weakened governmental credibility. The vaccine hesitancy resulted partly from distrust in the Hong Kong government, together with a vaccine complacency arising from Hong Kong's successful pandemic control performance in the previous waves of pandemic, left large number of Hong Kong residents unvaccinated, even more so among the elderly. Compared with China, Hong Kong lacked prerequisites of combating the Omicron variant with zero COVID strategy. Trailing behind Singapore in terms of vaccination rates, Hong Kong witnessed the fifth wave of the pandemic turned into a public health fiasco. In the words of Carrie

[116] Zheng et al. (2021).

[117] Wong et al. (2021, November 25).

Lam: "It's clear now that most foreign countries have a set of theories and measures against the pandemic, while our country has a different set...Hong Kong is caught in the middle."[118] That Hong Kong was caught in the middle has made Hong Kong the worst COVID scenario in the world.

4 Epilogue

The fifth wave of pandemic, deteriorated by Hong Kong government's doggedly adherence to zero COVID strategy, has hit Hong Kong hard not just in public health sphere but also in economy. The city has reported an 8.3% contraction in the first quarter of 2022, in contrast to over 3% growth in Singapore, South Korea, and Singapore; its travel restrictions on arriving passengers and inbound flights had only triggered an outflow of residents and expats escaping from the city.[119] Caught in between zero COVID and "live with COVID," Hong Kong did not succeeded in re-opening its border with Mainland China; even more severe was that as an international hub of aviation, business, and commerce, Hong Kong also isolated itself from the rest of the world.

Evidenced by the Shanghai lockdown and the Hong Kong debacle, a zero COVID strategy's effectiveness has been called into question concerning the high infection rates of the Omicron variants, the availability of vaccines, and the agreed-upon biological trait that most Omicron cases were mild or asymptomatic. Enlarging vaccination coverage and resuming economic activities are now priorities for previous zero COVID countries. In view of that the timely vaccination coverage reduced severe and death cases and greatly helped in making the pandemic a treatable and mild disease, the vaccination in Hong Kong became accelerated from February 2022. Organizations like the University of Hong Kong, mirroring their counterparts in Singapore, required "Vaccine Pass" for all employees and visitors.[120] The Hong Kong SAR government did not seek further and stricter restrictions. Repercussions of Hong Kong's lesson, such as the low vaccination rates and human tragedies, were also felt in Mainland China, the world's last zero COVID holdout. Since April 2022, China has shifted the focus of its vaccination campaign from essential workers to the elderly population.[121] Possibly perturbed by the Shanghai lockdown, the selection of large-scale city-wide lockdowns was treated with caution in China. In Beijing, there were eruption of cases across the metropolis since April 2022. No large-scale city-wide lockdown was imposed as of writing.

On 25 June, during the 12th Convention of the Chinese Communist Party delegates in Shanghai, the municipal Party Secretary Li Qiang declared that Shanghai had

---

[118] The Standard (2022).

[119] Riordan and Lin (2020).

[120] "HKU Vaccine Pass."

[121] Xie, "Learning from Hong Kong, China Targets Elderly in Covid-19 Vaccine Drive."

won the battle of defending Shanghai.[122] However, under the zero COVID arrangement, China resorted to keeping the coronavirus out rather than to vaccinations for population immunity, it raises the question if a Chinese city is ready for the next wave of pandemic which is all but inevitable. In one word, by being the world's last holdout of zero COVID, China seems determined to confront the pandemic by physical approaches: mass testing, restriction of human movement, and imposition of lockdowns. This physical approach turned out very effective in the first two years in Chinese social settings. Singapore adopted the same physical approach at the beginning months of the pandemic. However, in view of the high infection rates of the Delta variant which made the physical approach more and more ineffective, Singapore switched from this physical approach to confront the pandemic by medical (pharmaceutical) approach. The effectiveness of the physical approach, demonstrated by the Hong Kong fiasco and the Shanghai lockdown, was thus questioned.

The stories of navigating the pandemic in China, Singapore, and Hong Kong have reaffirmed that there is no "best," "eternally effective," nor "one-size-fits-all" pandemic control strategy. When a specific policy is enshrined into a doctrine, it compromised effectiveness and creates only chaos, uncertainties, and disasters. Each pandemic control strategy must be rooted in various prerequisites to be effective and take traits of variants and the balance of economic costs and human lives lost into consideration.

References

Abdullah, W. J., & Kim, S. (2020). Singapore's responses to the COVID-19 outbreak: A critical assessment. *American Review of Public Administration, 50*(6–7), 770–776. https://doi.org/.o0r.g1/107.171/0772/70527057047042002099442245 4

"Active epidemiological investigation on SARS-CoV-2 infection caused by Omicron Variant (Pango Lineage B.1.1.529) in Japan: Preliminary report on infectious period." National Institute of Infectious Diseases Disease Control and Prevention Center, National Center for Global Health and Medicine, January 5, 2022. https://www.niid.go.jp/niid/en/2019-ncov-e/10884-covid19-66-en.html

Al Kuwari, H. M., Abdul Rahim, H. F., Abu-Raddad, L. J., Abou-Samra, A.-B., Al Kanaani, Z., Al Khal, A., Al Kuwari, E., et al. (2020, 28 February–18 April). Epidemiological investigation of the first 5685 cases of SARS-CoV-2 infection in Qatar. *BMJ Open, 10*(10), e040428. https://doi.org/10.1136/bmjopen-2020-040428

Au, B. (2022, March 11). Carrie Lam admits 'not enough' done to vaccinate elderly hongkongers from Covid-19. *South China Morning Post*. https://www.scmp.com/video/coronavirus/3170182/carrie-lam-admits-not-enough-done-vaccinate-elderly-hongkongers-covid-19

Baharudin, H. (2021, November 29). Singapore tracking Omicron Covid-19 variant closely, may be forced to take steps back: PM Lee. *The Straits Times*. https://www.straitstimes.com/singapore/pap-conference-spore-tracking-omicron-covid-19-variant-closely-may-well-be-forced-to-roll

Business Standard. (2022, April 12). Singapore PM says his Govt 'did not get every call right' during Covid-19. https://www.business-standard.com/article/pti-stories/singapore-pm-lee-says-his-govt-did-not-get-every-call-right-during-covid-19-crisis-but-key-is-to-keep-learning-122041200812_1.html

[122] Xinmin (2022).

Chan, R. K. H. (2021, July). Tackling COVID-19 risk in Hong Kong: Examining distrust, compliance and risk management. *Current Sociology, 69*(4), 547–565. https://doi.org/10.1177/0011392121990026

ChannelNewsAsia. (2021, February 17). *Singapore receives its first shipment of Moderna's COVID-19 vaccine*. https://www.channelnewsasia.com/singapore/moderna-covid-19-vaccine-first-shipment-arrives-sia-singapore-356561

Chen, L. (2022, April 26). Relief, revival as Singapore scraps its COVID curbs. *Reuters*. https://www.reuters.com/world/the-great-reboot/relief-revival-singapore-scraps-its-covid-curbs-2022-04-26/

Chen, D., Song, F., Tang, L., Zhang, H., Shao, J., Qiu, R., Wang, X., & Ye, Z. (2020, September). Quarantine experience of close contacts of COVID-19 patients in China: A qualitative descriptive study. *General Hospital Psychiatry, 66*, 81–88. https://doi.org/10.1016/j.genhosppsych.2020.07.006

Cheng, L., & Cheung, E. (2022, March 5). Hong Kong now has the world's highest Covid-19 death rate. What happened, and how can this be fixed? *South China Morning Post*. https://www.scmp.com/news/hong-kong/health-environment/article/3169331/hong-kong-now-has-worlds-highest-covid-19-death

Cheung, R. (2022, March 9). In 'Zero COVID' Hong Kong, deaths smash global records. *Al Jazeera*. https://www.aljazeera.com/news/2022/3/9/in-zero-covid-hong-kong-deaths-smash-global-records

China.org.cn. (2014, January 17). Zhonggong Zhongyang Guanyu Quanmian Shenhua Gaige Ruogan Zhongda Wenti de Jueding 中共中央关于全面深化改革若干重大问题的决定 [Decision of the CCCPC on some major issues concerning comprehensively deepening the reform]. http://www.china.org.cn/chinese/2014-01/17/content_31226494_13.htm

Chung, I. (2022, March 14). Omicron variant vs. Delta: What are the differences? MedTecs. https://www.medtecs.com/omicron-vs-delta/#:~:text=Omicron%20variant%20spreads%20about%2070%20times%20faster%20than,can%20transmit%20faster%20among%20people%20than%20previous%20variants

Co, C. (2021, October 9). Unvaccinated people not allowed to dine in at hawker centres, enter shopping malls from Oct 13. ChannelNewsAsia. https://www.channelnewsasia.com/singapore/covid-19-no-dine-unvaccinated-hawker-coffee-shop-shopping-malls-oct-13-2232601

Das, M. (2022, April 4). COVID-19 and the elderlies: How safe are Hong Kong's care homes? *Frontiers in Public Health, 10*, 883472. https://doi.org/10.3389/fpubh.2022.883472

Davidson, H. (2021, May 25). Covid vaccine hesitancy could see Hong Kong throw away doses. *The Guardian*. https://www.theguardian.com/world/2021/may/25/covid-vaccine-hesitancy-could-see-hong-kong-throw-away-millions-of-doses

Davidson, H. (2022, February 14). Covid 'Onslaught' has overwhelmed Hong Kong's capacity, leader says. *South China Morning Post*. https://www.theguardian.com/world/2022/feb/14/covid-onslaught-has-overwhelmed-hong-kongs-capacity-as-cases-surge-leader-says

DW News. (2022, January 21). Henan Xianzhang Jinggao Wu 'Eyi Fanxiang' Beipi Guodu Fangyi 河南县长警告勿'恶意返乡' 被批过度防疫 [One Henan magistrate warns of 'malicious homecoming,' criticized for excessive pandemic control]. https://www.dw.com/zh/河南县长警告勿恶意返乡-被批过度防疫/a-60510308

Einhorn, B. (2022, March 3). *Hong Kong's nursing homes are unvaccinated hotbeds of Covid*. Bloomberg. https://www.bloomberg.com/news/articles/2022-03-03/hong-kong-s-nursing-homes-are-unvaccinated-hotbeds-of-covid

France 24. (2021, August 29). *Seventh consecutive weekly protest against France's Covid-19 health pass*. https://www.france24.com/en/france/20210829-seventh-consecutive-weekly-protest-against-france-s-covid-19-health-pass

France 24. (2021, September 10). *Thousands protest in Rome against Covid health pass*. https://www.france24.com/en/live-news/20211009-thousands-protest-in-rome-against-covid-health-pass

Gan, K. Y., Wong, L., & Kung, O. Y. (2021, June 24). Living normally, with Covid-19: Task force ministers on how s'pore is drawing road map for new normal. *The Straits Times*. https://www.straitstimes.com/opinion/living-normally-with-covid-19

Gorny, A. W., Bagdasarian, N., Koh, A. H. K., Lim, Y. C., Ong, J. S. M., Ng, B. S. W., Hooi, B., et al. (2021, February). SARS-CoV-2 in migrant worker dormitories: Geospatial epidemiology supporting outbreak management. *International Journal of Infectious Diseases, 103*, 389–394.https://doi.org/10.1016/j.ijid.2020.11.148

Gov.sg. (2020, April 11). *What you can and cannot do during the circuit breaker period*. https://www.gov.sg/article/what-you-can-and-cannot-do-during-the-circuit-breaker-period

Gov.sg. (2020, April 21). *Circuit breaker extension and tighter measures: What you need to know*. https://www.gov.sg/article/circuit-breaker-extension-and-tighter-measures-what-you-need-to-know

Gov.sg. (2020, December 23). *How Singapore is ensuring access to COVID-19 vaccines*. https://www.gov.sg/article/how-singapore-is-ensuring-access-to-covid-19-vaccines

Gov.sg. (2022, January 21). *Updates to safe management measures from 1 Feb 2022*. https://www.gov.sg/article/updates-to-safe-management-measures-from-1-jan-2022

Gov.uk. (2021, December 8). *UK marks one year since deploying world's first COVID-19 vaccine*. https://www.gov.uk/government/news/uk-marks-one-year-since-deploying-worlds-first-covid-19-vaccine

Guanghan.gov.cn. (2022, April 11). Jingzhun Fangkong Yiqing Shouhu Shimin Anquan Xiangcun Wanggeyuan: Zhimi 'Fanghuwang' Zhulao 'Fangkongqiang' 精准防控疫情 守护市民安全 乡村网格员: 织密 '防护网' 筑牢 '防控墙' [Precise on pandemic prevention, guard residents' safety, grid organiser in villages: Strenghten the net and wall of pandemic prevention]. http://www.guanghan.gov.cn/gk/xxgk/ahsb/ssxw/1620962.htm

Guo, R. (2021, June 7). Coronavirus: 18 million tests in three days as Guangzhou tries to stem spread in latest outbreak. *South China Morning Post*. https://www.scmp.com/news/china/science/article/3136378/coronavirus-18-million-tests-three-days-guangzhou-tries-stem

Hancock, T. (2022, March 29). China lockdowns cost at least $46 billion a month, academic says. BNN Bloomberg. https://www.bloombergquint.com/china/china-lockdowns-cost-at-least-46-billion-a-month-academic-says

Hartley, K., & Jarvis, D. S. L. (2020, July 2). Policymaking in a low-trust state: Legitimacy, state capacity, and responses to COVID-19 in Hong Kong. *Policy and Society, 39*(3), 403–423. https://doi.org/10.1080/14494035.2020.1783791

Hassan, A. (2022, March 14). Hong Kong lacks the resources for a China-style pandemic response, the city's leader says. *The New York Times*. https://www.nytimes.com/2022/03/14/world/asia/hong-kong-covid.html

Hong Kong Free Press. (2022, February 22). *Covid-19: Hong Kong reports record high of 32 daily deaths as mainland experts urge authorities to release more data*. https://hongkongfp.com/2022/02/22/covid-19-hong-kong-reports-record-high-of-32-daily-deaths-as-mainland-experts-urge-authorities-to-release-more-data/

Hong Kong Free Press. (2022, March 17). *Morgues overflowing as Hong Kong suffers deadly Covid wave, only 300 coffins remaining in city*. https://hongkongfp.com/2022/03/17/morgues-overflowing-as-hong-kong-suffers-deadly-covid-wave/

Huang, Z., & Feng, Z. (2021, November). Public health and private life under COVID-19 vaccination policies in China: A legal analysis. *Risk Management and Healthcare Policy, 14*, 4627–4638.https://doi.org/10.2147/RMHP.S336434

Khan, N. (2022, March 8). Hong Kong's Covid-19 death rate is the world's highest because of unvaccinated elderly. *The Wall Street Journal*. https://www.wsj.com/articles/unvaccinated-elderly-send-hong-kongs-covid-19-death-rate-to-worlds-highest-11646748091

Li, T., & Guo, Y. (April 10, 2022). Wanggeyuan Yong 'Tiejiaoban' Paochu Jingzhun Jiceng Shuju 网格员用 '铁脚板' 跑出精准基层数据 [Grid organizers master precise grass-roots level data with their iron feet]. *Taiyuan Ribao*. http://www.tynews.com.cn/system/2022/04/10/030435182.shtml

Li, V. Q. T., Ma, L., & Wu, X. (2022, January 28). COVID-19, policy change, and post-pandemic data governance: A case analysis of contact tracing applications in East Asia. *Policy and Society, 41*(1), 01–14. https://doi.org/10.1093/polsoc/puab019

Linghu, Q. (2022, March 11). Jilin Nongye Keji Xueyuan Zhuanyun Xuesheng 6556 Ren, Gei Fangyi Tichu Xinkaoyan 吉林农业科技学院转运学生 6556 人, 给防疫提出新考验 [Jilin Agricultural Science and Technology University transfers 6556 people, posing new tests for pandemic control]. Sohu. https://www.sohu.com/a/528977760_665455

Liu, C. (2022, March 13). At least 26 officials dismissed for poor performance in dealing with COVID-19. *Global Times*. https://www.globaltimes.cn/page/202203/1254740.shtml

Mahtani, S., & Yu, T. (2022, February 16). For two years, Hong Kong held off the pandemic. Then, everything fell apart. *The Washington Post*. https://www.washingtonpost.com/world/2022/02/16/hong-kong-coronavirus-surge-cases/

Marlow, I., & Tam, F. (2021, May 10). *Unused shots pile up as mistrust mars Hong Kong vaccinations*. Bloomberg. https://www.bloomberg.com/news/articles/2021-05-09/unused-shots-pile-up-as-mistrust-blights-hong-kong-vaccine-drive#xj4y7vzkg?leadSource=uverify%20wall

McLaughlin, T. (2021, April 1). The place with surprisingly high vaccine hesitancy. *The Atlantic*. https://www.theatlantic.com/international/archive/2021/04/hong-kong-trust-vaccine/618469/

Ministry of Health. (n.d.-a). *Vaccination statistics*. https://www.moh.gov.sg/covid-19/vaccination/statistics

Ministry of Health. (n.d.-b). *COVID-19 statistics*.

Ministry of Health. (2022, April 22). *Further easing of community and border measures*. https://www.moh.gov.sg/news-highlights/details/further-easing-of-community-and-border-measures

Ministry of Manpower. (2022, April 25). *Updated advisory on COVID-19 vaccination at the workplace*. https://www.mom.gov.sg/covid-19/advisory-on-covid-19-vaccination-in-employment-settings

National University of Singapore. (2021, December 27). *Circular 40 COVID-19: Update on vaccination-differentiated safe management measures*. https://emergency.nus.edu.sg/wp-content/uploads/2021/12/2021-1227-OSHE-COVID-19-Circular-40-11.47.pdf

National University of Singapore. (2022, January 31). *Circular 2 COVID-19: Update on vaccinated-differentiated safe management measures (VDS) for CET students*. https://emergency.nus.edu.sg/wp-content/uploads/2022/01/2022-0131-ORMC-COVID-19-Circular-2-12.04.pdf

NBD. (2021, March 31). Fabu Jiezhong Xinguan Yimiao 'Wubu' Zhuyi Shixiang? Zhen Zhengfu Daoqian Le 发布接种新冠疫苗 '五不' 注意事项?镇政府道歉了 [Declaring 'five prohibitions' for those not taking COVID Vaccines? Government apologized]. http://www.nbd.com.cn/articles/2021-03-31/1681942.html

NDTV. (2021, December 28). *Covid vaccination mandatory in Singapore from February 2022*. https://www.ndtv.com/world-news/coronavirus-covid-vaccination-mandatory-in-singapore-from-february-2022-2673179

NetEase. (2021, July 22). Yimiao Jiangli Buduan Shengji, Cong Songjidan Dao Songqian, Songshouji, Ruci Neijuan Weinaban? 疫苗奖励不断升级, 从送鸡蛋到送钱、送手机, 如此内卷为哪般? [Vaccine incentives continue to level up, from getting eggs, getting mobile phones, to getting money]. https://www.163.com/dy/article/GFHCSMP105328YVR.html

New York Post. (2020, March 29). *Rhode Island begins door to door checks for new yorkers fleeing coronavirus*. https://nypost.com/2020/03/29/rhode-island-begins-door-to-door-checks-for-new-yorkers-fleeing-coronavirus/

"News at a Glance: Hong Kong's High COVID-19 Death Rate, Biden's 'Test and Treat,' and a ResearchGate Lawsuit." (2022, March 10). https://doi.org/10.1126/science.adb1942

Pannett, R. (2021, August 10). Anti-vaccine protesters tried to storm the BBC's offices. But they had the wrong address. *The Washington Post*. https://www.washingtonpost.com/world/2021/08/10/vaccine-protesters-bbc-london/

Paper.cn. (2021, March 31). Hainan Wanning Yi Zhenzhengfu Zhiqian: Zai Xuanchuan Jiezhong Xinguan Yimiao Shi Guoyu Jiandan Cubao 海南万宁一镇政府致歉: 在宣传接种新冠疫苗时

过于简单粗暴 [Town government in Wanning, Hainan apologized: We were too simplistic and gruff]. https://www.thepaper.cn/newsDetail_forward_11980019

Paper.cn. (2021, August 12). Yin Bixu Xiedai 48 Xiaoshinei Hesuan Zhengming, Zhengzhou Gongjiaoche Gaofengqi Jin 5 Ming Chengke 因必须携带 48 小时内核酸证明, 郑州公交车高峰期仅 5 名乘客 [Because of the 48-hour negative nucleic acid testing requirement, only 5 passengers showed up]. https://m.thepaper.cn/baijiahao_14010582

Poa, H. (2022, January 10). Response to adjournment motion on vaccination differentiated safe management measures. Ministry of Health. https://www.moh.gov.sg/news-highlights/details/response-to-adjournment-motion-on-vaccination-differentiated-safe-management-measures/

Prime Minister's Office Singapore. (2021, October 9). *Update on the COVID-19 situation in Singapore by PM Lee Hsien Loong on 9 October 2021.* A Singapore Government Agency Website. https://www.pmo.gov.sg/Newsroom/Update-on-the-COVID-19-Situation-In-Singapore-by-Prime-Minister-Lee-Hsien-Loong-on-9-October

Prime Minister's Office Singapore. (2021, November 17). PM Lee Hsien Loong at the Bloomberg New Economy Forum 2021. https://www.pmo.gov.sg/Newsroom/PM-Lee-Hsien-Loong-at-the-Bloomberg-New-Economy-Forum-2021

Prime Minister's Office Singapore. (2022, March 24). Speech by PM Lee Hsien Loong on COVID-19: A new phase on 24 Mar 2022. https://www.pmo.gov.sg/Newsroom/PM-Lee-Hsien-Loong-on-COVID-19-A-New-Phase-on-24-Mar-2022

Reuters. (2021, October 21). *Hundreds protest against Bulgaria's COVID health pass.* https://www.reuters.com/world/europe/hundreds-protest-against-bulgarias-covid-health-pass-2021-10-20/

Riordan, P., & Lin, A. (2020, May 3). Hong Kong loosens Covid rules after sharp economic contraction. *Financial Times.* https://www.ft.com/content/f369920b-1873-437a-a412-994bec50b8fb

Schnirring, L., & Soucheray, S. (2022, February 22). *Hong Kong to harness mass testing in Omicron battle.* Center for Infectious Disease Research and Policy, University of Minnesota. https://www.cidrap.umn.edu/news-perspective/2022/02/hong-kong-harness-mass-testing-omicron-battle

Shepherd, C. (2022, May 19). Shanghai faces mental health crisis as COVID lockdown drags on. *The Washington Post.* https://www.washingtonpost.com/world/2022/05/19/china-covid-lockdown-shanghai-mental-health-crisis/

Silver, A. (2022, March 4). 'COVID Zero' regions struggle with vaccine complacency. *Nature.* https://doi.org/10.1038/d41586-022-00554-0

Singapore Management University. (2022, January 17). Safe management measures on SMU facilities' Access. https://www.smuaa.org.sg/newinfo

Singman, B. (2020, June 25). Attorney Sues Cuomo over 'unconstitutional' 14-day quarantine for out-of-state visitors. *Foxnews.* https://www.foxnews.com/politics/attorney-sues-cuomo-over-unconstitutional-14-day-quarantine-for-out-of-state-visitors

Taylor, L. (2022, March 17). Covid-19: Hong Kong reports world's highest death rate as zero Covid strategy fails. *BMJ.* https://doi.org/10.1136/bmj.o707

Tencent. (2022, February 23). Xingchengma Shi Zenmme Zhidao Ni Quguo Naxie Difang? Guanji, Badiao Shoujika Rengyou Jilu? 行程码是怎么知道你去过哪些地方?关机、拔掉手机卡仍有记录? [How do travel codes detect which places you have been? Can turning off or taking out SIM erase records?]. https://new.qq.com/omn/20220223/20220223A0D6UA00.html

Teo, J. (2021, November 8). Singapore hospitals under significant pressure; two-thirds of ICU beds occupied. *The Straits Times.* https://www.straitstimes.com/singapore/health/singapore-hospitals-under-significant-pressure-two-thirds-of-covid-19-icu-beds

The Pioneer. (2021, December 14). Booster shots of Covid vaccine will be key part of Singapore's strategy to fight Omicron: PM Lee. https://www.dailypioneer.com/2021/world/booster-shots-of-covid-vaccine-will-be-key-part-of-singapore-s-strategy-to-fight-omicron--pm-lee.html

The Standard. (2022, June 9). Carrie Lam rules out mainland China border reopening any time soon. https://www.thestandard.com.hk/breaking-news/section/4/191014/Carrie-Lam-rules-out-mainland-China-border-reopening-any-time-soon%C2%A0

The State Council of the People's Republic of China. (2022, April 18). China nucleic acid testing capacity at 51.65m samples a day.

The University of Hong Kong. (2022, September 8). *HKU Vaccine Pass*. https://covid19.hku.hk/control/hku-vaccine-pass/

Tsang, S. J. (2022, August). Predicting COVID-19 vaccine hesitancy in Hong Kong: Vaccine knowledge, risks from coronavirus, and risks and benefits of vaccination. *Vaccine: X, 11*, 100164. https://doi.org/10.1016/j.jvacx.2022.100164

U.S. Department of Health & Human Services. (2022, August 31). *COVID-19 vaccines*. https://www.hhs.gov/coronavirus/covid-19-vaccines/index.html

Wain, R., Insall, L., & Sleat, D. (2022, July 29). *Pressing go on mass testing*. Tony Blair Institute for Global Change.

Wang, Z. (2020, December 28). Wanggehua Shehui Zhili Fangshi de Lishixing Ji Benzhi 网格化社会治理方式的历史性及本质 [The historicity of grid community and its nature]. CE. http://views.ce.cn/view/ent/202012/28/t20201228_36163980.shtml

Weixin. (2022, March 14). Meihekou Shi Xinguan Feiyan Yiqing Gonggao [2022 Di Qi Hao] 梅河口市新冠肺炎疫情公告 [2022第7号] [Public announcement regarding coronavirus pandemic in Meihekou City, 7th of 2022]. https://mp.weixin.qq.com/s/NipLuuJz17d5y44gwCnRfg

Weixin. (2022, April 2). Xingtai Shi Nanhe Qu Fabu Gonggao Quanyu Shishi Fengkong Guanli 邢台市南和区发布公告全域实施封控管理 [Nanhe District, Xingtai issued public announcement imposing district-wide lockdown]. https://mp.weixin.qq.com/s/fGyHZ_Qaw6HdqKiWdYZcDA

Wenweipo. (2022, March 6). Qingyi Fangcang 5tian Zhuren Buzu 7% 青衣方舱5天住人不足7% [Tsing Yi cabin hospitals had an occupation rate of less than 7% after 5 Days]. https://www.wenweipo.com/a/202203/06/AP6223f67ee4b036dce99b2942.html

Wenzhou xinwen. (2021, December 17). Binjiang Jiedao Zhineng Menci 'Shanggang' Jujia Geli Duixiang Yikaimen, Houtai Jiufa 'Jingbao' 滨江街道智能门磁'上岗' 居家隔离对象一开门，后台就发'警报' [Intelligent door sensor 'came into effective' in Binjiang Street, alarm will automatically ring when those under house quarantine open the door]. http://news.66wz.com/system/2021/12/17/105425792.shtml

Wong, W., & Wu, A. M. (2021, December 6). State or civil society—What matters in fighting COVID-19? A comparative analysis of Hong Kong and Singapore. *Journal of Comparative Policy Analysis: Research and Practice*, 1–18. https://doi.org/10.1080/13876988.2021.1978819

Wong, J., Chaw, L., Koh, W. C., Alikhan, M. F., Jamaludin, S. A., Poh, W. W. P., & Naing, L. (2020, October 7). Epidemiological investigation of the first 135 COVID-19 cases in Brunei: Implications for surveillance, control, and travel restrictions. *The American Journal of Tropical Medicine and Hygiene, 103*(4), 1608–1613. https://doi.org/10.4269/ajtmh.20-0771

Wong, S. Y. S., Kwok, K. O., & Chan, F. K. L. (2020, May 11). What can countries learn from Hong Kong's response to the COVID-19 pandemic? *Canadian Medical Association Journal, 192*(19), E511–E515. https://doi.org/10.1503/cmaj.200563

Wong, O., Tsang, D., & Cheung, E. (2021, November 25). Hong Kong has met 'basic requirements' for reopening border with mainland China but more remains to be done. *South China Morning Post*. https://www.scmp.com/news/hong-kong/article/3157319/hong-kong-mainland-chinese-officials-hammer-out-final-details-cross

Xiao, F. (2022, March 18). Tamen Shi Wanggeyuan, Haishi Fuwuyuan, Paichayuan, Xuanchuanyuan...他们是网格员，还是服务员、排查员、宣传员...... [They are grid organisers, they are also servants, investigators, propagandists...]." Dazhong Ribao. http://124.133.228.83/articleContent/2205_995222.html

Xie, E. (2022, April 12). Learning from Hong Kong, China targets elderly in Covid-19 vaccine drive. *South China Morning Post*. https://www.scmp.com/news/china/science/article/3173891/learning-hong-kong-china-targets-elderly-covid-19-vaccine-drive

Xing, Y., Wong, G. W. K., Ni, W., Hu, X., & Xing, Q. (2020, December 3). Rapid response to an outbreak in Qingdao, China. *New England Journal of Medicine, 383*(23), e129. https://doi.org/10.1056/NEJMc2032361

Xinhua. (2022, February 22). *Mobile cabin COVID-19 hospitals supported by Mainland Being Constructed in Hong Kong*. https://english.news.cn/20220222/3e5e37b66d6f426499fe1b8d7df71c47/c.html

Xinmin. (2022, June 25). "Shanghaishi Di Shier Ci Dangdaihui Kaimu! Li Qiang: Hongyang Weida Jiandang Jingshen Jianxing Renmin Chengshi Linian 上海市第十二次党代会开幕!李强: 弘扬伟大建党精神 践行人民城市理念 [Shanghai Municipal 12th Party Delegates convention opens Li Qiang: Propagate the great "establish the Party Spirit" practice the ideal of building city for the people]. http://news.xinmin.cn/2022/06/25/32190122.html

Yan, Y., Pan, H., Shao, N., Xuan, Y., Wang, S., Li, W., Li, X., et al. (2020, December). COVID-19 in Singapore: Another story of success. *International Journal of Mathematics for Industry, 12*(01), 2050001. https://doi.org/10.1142/S266133522050001X

Yeo, R., & Ng, K. (2022, February 28). Bodies pile up at hospitals and mortuaries struggle to find space as Covid-19 deaths climb in Hong Kong. *South China Morning Post*. https://www.scmp.com/news/hong-kong/health-environment/article/3168601/bodies-pile-hospitals-and-mortuaries-struggle

Zheng, W., Cheung, E., Lam, N., & Tsang, D. (2021, November 22). Coronavirus: Tighten airport measures, Beijing delegation urges Hong Kong in inspection visit over border reopening. *South China Morning Post*. https://www.scmp.com/news/hong-kong/health-environment/article/3156917/coronavirus-hong-kongs-contact-tracing-efforts?utm_source=Yahoo&utm_medium=partner&utm_content=3157319&utm_campaign=contentexchange

Simon X. B. Zhao is Professor and Associate Dean at the Faculty of Humanities and Social Sciences BNU-HKBU United International College (UIC). He used to teach in the University of Hong Kong as a founding Director of International Center for China Development Studies and specializing at urban, regional and global studies and publishing in the broad areas of urban regional studies, geopolitical and economic development, spatial and global transformation, geography of international trade and finance, and development of international financial centers (IFCs). He is an editorial board member for four international journals and published more than 150 papers in international refereed journals, mainly SSCI/SCI/CSSCI Citation Journals, book chapters and consultative reports etc. Simon Zhao was a Member of the Central Policy Unit, HKSAR Government (Pan-PRD Panel) and chaired and co-lead many substantial academic research projects, providing widely consultancy services to China national, provincial, and local governments as well as private sector. His latest edited volume *COVID-19 Pandemic, Crises Response and the Changing World: Perspectives in Humanities and Social Sciences* (by Springer Nature, 2021) is well received globally.

Bo Yan received PhD in History from Nanyang Technological University, Singapore. He is a Research Fellow in the Faculty of Humanities and Social Sciences. Yan's main research interests include modern Chinese history, postwar United States, and modern Southeast Asian history, and the history of the global 1960s. He has published in the *Journal of Asian Studies, Twentieth Century China*, and others. Yan's PhD dissertation was listed as the International Convention of Asia Scholars (ICAS) Book Prize 2021-Best Dissertation in Humanities. Bo Yan is currently conducting studies and research on global pandemic of COVID-19.

Yutong Chen currently studies at Beijing Normal University-Hong Kong Baptist University United International College, where she majors in English Language and Literature studies with a particular focus on the representations of China in the 18th century British periodicals. Chen's areas of interest include comparative race and empire studies, postcolonial literature, and Asian American diaspora studies.

Economic Considerations Behind the COVID-19 Pandemic Control Policies

Simon X. B. Zhao, Bo Yan, Yu Liu, and Jiaqi Liu

1 Introduction

Since the outbreak of coronavirus cases at the turn of 2020, there gradually emerged two major pandemic combatting policies: "zero Covid" policy and "live with Covid" policy. Many previous studies have attributed differentiated paths each country adopts to their political systems.[1] Their main ideas are as follows: as China is often considered an authoritarian state, therefore, China adopts the zero Covid policy; Singapore, less authoritarian so, adopts "zero Covid" policy at a lesser degree; democratic countries in the West generally pursue to "live with Covid," while countries Sweden were most relaxed because of their social democratic institutions.[2] Such a dichotomy did not fit into realities in many countries. For example, Austria,[3] Australia, Canada,[4] and Taiwan[5] are indisputable democratic societies, yet these societies have adopted some of the most stringent control measures in the world.

This chapter does not intend to challenge political determinant in these already lengthy analyses. However, it seeks to address the economic dimension in representative pandemic control policies and strives to see what economic costs/benefits lie beneath the mentality of pandemic control policies. There exist a few research articles observing economic costs regarding the pandemic, however, they are mostly

[1] For example, see Martínez-Córdoba et al. (2021), Arlina Dewi et al. (2020), Nicholas Stowell et al. (2022), Scott L. Greer et al. (2020).

[2] Staffan Andersson et al. (2022).

[3] Jonathan Tirone and Naomi Kresge (2021).

[4] Tamer Oraby et al. (2021).

[5] Focus Taiwan: CNA English News (2021).

S. X. B. Zhao (✉) · B. Yan · Y. Liu · J. Liu
Beijing Normal University-Hong Kong Baptist University,
United International College, Zhuhai, Guangdong, China
e-mail: sxzhao@uic.edu.cn

© The Author(s), under exclusive license to Springer Nature Singapore Pte Ltd. 2023

S. X. B. Zhao et al. (eds.), *Comparative Studies on Pandemic Control Policies and the Resilience of Society*, https://doi.org/10.1007/978-981-19-9993-2_3

within the first year of the pandemic when the balance of human lives lost and economic costs were not obvious.[6] This chapter conducts an original set of research by firstly dividing the pandemic from 2020 to 2022 into two phases in its examination of pandemic control navigations in China, Australia, Singapore, and the United States; and secondly following closely Shanghai, a focal point in pandemic control approaches comparison and a timely and representative case city struggling between public health and socio-economic development. This chapter has referred to various data and put Shanghai in comparison in terms of lives lost and economic downswing. On this basis, this chapter highlights that when economic costs are considered compared to infected cases or human lives lost (which is often considered the sole standard in measuring the effectiveness of pandemic control policies), it will trigger serious considerations whether there exists a "best" or "one-size-fits-all" policy, or a pandemic policy that could be effective at all stages.

2 Economic Determinants behind Countries' Selection of Pandemic Control Policies

This section will observe three zero COVID pursuing nations in their first two years of fighting the pandemic: China, Australia, and Singapore. China is a prime example and the last holdout of a stringent zero Covid policy. Australia is one of the few Western countries that imposed large-scale lockdown and zero COVID policy (throughout most of 2020 and 2021). Singapore is one example that swiftly transited from a zero COVID country to live with Covid. By delineating their navigation through the pandemic, this section would suggest that economic considerations, side by side with considerations over infected case and death numbers, play a significant role in determining an individual country's pandemic control policy. This was evident in China's insistence on zero COVID, Australia's denial of a "trade-off" assumption, and Singapore's quick switch to live with COVID.

2.1 The Economic Consideration behind China's Insistence on Zero COVID Strategy in the First Two Years of the Pandemic

After hit by the very early coronavirus outbreak in Wuhan, China's quick suppression measures, such as mass testing and lockdowns, demonstrated an indisputable effectiveness in not only curbing the virus infection but also in resuming industrial production. Against a population of 1.4 billion, China has reported just 120,000 cases throughout the first two years of the pandemic.[7] Moreover, China's zero COVID

[6] Joanna Sokolowska and Tomasz Zaleskiewicz (2020).

[7] Our world in data.

strategy proved economically resilient in 2020 and 2021. The city of Wuhan had undergone a strict lockdown in early 2020. Yet, the Wuhan lockdown did not lead to an economic recession. Instead, it was estimated to save about 20,000 lives and save nearly 30 billion RMB on medical costs, resulting in long-term benefits on the society.[8] Compared with major economies worldwide, China's economy has reported a 3.2% increase, becoming the only major economy with net growth in 2020.

The low number of infected cases and the distinctive economic performance validated China's zero COVID strategy. There are analyses of how China COULD manage to contain the waves of outbreak with its staunch insistence on zero COVID policy (mainly from the perspective of political systems).[9] However, few studies have approached the question WHY China HAS TO adopt this zero COVID policy. Some analysts who would state that China's insistence on zero COVID strategy is due to China's limited medical resources. For example, one report estimated that once China "coexists" with Covid-19 like the United States did, there would be estimated daily new cases of 637,155 and daily severe cases of 22,364 (Fig. 1).[10]

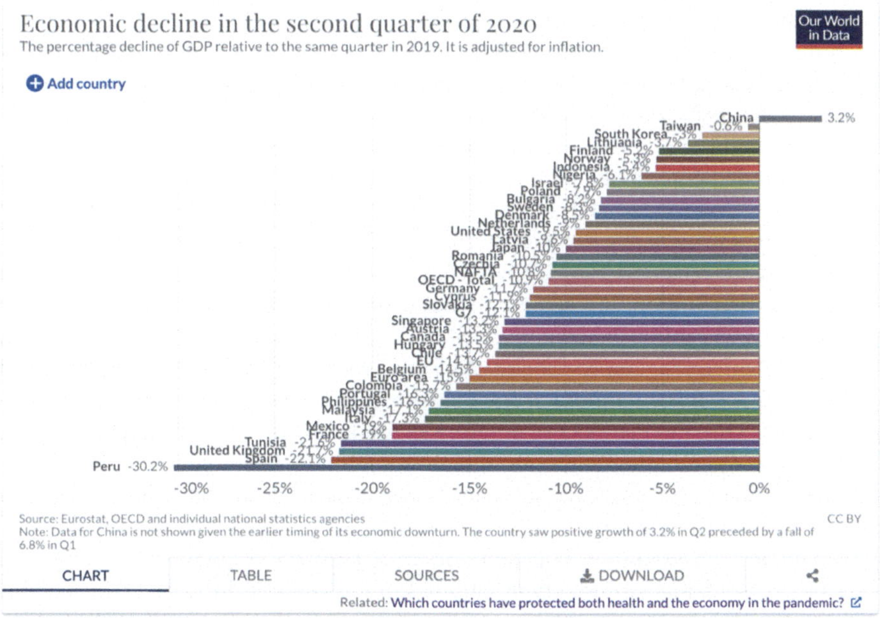

Fig. 1 China's quick effectiveness of combating the pandemic was evidenced by its economic performance in 2020[11]

---

[8] Shibing You et al. (2020).

[9] Zhuoran Li (2022).

[10] Yuan Zhang et al. (2021).

[11] From Our World in Data.

This section would suggest that China's seemingly draconian restriction measures under the zero Covid strategy can also be explained by its economic and industrial status. China has a giant domestic economy, giving itself an edge in maintaining its economic activities within its own proper without having to resume the economic production with border widely open. More importantly, China's insistence on the zero COVID policy was the result of the dominance of manufacturing in China's economy.

China's labour-intensive manufacturing asked for an elimination policy when manufacturing factories were hyper-prone to infection and vaccinations were unavailable. The manufacturing factories were highly **concentrated**, highly **closed**, and if hit by the virus, highly **contagious**. The three Cs had rendered manufacturing factories clusters of coronavirus outbreak throughout the world. For example, more than 200 people tested positive for Covid-19 at the KP Textile factory at Guatemala.[12] 90% of big United States meat plants had Covid-19 cases in the first year of the pandemic.[13] Manufacturing accounted for around 40% of China's GDP as of 2020,[14] and 95% of China's exports.[15] To avoid economic downturn caused by infections of workers in manufacturing industry thus constituted an economic consideration in China's decision-making on pandemic control policy.

China's *Communiqué on the Fourth National Economic Census* (No. 2) issued in 2019 presented statistics that might further explain China's economic motives undergirding a stringent zero COVID policy. By the end of 2018, there were 18.570 million corporate enterprises in the secondary and tertiary industry. Of the total, domestic-funded enterprises accounted for 98.8%, enterprises funded by investors from Hong Kong, Macao and Taiwan for 0.6% and foreign-funded enterprises for 0.6%. Among the domestic funded enterprises, state owned enterprises took up 0.4% and private enterprises 84.1%.[16] When it came to the number of persons employed by status of registration, the same national economic census reported that private enterprises (2.912 million) had employed 59.794 million employees.[17] On average, each private enterprise employed only 20.5 persons. Faced with the risk of infection clusters, shutdown, or temporary out of work, these enterprises would be almost indefensible. The small number of enterprises funded by investors from Hong Kong, Macao, Taiwan, and other countries also meant China's less urgent need for cross-border travelling. A closure of national border could be implemented at limited expense of China's economy.

In comparison, countries not adopting zero COVID strategy in the first two years of the pandemic suffered both fatalities and recession. In April 2021, Institut économique Molinari, a research institute located in Paris, France, published its report suggesting that the zero COVID strategy protected both people and economies more

---

[12] Business and Human Rights Resource Centre (2020).

[13] Leah Douglas (2022).

[14] Sean Ross (2022).

[15] Nicholas R. Lardy and Tianlei Huang (2020).

[16] National Bureau of Statistics of China (2019a).

[17] National Bureau of Statistics of China (2019b).

effectively. Taking France (not adopting zero COVID policy) as an example, the researchers suggested that the deaths per million people were 42 times higher in France, indicating 63,000 preventable deaths by the end of 2020 or 86,000 as of March 2021. The economy of France also suffered significantly due to not adopting a zero COVID strategy. The researchers' data was that the GDP declined five times more in France, representing lost earnings of 2,200 euros per French citizen.[18] The same report propounded that the zero COVID policy benefited the economy in both short and long term. Specifically, the researchers pointed out that zero COVID countries experienced a less severe economy decline than countries living with Covid: −4.5% versus −11.7% for the second quarter of 2020 and −1.2% versus −3.3% for the fourth quarter of 2020. The researchers also presented that the zero COVID strategy ensured better economic activities' stability and controlled uncertainty, measured by number of google searches of key terms like "workplace traffic" and "restaurants."[19]

Proponents of "live with Covid" would support the idea that living with COVID could be seen as a "trade-off" for economic continuity and recovery.[20] However, data-based research did not champion such a statement. In November 2020, the Institute for New Economic Thinking published their studies on whether saving the economy had to come at the expense of human lives. Titled "To save the economy, save people first," their report debunked the "trade-off" theory by presenting that lockdowns and other restriction measures did work to both save lives and recover the economy.[21] Citing the IMF's report "Fiscal policies database in response to Covid-19," the researchers showed how well each country had protected its citizens versus the total cost of doing so (Fig. 2).

As seen from the chart above, those countries that sacrificed economy to save lives, meaning that they mostly adopted a zero COVID policy in 2020, had saved both lives as well as livelihoods. Those countries pursuing a "trade-off," namely to sacrifice lives to save livelihoods, failed to achieve either.

Similar results were observed by Joe Hasell's study published on *Our World in Data*. Hasell presented that contrary to the idea of a trade-off, countries that suffered the most severe economic downturns, like Peru, Spain, and the UK, were generally among the countries with the highest Covid-19 death rate. The reverse was also true: that countries with lowest death rate tended to have less severe economic downturns.[22]

---

[18] Cécile Philippe and Nicolas Marques (2021).

[19] Philippe and Marques (2021).

[20] For example, see Ali Zeytoon-Nejad and Tanzid Hasnain (2021), Shalini Sarin Jain et al. (2022), Claudia F. Nisa et al. (2021).

[21] Philip Alvelda et al. (2020).

[22] Our World in Data.

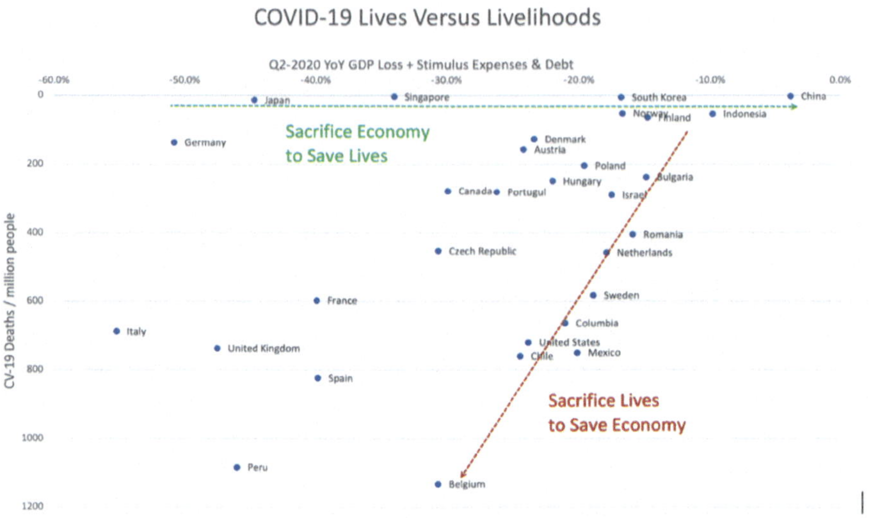

Fig. 2 COVID-19 Lives versus Livelihoods (From the Institute for New Economic Thinking)[23]

2.2 *The Economic Rationale behind Australia's Nation-Wide Lockdowns in 2020 and 2021*

The very idea that debunked the "trade-off" theory and supported saving lives first became theoretical underpinning of Australia's lockdown measures in response to the coronavirus pandemic in 2020–2021. As "one of the world's strictest virus control regimes,"[24] Australia imposed harsh lockdowns upon the nation's territories. For example, in the city of Melbourne, Australia has imposed the world's longest COVID related lockdowns, aiming to sacrifice temporary economy to save human lives. As of October 2021, Melbourne has experienced six lockdowns totalling 262 days since March 2020.[25] Intermittently, Sydney also underwent lockdowns lasting over 100 days.[26] It was estimated that it cost Melbourne $100 million per day due to lockdown[27] and in Greater Sydney area, a day of lockdown would cost $143 in lost activity and wipe more than $2 billion from GDP.[28]

The economic costs ran high though, researchers still supported that elimination (or a zero COVID) strategy fared better than suppression (or living with Covid) strategy.[29] One such study was conducted by Tom Kompas et al. (2021). By using

---

[23] Alvelda et al. (2020).

[24] Ben Westcott (2022).

[25] Al Jazeera (2021).

[26] BBC (2021).

[27] David Crowe et al. (2021).

[28] Ronald Mizen (2021).

[29] Tony Blakely et al. (2021).

a value of a statistical life year (VSLY) and an age-adjusted value of statistical life (A-VSL), the researchers estimated the human fatalities and economic costs caused by the early suppression methods. The researchers concluded that early rather than delayed mandated suppression imposed much lower economy and human lives lost, and the "go early, go hard" strategy, namely, the essence of zero COVID strategy, resulted in the lowest estimated public health and economy costs.[30]

Tony Blakely et al. (2021) used an agent-based model to estimate daily COVID infection rates and time and a proportional multistate lifetable model to estimate long-run health impacts (health adjusted life years [HALYs] arising from SARS-CoV-2) and costs (health systems, and health system plus GDP). Their findings have demonstrated that aggressive elimination (namely, zero COVID measures) resulted in the best performance of pandemic control in terms of numbers of infections and days of restriction; while moderate elimination (less restrictive zero COVID measures) would affect the economy to the least degree.[31] Tony Blakely et al. (2021) thus argued that from a health system perspective, an aggressive elimination was the optimal approach followed by a moderate elimination approach, while a moderate elimination approach is the optimal approach when it comes to social economy, followed by an aggressive elimination approach.[32]

Tony Blakely et al. (2021) built their research upon an analysis published in the *Lancet*. Miquel Oliu-Barton et al. (2021) argued that the trade-off, namely, the balance between public health, economic growth, democratic solidarity, and civil liberty, did not need to be an issue in the responses to COVID. Their research illustrates that countries that adopted elimination actions did fare better than countries that did not.[33] For example, measuring the COVID deaths per one million population among the OECD (Organisation for Economic Co-operation and Development) countries, the five countries that adopted an elimination approach (Australia, Iceland, Japan, New Zealand, and South Korea) have reported 25 times lower than other OECD countries that did not pursue elimination.[34] Researchers also considered Australia's pursuit of elimination, or a zero COVID strategy the reason behind Australia's "less economic scarring, and a stronger recovery than any other OECD country apart from South Korea."[35] In one word, a zero COVID strategy was the best response that Australia had to offer when trying to balance both human lives lost and economic costs.

It should be noted that the effectiveness of such zero COVID strategy, particularly in terms of economy, is temporal and subject to biological traits of variants of the virus. Various research and analyses, whether they are favouring a zero COVID or "live with COVID" strategy, often fail to examine the validity and effectiveness of the

[30] Tom Kompas et al. (2021).

[31] Blakely et al. (2021).

[32] Blakely et al. (2021).

[33] Miquel Oliu-Barton et al. (2021).

[34] Oliu-Barton et al. (2021).

[35] Patrick Abraham et al. (2021).

strategy against the biological traits of COVID-19, in other words, the differentiated variants that dominated the COVID pandemic.

The requirement of stringent lockdowns, one measure within the zero COVID strategy, lies in the fact that in the first one year and half, the original variant was the dominant variant of the coronavirus. The original variant of the coronavirus, compared with the two following major variants, the Delta and Omicron variants, has recorded a lower transmissibility rate but a higher death rate. These traits thus called for a zero COVID strategy to protect lives, especially in the context that vaccines were not available then. However, when the pandemic has entered the stage of the Delta and Omicron variants domination, a zero COVID strategy is then subject to review and reconsideration. The far higher infection rates of the Delta and Omicron variants have rendered key measures under zero COVID strategy, such as mass testing and large-scale lockdowns more costly and less effective. The availability of vaccines has greatly reduced COVID-related death and severe cases. In this context, a continuation of stringent zero COVID strategy, oftentimes a physical confrontation against the virus by restricting human movements, lockdowns, mass testing, was called into question.

2.3 Singapore's Quick Switch from Zero COVID Strategy to Live with COVID in 2021

Among all zero COVID pursuing countries, Singapore was one of the first to switch from zero COVID strategy to live with COVID. Singapore's decision was based on the biological traits of different coronavirus variants, the availability of vaccines, and the economic consideration.

Singapore did adopt a zero COVID strategy in the first and half year of the pandemic. Various restriction measures such as "circuit breaker," limitation of the size of groups dining in restaurants, and closure of national border have been the subjects of multiple research literature. The pandemic efforts of Singapore, however, did not stop at making these exceptional measures permanent.

This chapter looks at Singapore's quick switch from an economic angle. As a pure trade-oriented nation, Singapore could not afford a disconnection from foreign trade and foreign tourists while combating the spread of the virus. In terms of trade-to-GDP ratio, Singapore stands at a value of 321, meaning that international trade of Singapore is three times the amount of the tiny island-nation's gross domestic product.[36] As the regional aviation hub, Singapore's aviation industry stood as a pillar component in its economy. In 2018, 18.5 million foreign visitors came to Singapore with a total of 36.1 million passengers were carried through the air transport sector. Directly and indirectly, the aviation sector had generated more than 375,000 jobs in 2019 and a 36.6 billion US dollar contribution to the GDP of Singapore.[37] IATA, the International Air

[36] R. Hirschmann (2021).

[37] Easwaramoorthy Rangaswamy et al. (2022).

Transport Association, has provided a more detailed breakdown of the contribution of aviation sector to the economy of Singapore. Singapore's aviation sector, such as airlines, airport operators, airport on-site enterprises, restaurants, retails, aircraft manufacturers, and air navigation service providers has absorbed 119,000 jobs. In addition, the supply chain in air transport, such as goods and services providers to the airport and airlines, as well as local suppliers, have absorbed 78,000 jobs. Besides, the IATA has analysed that the air transport sector also supported a further 26,000 jobs "through the wages it pays its employees, some or all of which are subsequently spent on consumer goods and services." Another 152,000 jobs were generated through tourism, including foreign tourists arriving in Singapore via air. In total, 375,000 jobs in Singapore were supported by the air transport sector (as of 2016), which made up of roughly 6% of the total residents of Singapore.[38] Among the 36.6 billion US dollars generated, 13.3 billion US dollars were in direct employment, 6.6 billion US dollars were in supply chain jobs, 2.2 billion US dollars were in employment spending, and 14.5 billion US dollars were in tourism.[39] 11.8% of Singapore's GDP was supported by air transport and foreign tourists arriving by air.[40]

Heavily dependent upon international travel activities, Singapore's aviation sector was severely hit by various restriction measures imposed as response to the COVID-19 pandemic. Reported by ChannelNewsAsia, as of February 2021, the total passengers' movements were only 2% of pre-COVID levels.[41] What was even more disturbing to the economy of Singapore was the impact of such restriction measures upon international commerce and trade connected with Singapore. As 37,400 international companies based their operations out of Singapore and 7,000 of which were multinational corporations,[42] Singapore became home to the largest number of headquarter jobs from global Fortune 500 companies compared with other key Asian hubs.[43] The pandemic, as well as subsequent restriction measures, sent Singapore's economy to its worst ever recession in 2020. Although slightly better than an official forecast of 6–6.5% contraction, the economy of Singapore recorded a 5.8% contraction in 2020.

The economic disruption and the urgency to resume international connection, side by side with the evolving variant of the virus, validated Singapore's quick decision to shift from zero COVID strategy to live with COVID. In October 2021, marked by the speech by Prime Minister Lee Hsien Loong, Singapore officially transited to live with the COVID. PM Lee explained such motive by two reasons. The very first was the biological evolvement of virus variants. The high infection rate of the Delta variant was times higher than the original variant, rendering stringent restriction measures less effective. Therefore, Singapore concluded that a zero COVID strategy

---

[38] IATA (n.d.).

[39] IATA (n.d.).

[40] IATA (n.d.).

[41] Faizal Yahya (2021).

[42] Business Times (2018).

[43] The Singapore Economic Development Board (2022).

was no longer feasible.[44] Yet, the Delta variant had reported a lower rate of severe and death cases. By the time of Lee's speech, Singapore had achieved a high vaccination coverage across the nation,[45] preparing Singapore medically to live with COVID.

Another reason was economic consideration. To directly quote PM Lee's words:

> Singapore cannot stay locked down and closed off indefinitely. It would not work, and it would be very costly. We would be unable to resume our lives, participate in social activities, open our borders, and revive our economy. Each time we tighten up. Businesses are further disrupted. Workers lose jobs. Children are deprived of a proper childhood and school life. Families are separated for even longer. Especially families with loved ones overseas and extended families who have not been able to come together. All these cause psychological and emotional strain, and mental fatigue. For Singaporeans and for everyone else here with us, including our migrant workers.[46]

At the initial stage of the pandemic, saving human lives has outweighed temporary economic growth in the mindset of Singapore's leaders. Then, at the stage with more infectious variants and less fatal cases, the economic consideration was given a more prominent place in the pandemic control efforts and the changed pandemic control approach.

On 16 February 2021, in view of the adverse impact of the COVID-19 pandemic upon its aviation sector and the urgency to cement Singapore's status as the regional aviation hub, Singapore provided special grant to the aviation sector. This resilience package included an allocation of 870 million Singapore dollars to support the aviation sector, 10% landing charge rebate for all scheduled passenger flights landing in Singapore for airlines, and a 50% rebate on rental paid for ground handling companies' lounges and offices within Changi Airport and Seletar Airport terminal buildings.[47] Heng Swee Keat, Deputy Prime Minister and Minister of Finance, announced such decision in his budget speech to the parliament that the support was to secure Singapore's "position as a key aviation hub and maintain Changi's position as a safe, trusted, and well-connected airport."[48]

After switching to live with COVID, Singapore gradually eased travel restrictions imposed upon international travellers. Right after PM Lee's speech, Singapore launched quarantine-free travel arrangements with 10 countries following an over 20-month closure of national border. During the border closure, foreign travellers into Singapore were either required to quarantine at a hotel at travellers' own expenses or isolate at home. The re-opening of Singapore to vaccinated travellers from these VTL (vaccinated travel lanes) countries was met with economic responses. According to a Financial Times article, shares in Singapore Airlines rallied to their highest levels since April 2021, up 8% since the launch.[49] From 1 April 2022, Singapore eased its travel restrictions by allowing fully vaccinated travellers under the vaccinated

---

[44] Prime Minister's Office Singapore (2021).

[45] See chapter "Between zero COVID and 'Live with COVID'".

[46] Prime Minister's Office Singapore (2021).

[47] Yahya (2021).

[48] Ministry of Finance Singapore (2021).

[49] Mercedes Ruehl (2021).

travel framework (VTF) to enter Singapore without stay-home notice and on-arrival tests, only requiring them of pre-departure test 2 days prior to departure. On 26 April 2022, the pre-departure test requirement was dropped. Border control measures were further eased.

Singapore's quick and timely switch from zero COVID to live with COVID resulted in a recovery of the pandemic-hit economy. In 2021, the Singapore economy grew by 7.2%, rebounding from a 5.4% contraction in 2020.[50] The economic recovery accelerated in the final three months of 2021, following the ease of virus restrictions. This 7.2% recovery rebounded to fastest in more than a decade.[51] In 2022, even facing the pressure of inflation, the recovery continues. The first quarter of 2022 achieved 4% GDP growth, and the second quarter of 2022 was up 4.8%.[52] When it comes to Singapore's aviation industry, a 28 July 2022 news release from Singapore Airlines (SIA) has reported that a record first-quarter opening of $556 million (Singapore Dollar), which is also the second-highest quarterly profit in SIA's history. This came after Singapore fully opened its border in April 2022. In terms of number of passengers, Singapore Airlines and Scoot (a low-cost arm of the SIA group) carried 5.1 million passengers during the quarter, up 158.2% from the previous quarter. SIA's quarterly revenue per available seat-kilometre was 10.2%, a record for the full-service airline. Passenger flown revenue rose $1,456 million (+119.3%) quarter-on-quarter to $2,676 million on the back of a 126.7% growth in traffic. Passenger load factor rose 34.1% points to 79.0%, the highest since the onset of the pandemic, as the traffic growth outpaced the capacity expansion of 28.9%. The $556 million operating profit is also a $623 million improvement from the $67 million loss in the previous quarter, and on a year-on-year basis, this was better by $830 million from the previous year's opening loss of $274 million.[53] Notably, the manufacturing output has grown by 13.8% in May 2022,[54] in contrast to China's economic downswing at the same period, which the chapter will look into below.

3 Economic (Re)considerations in the Era of Omicron

This section observes economic performance of countries in the era of Omicron. The countries being analysed are China and the United States. These two countries are of special value in an observation and comparison of economic performances in relation to the "zero COVID" and "live with COVID" dichotomy. China remains as the world's last major zero COVID pursuing nation. The United States, with the responsibility of handling the coronavirus crisis left to the 50 states, saw two approaches

[50] Kentaro Iwamoto (2022).

[51] De Wei Low (2022).

[52] Rajiv Biswas (2022).

[53] Singapore Airlines (2022).

[54] Biswas (2022).

to the pandemic, namely closure and opening side by side in one country. The two countries' economic performances tied to pandemic control policies have further corroborated the complexity of economic considerations in the era of Omicron.

3.1 *Economic Decline following China's Lockdowns in 2022*

Following the step of Singapore, many previous zero COVID nations, such as Australia,[55] New Zealand,[56] and South Korea,[57] have pivoted to live with COVID. China has thus become the world's last strong hold-out of zero COVID pursuit.

As the aim of zero COVID strategy was the elimination of infected cases, therefore, when measured by the number of cases, China's zero COVID strategy proved to be very effective in the first two years of the pandemic. As of writing (July 2022), China has recorded less than 900,000 confirmed cases.[58] However, the zero COVID strategy that China has adopted in combating the pandemic came under growing scrutiny and even criticism when the Omicron variant has become the dominant variant. As happened in previous zero COVID countries, the highly infectious Omicron variant has rendered restriction measures more and more costly, the lower rate of severe and death cases caused social and psychological[59] fatigue over draconian measures. The effectiveness of China's zero COVID strategy in the era of Omicron was called into further question when an economic performance standard was introduced to examine the effectiveness of the strategy. Following a controversial lockdown in the economic powerhouse of Shanghai, China's economy was sharply contracted.

An estimation in a KPMG report monitoring China's economy by May 2022 suggested that there were 41 cities containing risk area(s) in April over 10 days, representing about 21.5% of China's population or 25.7% of GDP.[60] In April 2022, economists from Hong Kong and Mainland China estimated economic cost of lockdowns in China in response to the pandemic. By comparing 1.8 million city-to-city truck flow data in cities with and without lockdowns, the economists estimated that a one-month full-scale lockdown would imply a decline in the real income in affected cities. When it came to more exact numbers, the economists appraised that if one-month full-scale lockdown would be imposed on China's four largest cities (Shanghai, Beijing, Guangzhou, and Shenzhen), the four cities would lose their real income by 61% and the national income by 8.6%.[61]

Among the four cities, Shanghai experienced since the late February 2022 to June the most enduring, stringent, and consequential city-wide COVID-related lockdown

[55] Reuters (2022).

[56] Praveen Menon (2021).

[57] The New York Times (2022).

[58] From Our World in Data.

[59] The Lancet (2022).

[60] KPMG (2022).

[61] Jingjing Chen et al. (2022).

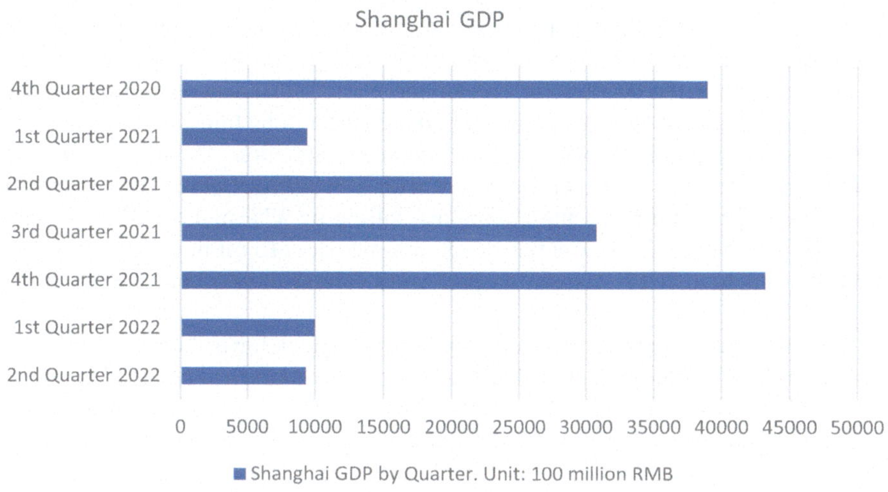

Fig. 3 Shanghai's GDP by quarter from the end of 2020 to 2nd quarter of 2022. Data is from the National Bureau of Statistics of China. Graph is made by authors

after Wuhan in China. While partly ongoing, the Shanghai lockdown's economic impact could be sensed from an assortment of detailed data. On 15 July 2022, the National Bureau of Statistics released China's economic data of the second quarter of 2022. Shanghai was reported with a sharp decrease by 13.7% in its GDP of the second quarter and a 5.7% contraction. Recording a GDP number at 933.9 billion (CNY), Shanghai's GDP fell not just short of 1trillion[62] from the first quarter of 2022 but witnessed an even apparent disparity of 4.32 trillion from the fourth quarter of 2021. Compared by sheer number, the GDP of Shanghai in the second quarter of 2022 was only half of the second quarter of 2021.[63] The economic contraction surrendered Shanghai's title as China's No. 1 economic metropolis to Beijing, though the latter also suffered a 3% contraction in the second quarter of 2022.[64] It should be noted that official data usually tended to be more "conservative" than actual economic performance, therefore, Shanghai's real economic data was likely to be even worse than officially recognized (Fig. 3).

Various industries in China have been negatively affected by the lockdowns. For example, the aviation in China was one industry that was severely affected due to lockdowns, health/travel codes restrictions, and local quarantine requirements. As Shanghai and Beijing stood as aviation hubs in China, the Shanghai lockdown as well as Beijing's harsh policies regulating inbound travellers severely impacted the aviation industry in China. A July 2022 news conference by Civil Aviation Administration of China (CAAC) stated that travel hours of all industries in the first half of 2022 had dropped from the levels of 2020 and 2019. It should be noted that when it

---

[62] People.cn (2022a).

[63] National Data, National Bureau of Statistics of China (n.d.).

[64] Dandan Meng (2022).

came to general aviation (all civil aviation aircraft operations except for commercial air transport or aerial work), there was an increase.[65] Therefore, the sharp drop was the direct result of the decrease in human and cargo movement. According to the CAAC announcement, the travel hours for all industries, total transport turnover, and passenger, cargo, and flight volume all witnessed contraction. The travel hours for all industries stood at 3.099 million hours with a 38.4% decrease. Total transport turnover was 29.34 billion tonne-kilometres, falling by 37% and equalling to only 46.7% of the level of 2019. The passenger volume was 118 million visits dropping by 51.9% (36.7% of 2019). Cargo transport saw a volume of 3.077 million, declining by 17.9%.[66]

Other industries were hit as well. The stringent lockdown, human movement restrictions, and income deduction due to economic disruption had pummelled manufacturing,[67] tourism,[68] catering,[69] as well as leisure and entertainment industries.[70] A spill-over results of the lockdown-led economic disruption was the unemployment, particularly the youth unemployment in China in 2022. On 15 June 2022, during a news conference, Fu Linghui, spokesperson of the National Bureau of Statistics, acknowledged that because of the effects of pandemic, the difficult situations of enterprises, and the shrunk capacity for enterprises to absorb employment, China's youth (16–24) unemployment rates had reached 18.4% in May, 0.2% up from April.[71] China's youth unemployment rate was faring far worse than youth unemployment rates in other major economies, in comparison, youth unemployment rate was 8.1% in the United States, 13.3% in the European Union, 3.8% in Japan, and 6.9% in South Korea.[72] In July, the youth unemployment rate deteriorated and soared to 19.3%.[73] Again, as official data tended to be more conservative, it could be estimated that actual youth unemployment could go even higher.

The economic downturn has been sensed by the Chinese authorities. In late May, Li Keqiang held a massive videoconference on the economy. As reported, Li's videoconference had summoned 100,000 attendees comprised of provincial, municipal, and county officials across China. The sheer number and scale led even the Global Times, a newspaper with governmental background to agree that the meeting was "unprecedented."[74] During the conference, Li was reported to warn that the economic difficulties, in some areas and to a certain extent, were even greater than the severe

[65] People.cn (2022b).

[66] People.cn (2022b).

[67] Jonathan Cheng (2022).

[68] BBC (2022).

[69] Informa Markets (n.d.).

[70] Kyle Mullin (2022).

[71] National Bureau of Statistics (2022).

[72] Leona Ying Liu (2022).

[73] Luna Sun (2022).

[74] GT Staff Reporters (2022).

shock of the pandemic in 2020.[75] By highlighting the gravity of the economic situation, Li emphasized on "stabilizing the economy base" (稳住经济大盘),[76] rather than an usual official narrative of "stabilizing economic growth" (稳增长), another implication that China's overall economy did not just achieve growth, but faced the danger of an unstable base. Also, Li emphasized to orchestrate the pandemic control and socioeconomic development in a highly effective manner (高效统筹疫情防控和经济社会发展). Han Yong Hong, Associate Editor of Singapore's *Zaobao*, analysed this statement as a warning from Li against concentrating on one issue (namely, the draconian pandemic control measures) at the expense of other priorities and implementing one-size-fits-all measures.[77]

3.2 *Closure versus Opening? Two Paths in Two Americas*

The United States became the worst-hit country by the COVID-19 pandemic. However, the simple statement veiled the fact that the 50 states the District of Columbia could implement varying pandemic control measures. It might not be accurate to argue that zero COVID strategy and "live with COVID" strategy were struggling among the states. There was indeed a divergence between closure and opening strategies in the United States, which resembled the measures taken under zero COVID and "live with COVID" strategies.

In *A Final Report Card on the States' Response to COVID-19*, a working paper to the National Bureau of Economic Research of the United States, researchers measured and compared the different economic and health trajectories across the 50 states and DC. The researchers approached the measurement and comparison from three categories: the economy, education, and mortality. They measured the economic performance by unemployment rates and GDP after industry composition adjustment. They used the Burbio cumulative in-person instruction percentage for the complete 2020–2021 school year to examine education performance. With regards to mortality, the researchers calculated COVID-associated deaths reported to the US Centers for Disease Control and Prevention (CDC) and all-cause excess mortality.[78]

It was understood that the economy and education were positively correlated, as states more willing to resume business and economic activities were more willing to resume in-person education. However, the researchers found that there was virtually no relation between economy scores and mortality, meaning that states trying to fight the pandemic by restricting economic activities did not significantly improve their health performance.

When it came to specific results, the states of New Jersey, New York, California, and the District of Columbia, all resorting to strict closure and lockdown measures,

[75] Evelyn Cheng (2022).

[76] GOV.cn (2022b).

[77] Yong Hong Han (2022).

[78] Phil Kerpen et al. (2022).

were among the worst performing states in all categories. They scored lowest in terms of normalizing economic activities while maintaining highest mortality rates. While states like Florida, Utah, Nebraska, Montana, South Dakota, all sticking to opening and easing COVID restrictions, ranking highest in navigating the pandemic.

The case of Hawaii needed special attention. Hawaii ranked first in health performance. The state had recorded the lowest COVID-related deaths per 100,000 people and lowest average mortality rates. Geographical location of Hawaii was believed to be attributable to Hawaii's low number of infected cases.[79] Nevertheless, Hawaii's low number of cases resulted from that the state had one of the strictest sets of control measures among the 50 states. Up to 26 March 2022, unvaccinated or those without a negative test from other places in the United States were still required to quarantine for five days. Hawaii was also the last state in the US to maintain pandemic safety measures such as wearing masks indoors.[80] Such strict set of measures, however, came with Hawaii ranking the last in economic performance. Its −6.3% GDP contraction was the second biggest among the states after Connecticut, and its unemployment rate was just above New York. Resembling China, Hawaii had also observed an obvious negative correlation of strict pandemic control measures and economic performance.

3.3 China's Pandemic Control Measures Adjustment: In View of Dire Economy?

Faced with a dire economic prospect, there emerged signs that governments at varied levels in China had slightly retreated from their preceding draconian and strict control measures.

Large-scale of mass COVID testing, one of the two key requirements under China's zero COVID strategy planning, has faced rendered multiple cities across China face financial challenges as local fiscal revenues were struggling. The National Health Care Security Administration of China announced following Li's meeting that massive testing fees must be borne by local authorities[81] and local basic health insurance should no longer cover the cost of massive testing but only the test bills for insured patients seeking medical treatment in a hospital.[82] The massive testing fees used to be supported by local fiscal revenues. However, the local fiscal revenues were severely contracted following the zero COVID strategy and the disruption of normal economic activities. Since May 2022, news about the shrinkages in local fiscal revenues across China have made headlines.[83] Of the 22 provinces that announced its

[79] Alejandro De La Garza (2020).
[80] Jennifer Sinco Kelleher (2022).
[81] GOV.cn (2022a).
[82] Kang Jia and Chunzhi Ou (2022).
[83] 163 (2022).

fiscal balance, more than 10 have reported fiscal revenues decrease. In the pandemic-hit province of Jilin, the public budget revenues saw a 44.2% drop.[84] It was reported that Shenzhen had witnessed a 44% of decrease in fiscal revenue in April 2022. Nanning, Guangxi suffered a 6% decrease, Changzhou, Jiangsu a 4.5%, Zhuzhou, Hunan a 4.9% decrease.[85]

Langzhong, Sichuan Province announced on 28 May 2022 that residents were to pay at their expenses for mandatory mass testing.[86] One day later, facing the public discontent, the authority changed the requirement from mandatory mass testing to voluntary testing. Analysts have attributed the Langzhong episode to the financial drainage.[87]

Langzhong was not the only Chinese city that changed stance upon mass COVID testing in view of financial revenues stumble. Hefei, Anhui Province, announce the suspension of mass COVID testing on 21 June 2022, becoming the first provincial capital in China to do so.[88] Other cities in Anhui Province, such as Xuancheng, Lu'an, Huainan, and Huangshang, also suspended mass testing on a regular basis. In Jiangsu Province, Hai'an announced to suspend regular COVID testing on 10 June 2022; while Suzhou dropped the requirement of presenting negative COVID testing results.[89] Huizhou in Guangdong Province and Dezhou in Shandong Province also dropped requirements of negative COVID testing presentation.[90]

Another change in COVID prevention measures was reflected in one's travel code. As discussed in chapter "Between Zero COVID and 'Live with COVID'," a travel code detected one's travelling histories in the past 14 days and whether one's located area reported confirmed COVID cases. A travel code presentation is oftentimes required to enter public domain or take public transport. On 29 June 2022, Ministry of Industry and Information Technology of China announced that "to orchestrate the pandemic control and socioeconomic development in a highly effective manner" and to "facilitate travelling," the asterisk on one's travel code would be dropped. On 8 July, it was announced that a travel code would only detect one's travel histories for the past 7 days, shortened from 14 previously. One aim for such change was believed to boost travel, tourism, and personal spending in stimulating the economy. It was reported that just half an hour after the news of dropping the asterisk was released, searches for air/train tickets and hotel booking went hiking on travel agency platforms.[91]

The quarantine requirements upon inbound travellers seeking entry to Mainland China also witnessed changes. On 28 June 2022, China announced to change the

84 Caijing (2022).

85 Sohu (2022a).

86 Sohu (2022b).

87 Sohu (2022b).

88 Jiemian (2022); Note that Hefei resumed mass testing when a returnee was detected positive on 27 June 2022. Lei Liang (2022).

89 Weixin (2022).

90 Simin Wu (2022).

91 Lu Wang (2022).

quarantine requirements from "14 days of quarantine +7 days of stay-at-home health monitoring" to "7 days of quarantine +3 days of stay-at-home health monitoring."[92] In addition, inbound travellers would be required to take only throat swab tests rather than the previous nostril swab.[93] It can be inferred that these measures were in line with Premier Li Keqiang's requirements of "stabilizing foreign capitals" at various occasions. In the wake of the slightly loosened immigration requirement, the market responded with the Shanghai Stock Exchange Composite Index jumping.[94]

4 Conclusion

Since the outbreak of coronavirus cases, there gradually emerged two major pandemic control approaches. One is a zero COVID strategy adopted by China, Australia, New Zealand, Singapore, and Vietnam. The other is to "live with COVID," followed by most European and North American countries. While much extant research literature attempted to measure effectiveness of the two approaches in individual countries mainly by the number of infected cases, this study turns to the economic dimension in an up-to-date assessment of a pandemic control policy.

A zero COVID strategy might be effective in both suppressing the spread of virus and economic recovery in the initial days of the pandemic. The cases of China, Australia, and Singapore had all attested this point. The effectiveness of such zero COVID strategy, however, faced growing challenges when more infectious variants dominated. While acknowledging the necessity of lockdowns in Australian cities, researchers have warned Australian government to "consider when and how to lift lockdown and other restrictions, as there is no doubt the cost of getting this wrong is very high."[95] Singapore declared its divergent path from zero COVID to live with COVID when hit by the Delta variant in October 2021. An April 2022 report from the United States even argued for the zero connection between the economy and health management. The assessments that states trying to curb virus infection by restricting economic activities had suffered the most fatalities had further corroborated the necessity of strategy shift.

When China remained as the sole holdout of zero COVID practices, it also faced mounting difficulties in maintaining various strict restriction measures in the era of Omicron. In the first place, the high infection rate of the Omicron variant has rendered the elimination of COVID cases more and more difficult and costly. Of the total confirmed cases in China (900,000 as of late July), over 700,000 were detected in 2022 when the Omicron variant has become the dominant variant.[96] As of July 2022, China has seen city- or province-wide lockdowns in various places, such as

[92] GOV.cn (2022c).

[93] GOV.cn (2022c).

[94] Chris Buckley (2022).

[95] Anton Pak et al. (2021).

[96] Our World in Data.

Xi'an, Shenzhen, Jilin Province, and Shanghai, all of which witnessed the worst scenario of COVID outbreak in two years. An internet meme went that the one-week lockdown in Shenzhen, China's southern economic powerhouse, has cost the municipality 60 billion RMB, only to find out 643 cases including 21% asymptomatic cases, 72% mild cases, 7% ordinary cases, and zero severe and death case. The 100 million cost per positive case was not scientifically validated though, however, it did reflect that tracing, locating, quarantining, and eliminating confirmed positive cases are increasingly costly in the era of Omicron.

Secondly, from an economic perspective, the two pillars of a zero COVID strategy, namely, the mass testing and strict lockdowns, lacked economic support. Instead, they have become pure economic consumptions. According to calculations by Soochow Securities Co., a well-known invest services company in China, had all the Chinese cities in tier 2 and above established regular COVID testing, the cost would be 1.7 trillion RMB every year, or 1.3% of China's nominal GDP for 2021.[97]

It is inferred that restrictions on human movement and suspension on economic activities will help curb the spread of virus. However, after two years of implementing stringent pandemic control measures in China, the questions now are *At what costs?* and *Can China's economy survive the pandemic control restrictions?* As discussed, China's economy experienced a downturn that puts the once effective zero COVID strategy into question. Particularly, the coronavirus caused mainly mild or no symptoms among infected cases and resulted in extremely low fatalities in China.[98] According to the National Health Commission of China, in 2021, the COVID-19 pandemic in China was responsible for only 2 deaths, with a death rate at 0.0001 per 100,000 people.[99] The very low death rates only add suspicion over the set of draconian measures in response to the COVID pandemic. The tricky balance of human lives lost and economic costs haunts China like it does in other countries. One thing is certain: economic costs are also important dimensions in measuring the effectiveness of a pandemic control policy. When it comes to economic costs, a once indisputably effective control policy will be subject to review, question, and reconsideration. It reaffirms that there might not exist a "best," "universal," "one-size-fits-all," "contextless" pandemic control policy, or a control policy effective throughout every stage of the pandemic.

5 Limitations of the Research

This study is a tentative exploration of economic costs considerations behind adoptions of COVID-19 pandemic control policies. It should be noted that this study has some limitations. The first is that the COVID-19 pandemic is still ongoing and developing, with variants still evolving. Therefore, the tricky balance of human lives lost

[97] Chuan Tao et al. (2022).

[98] National Bureau of Disease Control and Prevention (2022).

[99] National Bureau of Disease Control and Prevention (2022).

and economic costs are subject to new development of the pandemic and new conditions. The second limitation of this research is that the research resorts to reviewing existing research literature as its main methodology except for the original research for the Shanghai part. It is common practice for analysts to refer to large volumes of media report in view of the pandemic though, it is still expected that more empirical studies could emerge to come up with more exact number of the economic costs versus economic gains through the implementation of a pandemic control policy.

Index	The first half of 2022		
	Whole residents	Urban permanent residents	Rural permanent residents
Per capita disposable income (Chinese Yuan)	38,996	40,950	21,434
Per capita consumption expenditure (Chinese Yuan)	21,028	22,043	11,901
Increase of disposable income (%)	−3.4	−3.3	−4.9
Increase of consumption expenditure (%)	−11.1	−11.2	−9.3

References

Abraham, P., Bablani, L., Carvalho, N., & Blakely, T. (2021, June 30). Yes, lockdowns are costly. But the alternatives are worse. *The Conversation*. https://theconversation.com/yes-lockdowns-are-costly-but-the-alternatives-are-worse-163572

Al Jazeera. (2021, October 17). Australia: Melbourne to bring an end to world's longest lockdowns. *Al Jazeera*. https://www.aljazeera.com/news/2021/10/17/australias-melbourne-set-to-end-worlds-longest-lockdowns

Alvelda, P., Ferguson, T., & Mallery, J. C. (2020, November 18). To save the economy, save people first. *Institute for New Economic Thinking*. https://www.ineteconomics.org/perspectives/blog/to-save-the-economy-save-people-first

Andersson, S., Aylott, N., & Eriksson, J. (2020, May 9). Democracy and technocracy in Sweden's experience of the COVID-19 pandemic. *Frontiers in Political Science, 4*, 832518. https://doi.org/10.3389/fpos.2022.832518

BBC. (2021, October 11). Covid Australia: Sydney celebrates end of 107-day lockdown. *BBC*. https://www.bbc.com/news/world-australia-58866464

BBC. (2022, April 7). China lockdowns: Tourism spending plunges over key holiday. *BBC*. https://www.bbc.com/news/business-61019241

Biswas, R. (2022, July 15). Singapore economy continues on road to recovery. *IHS Markit*. https://ihsmarkit.com/research-analysis/singapore-economy-continues-on-road-to-recovery-July22.html

Blakely, T., Thompson, J., Bablani, L., Andersen, P., Ouakrim, D. A., Carvalho, N., Abraham, P., Boujaoude, M. A., Katar, A., Akpan, E., Wilson, N., Stevenson, M. (2021, January 13). Determining the optimal COVID-19 policy response using agent-based modelling linked to health and cost modelling: Case study for Victoria, Australia. *Epidemiology*. https://doi.org/10.1101/2021.01.11.21249630

Buckley, C. (2022, June 28). China halves quarantine time for international arrivals, cheering markets. *The New York Times*. https://www.nytimes.com/2022/06/28/world/asia/china-covid-quarantine.html

Business & Human Rights Resource Centre. (2020, August 7). Guatemala: Covid-19 outbreak at factories supplying major global garment brands reveal lack of adequate safety measures for workers. *Business & Human Rights Resource Centre*. https://www.business-humanrights.org/en/latest-news/guatemala-covid-19-outbreak-at-factories-supplying-major-global-garment-brands-reveal-lack-of-adequate-safety-measures-for-workers/

Business Times. (2018, March 30). Singapore can be Asia hub for MNCs as HK turns to China. *Business Times*. https://www.businesstimes.com.sg/government-economy/singapore-can-be-asia-hub-for-mncs-as-hk-turns-to-china

Caijing. (2022, June 24). (Duodi Qian 5 Yue Caizheng Shouru Shuju Chulu 10 Yu Sheng Yiban Gonggong Yusuan Shouru Xiahua) 多地前 5 月财政收入数据出炉 10 余省一般公共预算收入下滑 [Multiple places release financial revenues data for May, over 10 provinces saw drop]. http://m.caijing.com.cn/api/show?contentid=4870660

Chen, J., Chen, W., Liu, E., Luo, J., & Song, Z.(M.). (2022, April 3). The economic cost of locking down like China: Evidence from city-to-city truck flows. *The Chinese University of Hong Kong*. https://www.econ.cuhk.edu.hk/econ/images/Documents/Truck_Flow_and_COVID19_220315.pdf

Cheng, E. (2022, May 26). China holds an unprecedented, massive videoconference on the economy. *CNBC*. https://www.cnbc.com/2022a/05/26/china-holds-an-unprecedented-massive-videoconference-on-the-economy.html

Cheng, J. (2022, April 29). China's manufacturing activity pummeled by Covid restrictions. *The Wall Street Journal*. https://www.wsj.com/articles/chinas-manufacturing-activity-contracted-further-in-april-11651285563

Crowe, D., Ilanbey, S., & Bonyhady, N. (2021, March 30). 'Take responsibility': Feds Rebuff Victorian Plea for lockdown assistance. *The Sydney Morning Herald*. https://www.smh.com.au/politics/federal/take-responsibility-feds-rebuff-victorian-plea-for-lockdown-assistance-20210530-p57whj.html

De La Garza, A. (2020, November 25). Hawaii is riding out the COVID-19 storm. But geographic isolation isn't the blessing it may seem. *Time*. https://time.com/5915084/hawaii-covid-coronavirus/

Dewi, A., Nurmandi, A., Rochmawati, E., Purnomo. E. P., Rizqi. M. D., Azzahra, A., Benedictos, S., Suardi, W., & Dewi, D. T. K. (2020, November 7). Global policy responses to the COVID-19 pandemic: Proportionate adaptation and policy experimentation: A study of country policy response variation to the COVID-19 pandemic. *Health Promotion Perspectives, 10*(4), 359–365. https://doi.org/10.34172/hpp.2020.54

Douglas, L. (2022, January 15). Nearly 90% of big US meat plants had COVID-19 cases in pandemic's first year - data. *Reuters*. https://www.reuters.com/business/nearly-90-big-us-meat-plants-had-covid-19-cases-pandemics-first-year-data-2022-01-14/

Focus Taiwan: CNA English News. (2021, August 25). Strict border control measures to remain until vaccination rate rises: CECC. https://focustaiwan.tw/society/202108250022

GOV.cn. (2022a, May 26). (Guojia Yibaoju: Changtaihua Hesuan Jiance Feiyong You Gedi Zhengfu Chengdan) 国家医保局: 常态化核酸检测费用由各地政府承担 [National healthcare security administration: Normalized nucleic testing fees be born by local governments]. http://www.gov.cn/xinwen/2022a-05/26/content_5692466.htm

GOV.cn. (2022b, May 25). (Li Keqiang Zai Quanguo Wenzhu Jingji Dapan Dianshi Dianhua Huishang Zuo Zhongyao Jianghua Han Zheng Zhuchi) 李克强在全国稳住经济大盘电视电话会上作重要讲话 韩正主持 [Li Keqiang delivers important speech at teleconference on stablizing National Economic Base Han Zheng Presides]. http://www.gov.cn/xinwen/2022b-05/25/content_5692298.htm

GOV.cn. (2022c, June 28). (Mijie, Rujing Renyuan Geli Guankong Shijian Tiaowei 7+3, Chengli Zhengzhi 'Cengcengjiama' Wenti Zhuanban Kaitong Liuyanban...Zuixin Qingkuang

Tongbao!) 密接、入境人员隔离管控时间调为"7+3", 成立整治"层层加码"问题专班开通留言板……最新情况通报! [Close contacts, immigration quarantine adjusted to 7+3, 'levelled-up control' special committee established to receive comments...lates information notified]. http://www.gov.cn/fuwu/2022c-06/28/content_5698267.htm

Greer, S. L., King, E. J., da Fonseca, E. M., & Peralta-Santos, A. (2020, September 1). The comparative politics of COVID-19: The need to understand government responses. *Global Public Health, 15*(9), 1413–1416. https://doi.org/10.1080/17441692.2020.1783340

GT Staff Reporters. (2022, May 25). China holds unprecedented meeting on stabilizing economy with focus on policy implementation. *Global Times*. https://www.globaltimes.cn/page/202205/1266575.shtml

Han, Y. H. (2022, May 27). "Chinese Premier Li Keqiang's meeting of 100,000 attendees conveys sense of urgency on the economy. *ThinkChina*. https://www.thinkchina.sg/chinese-premier-li-keqiangs-meeting-100000-attendees-conveys-sense-urgency-economy

Hirschmann, R. (2021, November 1). Trade to gross domestic product (GDP) ratio of Singapore from 2011 to 2020. *Statista*. https://www.statista.com/statistics/1113850/singapore-trade-to-gdp-ratio/#statisticContainer

IATA. (n.d.). *The importance of air transport to Singapore*. https://www.iata.org/en/iata-repository/publications/economic-reports/singapore--value-of-aviation/

Informa Markets. (n.d.). Effects of COVID-19 on the Chinese F&B industry. https://www.figlobal.com/china/en/visit/news-and-updates/affects-coronavirus-chinese-food-beverage-industry.html

Iwamoto, K. (2022, January 3). Singapore GDP rebounds 7.2% in 2021 as omicron looms. *Nikkei Asia*. https://asia.nikkei.com/Economy/Singapore-GDP-rebounds-7.2-in-2021-as-omicron-looms

Jain, S. S., Jain, S. P., & Li, Y. J. (2022, March 30). Sustaining livelihoods or saving lives? Economic system justification in the time of COVID-19. *Journal of Business Ethics*. https://doi.org/10.1007/s10551-022-05091-4

Jia, K., & Ou, C. (2022, June 13). Regular nucleic acid tests in risky areas help recovery. *China Daily*. https://global.chinadaily.com.cn/a/202206/13/WS62a66e46a310fd2b29e6235b.html

Jiemian. (2022, June 22). (Hefei Zhihou, Haiyou Naxie Chengshi Zanting Changtaihua Hesuan Jiance?) 合肥之后, 还有哪些城市暂停常态化核酸检测? [After Hefei, which other cities have suspended normalized nucleic acid testing?]. https://m.jiemian.com/article/7631328.html

Kerpen, P., Moore, S., & Mulligan, C. B. (2022, April). *A final report card on the states' response to COVID-19*. National Bureau of Economic Research. https://www.nber.org/system/files/working_papers/w29928/w29928.pdf

Kompas, T., Quentin Grafton, R., Che, T. N., Chu, L., & Camac, J. (2021, June 4). Health and economic costs of early and delayed suppression and the unmitigated spread of COVID-19: The case of Australia. Edited by Brecht Devleesschauwer. *PLoS One, 16*(6), e0252400. https://doi.org/10.1371/journal.pone.0252400

KPMG. (May 2022). China economic monitor. *KPMG*. https://assets.kpmg/content/dam/kpmg/cn/pdf/en/2022/05/china-economic-monitor-q2-2022.pdf

Lardy, N. R., & Huang, T. (2020, April 15). Weak exports will not stall China's manufacturing recovery. *Peterson Institute for International Economics*. https://www.piie.com/blogs/china-economic-watch/weak-exports-will-not-stall-chinas-manufacturing-recovery

Li, Z. (2022, March 22). What keeps China's zero-COVID policy going. *The Diplomat*. https://thediplomat.com/2022/03/what-keeps-chinas-zero-covid-policy-going/

Liang, L. (2022, June 27). (Hefei Quanshi Kaizhan Xinyilun Quyu Hesuan Jiance) 合肥全市开展新一轮区域核酸检测 [Hefei launches new wave of citywide nucleic acid testing]. *Shanghai Securities News*. https://news.cnstock.com/news,bwkx-202206-4908941.htm

Liu, L. Y. (2022, July 18). As China's youth unemployment keeps rising, Beijing says regional governments need to step up. *South China Morning Post*. https://www.scmp.com/economy/china-economy/article/3185707/chinas-youth-unemployment-keeps-rising-beijing-says-regional

Low, D. W. (2022, January 3). Singapore growth rebounds to fastest in more than a decade. *Bloomberg*. https://www.bloomberg.com/news/articles/2022-01-03/singapore-s-recovery-caps-fastest-gdp-growth-in-more-than-decade#xj4y7vzkg

Martínez-Córdoba, P.-J., Benito, B., & García-Sánchez, I.-M. (2021). Efficiency in the governance of the Covid-19 pandemic: Political and territorial factors. *Globalization and Health, 17*(1), 113. https://doi.org/10.1186/s12992-021-00759-4

Meng, D. (2022, July 19). (Shanghai GDP Dierji Weisuo Jin 14%, Diyi Jingji Dashi Paiming Ranggei Beijing) 上海GDP第二季萎缩近14% 第一经济大市排名让给北京 [Shanghai's GDP contracted by 14%, surrendering the no.1 economic city to Beijing]. *Zaobao*. https://www.zaobao.com.sg/news/china/story20220719-1294194

Menon, P. (2021, November 22). New Zealand to end tough COVID curbs, adopt new virus-fighting system. *Reuters*. https://www.reuters.com/business/healthcare-pharmaceuticals/new-zealand-end-tough-covid-curbs-adopt-new-virus-fighting-system-next-week-2021-11-22/

Ministry of Finance Singapore. (2021, February 16). Budget 2021. *Ministry of Finance Singapore*. https://www.mof.gov.sg/singapore-budget/budget-2021

Mizen, R. (2021, June 27). NSW lockdown will deliver $2b economic hit. *Financial Review*. https://www.afr.com/policy/economy/nsw-lockdown-will-deliver-2b-economic-hit-20210627-p584m3

Mullin, K. (2022, July 14). For China's nightlife scene, 'zero covid' an unceasing ordeal. *Al Jazeera*. https://www.aljazeera.com/economy/2022/7/14/for-chinas-nightlife-scene-zero-covid-an-unceasing-ordeal

National Bureau of Disease Control and Prevention. (2022, April 22). (2021nian Quanguo Chuanranbing Yiqing Gaikuang) 2021年全国法定传染病疫情概况 [Overview of Infectious Diseases in China 2021]. *National Bureau of Disease Control and Prevention*. http://www.nhc.gov.cn/jkj/s3578/202204/4fd88a291d914abf8f7a91f6333567e1.shtml

National Bureau of Statistics. (2022, June 15). (Guojia Tongjiju Xinwen Fayanren Jiu 2022 Nian 5 Yuefen Guomin Jingji Yunxing Qingkuang Da Jizhe Wen) 国家统计局新闻发言人就 2022 年5月份国民经济运行情况答记者问 [Press conference by National Bureau of Statistics spokersperson on National Economy of May 2022]. http://www.stats.gov.cn/tjsj/sjjd/202206/t20220615_1858320.html

National Bureau of Statistics of China. (2019a, November 20). Communiqué on the fourth national economic census (no. 2). *National Bureau of Statistics of China*. http://www.stats.gov.cn/english/PressRelease/2019a11/t2019a1120_1710327.html

National Bureau of Statistics of China. (2019b, November 20). Communiqué on the fourth national economic census (no. 3). *National Bureau of Statistics of China*. http://www.stats.gov.cn/english/PressRelease/2019b11/t2019b1120_1710328.html

National Data, National Bureau of Statistics of China. (n.d.). GDP of Shanghai in the first quarter of 2022. https://data.stats.gov.cn/easyquery.htm?cn=E0102&zb=A0101&reg=310000&sj=2022A

NetEase. (2022, July 5). (DuodiCaizheng Shouru Hanjian Chuxian Xiahua) 多地财政收入罕见出现下滑 [Multiple Places Witness Rarely Seen Decline]. https://www.163.com/dy/article/HBGNT55F0525CHJG.html

Nisa, C. F., Bélanger, J. J., Faller, D. G., Buttrick, N. R., Mierau, J. O., Austin, M. M. K., Schumpe, B. M., Sasin, E. M., Agostini, M., Gützkow, B., Kreienkamp, J., Abakoumkin, G., Khaiyom, J. H. A., Ahmedi, V., Akkas, H., Almenara, C. A., Atta, M., Bagci, S. C., Basel, S., … Pontus Leander, N. (2021, December). "Lives versus livelihoods? Perceived economic risk has a stronger association with support for COVID-19 preventive measures than perceived health risk. *Scientific Reports, 11*(1), 9669. https://doi.org/10.1038/s41598-021-88314-4

Oliu-Barton, M., Pradelski, B. S. R, Aghion, P., Artus, P., Kickbusch, I., Lazarus, J. V., Sridhar, D., & Vanderslott, S. (2021, June). SARS-CoV-2 elimination, not mitigation, creates best outcomes for health, the economy, and civil liberties. *The Lancet, 397*(10291), 2234–2236. https://doi.org/10.1016/S0140-6736(21)00978-8

Oraby, T., Tyshenko, M. G., Maldonado, J. C., Vatcheva, K., Elsaadany, S., Alali, W. Q., Longenecker, J. C., & Al-Zoughool, M. (2021, December). Modeling the effect of lockdown timing as a COVID-19 control measure in countries with differing social contacts. *Scientific Reports, 11*(1), 3354. https://doi.org/10.1038/s41598-021-82873-2

Pak, A., Adegboye, O. A., & McBryde, E. S. (2021, December 2). Are we better-off? The benefits and costs of Australian COVID-19 lockdown." *Frontiers in Public Health, 9*, 798478. https://doi.org/10.3389/fpubh.2021.798478

People.cn. (2022a, July 15). (Guojia Tongjiju Gongbu 2022a Dierjidu Ji Shangbannian 31 Shengfen GDP Shuju) 国家统计局公布 2022a 年二季度及上半年 31 省份 GDP 数据 [National Bureau of Statistics of China announces GDP data of 31 provinces for the second quarter and first half of 2022a]. http://finance.people.com.cn/n1/2022/0715/c1004-32476572.html

People.cn. (2022b, July 12). (Minhangju: Shangbannian Quanhangye Wancheng Lvke Yunshuliang 1.18 Yi Renci) 民航局: 上半年全行业完成旅客运输量 1.18 亿人次 [CAAC: Total transport volume reached 118 million]. http://finance.people.com.cn/n1/2022b/0712/c1004-32473070.html

Philippe, C., & Marques, N. (2021, April). The zero COVID strategy protects people and economies more effectively. *Institut Économique Molinari*. https://www.institutmolinari.org/wp-content/uploads/2021/03/etude-zero-covid2021_en.pdf

Prime Minister's Office Singapore. (2021, October 9). Update on the COVID-19 situation in Singapore by PM Lee Hsien Loong on 9 October 2021. *A Singapore Government Agency Website*. https://www.pmo.gov.sg/Newsroom/Update-on-the-COVID-19-Situation-In-Singapore-by-Prime-Minister-Lee-Hsien-Loong-on-9-October

Rangaswamy, E., Hoe, W. H. K., & Nawaz, N. (2022, April 1). Impact of COVID-19 pandemic on Singapore's aviation industry. Preprint. In Review. https://doi.org/10.21203/rs.3.rs-1505332/v1

Reuters. (2020, March 12). Australia nears living with COVID like flu-PM Morrison. *Reuters*. https://www.reuters.com/world/asia-pacific/australia-nears-living-with-covid-like-flu-pm-morrison-2022-03-12/

Ross, S. (2022, January 13). The 3 industries driving China's economy. *Investopedia*. https://www.investopedia.com/articles/investing/091515/3-industries-driving-chinas-economy.asp

Ruehl, M. (2021, October 19). Singapore launches quarantine-free travel to 10 countries. *Financial Times*. https://www.ft.com/content/70cb075e-41e8-4edc-85a5-a14b70c8fb45

Sinco Kelleher, J. (2022, March 9). *Hawaii to lift last US state mask mandate by March 26*. Associated Press. https://apnews.com/article/covid-health-hawaii-david-ige-1d41dba1b322af8f7458fbb98c22549c

Singapore Airlines. (2022, July 28). SIA group posts first quarter operating profit of $556 million on surging passenger demand. *Singapore Airlines*. https://www.singaporeair.com/saar5/pdf/Investor-Relations/Financial-Results/SGXNET/bu-q1fy2223.pdf

Sohu. (2022a, May 13). (Shenzhen Caizheng Shouru Xiahua Yue 44%, 4 Yue Difang Caizheng Shouzhi Maodun Jiada) 深圳财政收入下滑约 44%, 4 月地方财政收支矛盾加大 [Shenzhen financial revenue drops by 44%, deficits for April worsened]. https://www.sohu.com/a/546651527_120179484

Sohu. (2022b, May 29). (Sichuan Langzhong Huiying Zifei Changtaihua Hesuan: Zhengfu Chengdan Buqi, an Tongzhibanshi) 四川阆中回应自费常态化核酸: 政府承担不起, 按通知办事 [Langzhong, Sichuan responds to normalized nucleic acid testing at one's own fee: The government could not afford it]. https://www.sohu.com/a/552217187_115479

Sokolowska, J., & Zaleskiewicz, T. (2020, October 27). Willingness to bear economic costs in the fight against the COVID-19 pandemic. *Frontiers in Psychology, 11*, 588910. https://doi.org/10.3389/fpsyg.2020.588910

Stowell, N., Immormino, J., Salazar, A. O., Zeng, Y., & Arbetman-Rabinowitz, M. (2022, January 2).Politics, policy, and pandemic control: A global perspective. *Canadian Foreign Policy Journal, 28*(1), 1–21. https://doi.org/10.1080/11926422.2021.2007966

Sun, L. (2022, July 15). China jobs: Youth unemployment hits record high in June – nearly 1 in 5 young people out of work. *South China Morning Post*. https://www.scmp.com/economy/china-economy/article/3185473/china-jobs-youth-unemployment-hits-record-high-june-nearly-1

Tao, C., Shao, Y., & Duan, M. (2022, May 10). (Changtaihua Hesuan Jiance de Jingjizhang) 常态化核酸检测的经济账 [Economic costs behind normalized nucleic acid testing]. *Economic View*. https://www.jwview.com/jingwei/html/m/05-10/481896.shtml

The Lancet. (2022, June). Mental health after China's prolonged lockdowns. *The Lancet, 399*(10342), 2167.https://doi.org/10.1016/S0140-6736(22)01051-0

The New York Times. (2022, April 25). Covid news: South Korea to end virtually all restrictions. *The New York Times*. https://www.nytimes.com/live/2022/04/14/world/covid-19-mandates-cases-vaccine

The Singapore Economic Development Board. (2022, July 1). Headquarters. *A Singapore Government Agency Website*. https://www.edb.gov.sg/en/our-industries/headquarters.html

Tirone, J., & Kresge, N. (2021, November 19). Austria enters full national lockdown, orders mandatory Covid vaccines. *Bloomberg*. https://www.bloomberg.com/news/articles/2021-11-19/austria-to-lock-down-in-bid-to-halt-record-covid-surge-reports

Wang, L. (2022, June 29). (Xingchengka Quxiao Xinghao: Jipiao Sousuoliang Shangzhang 180% Shuqi Huo Ying Chuxing Gaofeng) 行程卡取消星号: 机票搜索量上涨180% 暑假或迎出行高峰 [Travelling codes no longer show asterick: Air tickets search surge by 180%]. *36Kr*. https://36kr.com/p/1805829651088393

Weixin. (2020, June 25). (Duodi Hesuan Jiance Jiange Yanchang Huo Quxiao, Zhuanjia:) 多地核酸检测间隔延长或取消, 专家: 现阶段大规模核酸用处不大 [Multiple places loosened or cancelled nucleic acid testing, experts: Mass testing of limited use at this stage]. https://mp.weixin.qq.com/s/zGWhSHnNW1uY_se_uf2oiA

Westcott, B. (2022, February 18). Australia's last Covid-zero holdout finally reopens to world. *Bloomberg*. https://www.bloomberg.com/news/articles/2022-02-18/australia-s-last-covid-zero-holdout-finally-reopens-to-the-world#xj4y7vzkg?leadSource=uverify%20wall

Wu, S. (2022, June 21). (Hefei Zanting Changtaihua Hesuan Jiance, Haiyou Zhexie Diqu Dongtai Tiaozheng) 合肥暂停常态化核酸检测, 还有这些地区动态调整 [Hefei cancelled nornalized nucleic acid testing, other places have dynamic adjustments]. *Yicai*. https://www.yicai.com/news/101450902.html

Yahya, F. (2021, February 18). Commentary: This is why Singapore needs to save its airlines and aviation sector. *ChannelNewsAsia*. https://www.channelnewsasia.com/commentary/singapore-needs-to-save-airlines-aviation-budget-sia-support-359436

You, S., Wang, H., Zhang, M., Song, H., Xu, X., & Lai, Y. (2020). Assessment of monthly economic losses in Wuhan under the lockdown against COVID-19. *Humanities and Social SCiences Communications, 7*(1), 52. https://doi.org/10.1057/s41599-020-00545-4

Zeytoon-Nejad, A., & Hasnain, T. (2021). The coronavirus trade-off – life vs. economy: Handling the trade-off rationally and optimally. *Social Sciences & Humanities Open, 4*(1), 100215. https://doi.org/10.1016/j.ssaho.2021.100215

Zhang, Y., You, C., Gai, X., Zhou, X., School of Mathematical Sciences, Peking University, Beijing, China, Center for Statistical Sciences, Peking University, Beijing, China, Beijing International Center for Mathematical Research, Peking University, Beijing, China, and Department of Biostatistics, School of Public Health, Peking University, Beijing, China. (2021). On coexistence with COVID-19: Estimations and perspectives. *China CDC Weekly, 3*(50), 1057–1061. https://doi.org/10.46234/ccdcw2021.245

Simon X. B. Zhao is Professor and Associate Dean at the Faculty of Humanities and Social Sciences BNU-HKBU United International College (UIC). He used to teach in the University of Hong Kong as a founding Director of International Center for China Development Studies and specializing at urban, regional and global studies and publishing in the broad areas of urban regional studies, geopolitical and economic development, spatial and global transformation, geography of international trade and finance, and development of international financial centers (IFCs). He is an editorial board member for four international journals and published more than 150 papers in international refereed journals, mainly SSCI/SCI/CSSCI Citation Journals, book chapters and consultative reports etc. Simon Zhao was a Member of the Central Policy Unit, HKSAR Government (Pan-PRD Panel) and chaired and co-lead many substantial academic research projects, providing widely consultancy services to China national, provincial, and local governments as well as private sector. His latest edited volume COVID-19 Pandemic, Crises Response and the Changing World: Perspectives in Humanities and Social Sciences (by Springer) is well received globally.

Bo Yan received PhD in History from Nanyang Technological University, Singapore. He is a Research Fellow in the Faculty of Humanities and Social Sciences. Yan's main research interests include modern Chinese history, postwar United States, and modern Southeast Asian history, and the history of the global 1960s. He has published in the Journal of Asian Studies, Twentieth Century China, and others. Yan's PhD dissertation was listed as the International Convention of Asia Scholars (ICAS) Book Prize 2021-Best Dissertation in Humanities. Bo Yan is currently conducting studies and research on global pandemic of COVID-19.

Yu Liu studies English Language and Literature at BNU-HKBU UIC. Liu's main research interests include comparative literature and modern poetry. Liu's research projects relating to Fu Manchu and western popular cultures received financial support from the faculty. In addition to academic research, Liu is a prolific writer and poet. Liu's poet Good Night has won Excellence in Creative Writing Award in Elsewhere Issue 4.

Jiaqi Liu received a master's degree in Hospitality and Real Estate Management from The Chinese University of Hong Kong, and a bachelor's degree in Applied Economics (First Class Honours) from Beijing Normal University-Hong Kong Baptist University United International College (UIC). Her research interests include economic development, real estate, and urban renewal.

Social Resilience and COVID-19 Recovery: Evidence from 107 Countries

Fangxin Yi, Yangtian Feng, and Tao Li

1 Introduction

Since early 2020, the COVID-19 pandemic has become the most crucial public health catastrophe. According to the COVID-19 Dashboard of the Centre for Systems Science and Engineering (CSSE) of Johns Hopkins University (JHU), as of Feb 8, 2022, about 400 million confirmed cases and six million deaths have been reported globally. While public healthcare systems are under severe strain, governments across the world are dealing with serious disruptions and policy issues as a result of the pandemic and public confidence in governments and institutions is eroding. As a result, societies attempting to combat the COVID-19 pandemic have placed a greater emphasis on resilience as a policy aim or attribute.

Despite the growing interest in COVID-19 policy responses from researchers, social resilience has received less attention in COVID-19 studies and more attention has been paid to other effects and consequences of the virus. Much of the previous work has been based on single-case or small-n studies. Developing large-n comparative case studies of countries from various locations around the world has received little attention thus far. More samples are needed to examine how possible interactions of political conditions and other factors affect the evolving pandemic responses over time. As a result, this paper intends to conduct an empirical analysis between social resilience in terms of recovery and its causal factors to investigate which combination of factors are most influential in building a resilient society. Moreover, by structuring the panel data of 107 countries globally and making a comparative analysis, research will find out how could a country recover from the pandemic.

As important qualities of social resilience, we emphasise on demographic situation, government response, medical service, and resources. We applied four distinct

F. Yi (✉) · Y. Feng · T. Li
Beijing Normal University, Beijing, China
e-mail: francineyi@gmail.com

© The Author(s), under exclusive license to Springer Nature Singapore Pte Ltd. 2023
S. X. B. Zhao et al. (eds.), *Comparative Studies on Pandemic Control Policies and the Resilience of Society*, https://doi.org/10.1007/978-981-19-9993-2_4

combinations of social resilience across the 107 countries based on these criteria and our data analysis. The resilience of the society is not dependent on any single aspect or trait. Instead, it is a result of multiple variables, with different countries and cities demonstrating diverse degrees of resilience based on various combinations of elements or attributes. It is expected that by identifying these elements, we will be able to shed light on the qualities that may be required for resilient institutions and policy designs in the event of a worldwide pandemic. This, in turn, should encourage more research into policy design for a pandemic resilient society.

2 Literature Review

2.1 Concept of Social Resilience

Groups and agencies engaged in disaster alleviation recent years have emphasised more on achieving a disaster-resilient society. Resilience refers to an entity's ability return to its original state after exposure to external shocks and the concept has been widely applied to material world (Vasu, 2020). Since disruptions are unavoidable, and there are a variety of internal and external variables that contribute to disruptions in companies. When it comes to resilience, the main concerns are organizational vulnerability, disruptions, and complexities, as these are the main challenges that affect the system's resilience (Rai et al., 2021).

Various models have been developed in determining dimensions of resilience. In a disaster-related literature under the background of earthquake rescue, Bruneau et al. (2003) conceptualised resilience into four dimensions: technical, organizational, economic, and social, whereas the social dimension of resilience consists of measures specifically designed to lessen the extent to which earthquake-stricken communities and governmental jurisdictions suffer negative consequences due to the loss of critical services as a result of disaster. Stanton-Geddes et al. (2013) divided urban resilience into four parts: infrastructural resilience, institutional resilience, economic resilience and social resilience, of which social aspect specifically refers to the demographic structure, elements of organisation and social capital.

As for social resilience, which has long been an area of interest in understanding social recovery, there is still relatively little dedicated definition in the previous literature. For example, Keck and Sakdapolrak (2013) defined social resilience as "social entities and their abilities to tolerate, absorb, cope with and adjust to environmental and social threats of various kinds", while Obrist et al. (2010) presented a multi-dimensional concept of social resilience based on a 'Multi-layered social resilience framework'. Social resilience, according to this definition, can be attained if a society and its social actors have three capacities and qualities. Reactive capacity, proactive capacity, and positive results are the three.

Along with the definitions, there are various interpretations of social resilience in the existing theoretical literature. Some academics looked at social resilience via the lens of complex adaptive systems, while others looked at it through the lenses of social and health sciences. As a result, numerous indicators, and identities for displaying social resilience. Most of the literature coping with indicators of social resilience refers to natural disasters, like earthquakes or floods (Khoja et al., 2020).

2.2 Measuring Social Resilience under COVID Background: Selection of Indicators and Frameworks

According to Paton and Johnston (2006), the measure of social resilience aims to assess the capacity to adapt to challenges before and after a disaster. The major approach in measuring social resilience is to utilise various indicators and quantify the level of resilience, with further comparisons between communities, places and disasters (Cutter et al., 2008).

When it comes to the indicators for measurement of social resilience in the disaster context, previous literatures have developed a number of frameworks with different emphasis and scope to determine key characters. For example, Saja et al. (2019) has put forward a multi-layered structure, which includes dimension, characteristic and indicator, and a "5s" framework consisting social structure, social capital, social mechanism, social equity and diversity, and social belief is constructed in measuring social resilience.

It is worth noting that some of the previous models for social resilience measurement may not be perfectly applicable under the COVID background. In case of a pandemic crisis, some important factors such as social networks or social support, may have to be complemented by additional factors (Busic & Schubert, 2021), since the pandemic has brought a greater impact on public health and socio-economic aspect, instead of physical damage of housing and infrastructures. According to University of Oxford, policy responses to COVID pandemic can be classified as following 9 aspects: school and workplace closures; cancellation of public events and restrictions on public gatherings; closures of public transport; stay-at-home requirements; public information campaigns; restrictions on internal movements and international travel controls. As stated by Hynes et al. (2020), social resilience with COVID background entails identifying the individuals and activities that are most affected, determining how interventions to help them will affect others, and emphasizing that difficult trade-offs between health, economic, social, and other goals are unavoidable in the short term. In the long run, if policies are to be effective, they must respond to the systemic causes and repercussions of large shocks.

The COVID-19 crisis also demonstrates the importance of having resources on hand in case the system experiences unanticipated upheavals that prohibit it from running regularly. Despite this, social resilience has received less attention in COVID-related studies and more attention has been paid to other effects and consequences (Alizadeh & Sharifi, 2021). Aiming such circumstances, Yi et al. (2023) built a pandemic-oriented social resilience measurement framework. Four major factors affecting the recovery from COVID-19, which are robustness, preparedness, resource and social capital, and government response, were summarised, and three perspectives containing COVID recovery, economy recovery and future protection capacity were proposed to measure recovery outcome.

To make a conclusion, in analysing disaster recovery, most of the previous literature focused on two most essential views, that are economic and social perspectives, and are used to assess the region's recovery from the pandemic's impact in the past. Since COVID-19 has appeared in various variations, the world will be troubled by the pandemic continuously for a foreseeable long period of time. Because of that, indicator of future protection capability is suggested to assess the region's future risk resistance from the pandemic's destruction. Based on an in-depth review of previous resilience attributes and assessment models, this chapter will propose an analytical framework that integrates the various resilience indicators and applies this to the study of 107 countries, which includes their pandemic response outcome and comparative analysis.

3 Theoretical Framework

3.1 Analytical Framework

As mentioned previously, there are little widely accepted models and frameworks for social resilience measurement. Moreover, the uniqueness of COVID-19 as a public health catastrophe means some indicators in previous research models related to physical damage such as earthquakes and floods may not be applicable (Yi et al., 2023). In this research, we refer to the very fundamental concept of resilience, that is the ability to cope with external stresses and disturbances. When it applied country or society wide, it means certain aspects of a society must return to (or even exceed) the original state. Hence, it is assumed here that countries and regions with better recovery outcome are more resilient. This research proposes two major perspectives for recovery measurement: economy recovery and future protection capacity. The first is an indicator widely used in evaluating a region's recovery from disaster's devastation. The second indicator, future protection capacity, is proposed here because COVID has continuously appeared in different variants, and it is necessary to assess the resistance for future COVID waves. According to WHO, access to safe and effective vaccines is critical to ending the COVID-19 pandemic. A substantial number of studies demonstrated that when COVID-19 vaccine acceptance rate is below 60%, it would pose a serious problem for efforts to control the current COVID-19 pandemic, while high acceptance rates would help to achieve proper control (Sallam, 2021). Therefore, we take vaccine rate as the indicator for future protection capacity measurement.

For the factors influencing recovery, fundamentals factors like pre-disaster preparation and socio-economic conditions need to be considered. Based on previous literature, we summarised four major factors that affect COVID outcome: demographic character, government response, medical service, and resources. The relationship between demographic science and disaster has long been discussed. As for COVID-19, deaths have been observed to occur at older ages, the important role of demography is highlighted. Previous studies have revealed how the age structure of a society may affect transmission and fatality rates across countries (Dowd et al., 2020). In addition, government response is surely assessed because it undeniably intervened the COVID outcome and recovery. Major government responses include public health measures like closures and containment and economic measures like quantitative easing. As a public health event, medical service has an unignorable impact on COVID recovery. On the one hand, it must deal with new cases with proper treatment. On the other hand, the escalating cases will consume the limited medical resources excessively. Existing studies of multiple areas have proved emergency services, outpatient clinics and long-term care will all be disrupted by the pandemic (Chung et al., 2020; Katayama et al., 2020). Lastly, the global pandemic has seriously disrupted trades and logistics, and limitations and fragility of global supply chain has been exposed (Sarkis, 2021). Under such circumstances, shortage of daily necessities and panic buying can be observed in almost every country. Abundant reservation of resources is predominant for a COVID resilient society.

As for the detailed indicators for recovery influencing factors (Fig. 1), our efforts to determine country or region's demographic including collecting data on life expectancy, working age (15–64) population and morality rate. To measure the government response, we referred to the Oxford COVID-19 Government Response Tracker (OxCGRT) as our indicator, which consists of closures and containment, economic measures, health measures and vaccine policies. When it comes to judging a country's medical service, data of current health expenditure, number of hospital beds and doctors are taken into consideration. Lastly, to examine a country's resource, we collected data of its GDP, energy use and food production. All the data we collected in this research is from trustworthy databases, including World Bank, IMF, WHO and OxCGRT. The detailed measurement indicators and data sources are elaborated in Appendix 1.

4 Methodology

4.1 Resilience Measurement and Calibrations

In this research, we measure the resilience of a country from two aspects: economy recovery and future protection.

Economy: This research evaluates economy recovery through GDP growth rate. GDP is the most used indicator to measure the development status of a region. The pandemic has disrupted the normal economic order to varying degrees, so we obtained the actual GDP growth rate in 2020 from World Bank and the predicted

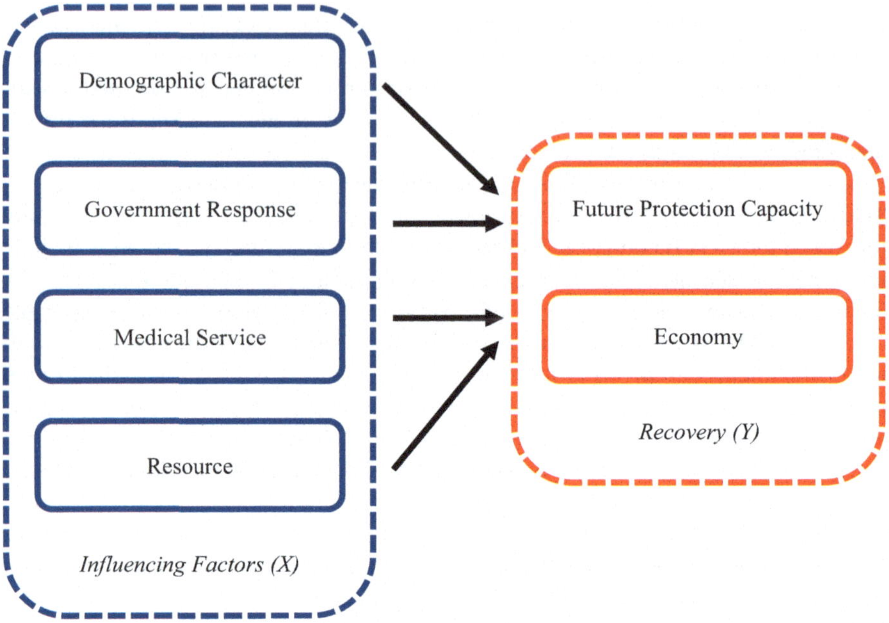

Fig. 1 Theoretical framework for factors influencing COVID recovery

GDP growth rate in 2021 from the IMF, and calculated the percentage of the city's GDP growth in 2021 compared to 2019.

Future Protection Capacity: At present, the coronavirus is still spreading wantonly, and the newly mutated strains are highly contagious, making the duration of the pandemic full of uncertainty. Immunization by vaccination is one of the best ways to avoid large-scale transmission again. Counties with higher vaccination rates can more easily return to normal life and restore economic development. We obtain total vaccinations per hundred people from WHO database.

Raw data collected is then needed to be calibrated to a number between 0 and 100 for analysis. For standardization, all calibrations follow the formula as shown below:

$$\text{Calibrateddata} = \frac{X - W_{min}}{W_{max} - W_{min}} \times 100$$

where X represents a given country's data, while Wmin and Wmax represent the world minimum and maximum for that set of data, world maximum and minimum are used to calibrate data so that countries not included in the list could also make use of the results obtained from this research. This calibration method is also applied in influencing factors standardization.

We propose an artificial criterion to classify all countries into three categories of resilience level based on the Y score. We believe that the higher the Y score, the stronger the resilience of the country, that is, the higher level and faster speed to recover from public crisis events. Our classification method is to classify the countries with Y score greater than 60 as high resilience level, a total of 31 countries; Countries with Y score between 30 and 60 are classified as the mediate resilience level, a total of 57 countries; Countries with Y values below 30 are classified as low resilience level.

4.2 Influencing Factors and Calibration

In this research, four influencing factors are set to evaluate the resilience of countries.

Demographic Character: Life expectancy at birth and age dependency ratio from World Bank and mortality rate from WHO are selected to evaluate the demographic character of a country. People in areas with high life expectancy and lower mortality rate are in better health, are more immune to pandemic and less likely to get infected. The age dependency ratio means that the non-working population is divided by the working population. If the number is high, it indicates that the elderly and children, who are vulnerable to infection, account for a large proportion in the country, which deserves government departments' attention.

Government Response: This research selects 16 indicators from Oxford's COVID policy tracker to evaluate whether the government has implemented effective measures during the pandemic. These measures include closure, economic and health aspects. The Oxford database counts the presence and frequency of government policies and give a score to each item.

Medical Level: Current Health Expenditure per Capita from World Bank and Hospital beds and Medical doctors (per 10,000 people) from WHO are selected to represent this indicator. Medical levels in countries are relatively fixed, there are no obvious fluctuations in the short term. If the spread of the virus is not controlled, sooner or later, medical resources will be depleted, resulting in humanitarian disaster and social instability. Therefore, for regions with insufficient medical resources, it is essential that they take the lead in controlling the pandemic and avoid the depletion of medical resources.

Resource: How a country supplies basic living materials is one of the important evaluation factors to measure the resilience. Food production index, GDP, Energy use from FAO is selected to reflect the basic supply capacity of the city. During the raging period of the virus, the production of agricultural products was negatively affected, and the import link was hindered. At the same time, fighting the pandemic means a lot of capital needs. The government needs not only to increase investment in medical care, but also to ensure the basic living needs of the public. Countries with adequate infrastructure and adequate energy supply are more resilient to disasters.

5 Overall Results

We rank all countries from top to bottom in terms of their Y recovery score. We consider that the higher the score is, the stronger the resilience level is, we give out the resilience rank of 107 countries. We divided 107 countries into three resilience level categories based on the size of their Y Recovery score range. The first resilience level classification, High resilience, of which Y Recovery score between 60 and 100, from high to low includes Ireland, Chile, China, Korea, Rep., Brunei Darussalam, Denmark, Vietnam, Singapore, Norway, Luxembourg, United Arab Emirates, Malta, Australia, Finland, Belgium, Uruguay, France, Iceland, Italy, Netherlands, Sweden, Cambodia, Germany, Portugal, Cyprus, Estonia, Israel, Austria, Lithuania, Brazil, United States. The second resilience level classification, Mediate resilience, of which Y Recovery score between 30 and 60, includes Bhutan, Peru, Spain, Saudi Arabia, El Salvador, Nicaragua, Bangladesh, Tajikistan, Dominican Republic, Slovenia, Seychelles, Japan, Colombia, Czech Republic, Latvia, Ecuador, Morocco, Thailand, Paraguay, Slovak Republic, Panama, Russian Federation, Belarus, Kazakhstan, Croatia, Pakistan, Azerbaijan, Guinea, Guatemala, Romania, Jordan, Mauritius, Tunisia, Benin, Comoros, Bolivia, Togo, Mozambique, Belize, Georgia, Niger, Burkina Faso, Bulgaria, Tanzania, Mauritania, Kenya, Botswana, Uganda, Senegal, Djibouti, Trinidad and Tobago, Cabo Verde, Bosnia and Herzegovina, South Africa, Moldova, Malawi, Liberia. The third resilience level classification, Low resilience, of which Y Recovery score between 0 and 30, includes Sierra Leone, Mali, Solomon Islands, Gabon, Fiji, Cameroon, Nigeria, Central African Republic, Barbados, Angola, Jamaica, Burundi, Bahamas, The, Suriname, Sudan, Kyrgyz Republic, Zambia, Haiti, Iraq. 31 countries are high resilience countries, 57 countries are mediate countries and 19 countries are low mediate countries.

We calculate the percentage of developed countries in each category. 54.8% of the countries in the High resilience category are developed countries, including Singapore, Luxembourg, Malta, Australia, Finland, Belgium, Iceland, Italy, Netherlands, Germany, Portugal, Cyprus, Estonia, Israel, Austria, Lithuania, United States. While merely 8.7% of the countries in the Mediate resilience category are developed countries, including Spain, Slovenia, Japan, Czech Republic, Latvia. And none of the countries in the Low resilience category are developed countries. Developed countries are generally characterized by a high human development index, per capita GROSS national product, industrialization level and quality of life. Therefore, they are more likely to recover from a sudden public crisis to the original level quickly and adjust themselves in this crisis to surpass the original level.

6 Conclusion and Discussion

With the emergence of new variants, the COVID epidemic is expected to continue in the short term. As more countries are giving up "Zero-COVID" policies and trying to live with the virus, it is undeniably important to seek a more resilient society which can recover faster from the pandemic and is capable to withstand the next outbreak. Social resilience and policy implication has been a major focus of research interest in the context of COVID and an overprovision of studies can be seen. Much of the existing literature aiming COVID recovery and resilience building, however, consist single-case or small-n studies, and little effort has been paid to develop a large-n comparison of multiple countries and regions. By assessing the recovery and resilience of 107 countries, this paper has sought to address this gap in the previous literature.

By collecting data of each country's economy and future protection capacity, we identified countries with better recovery outcome and resilience. Moreover, we figured out four combinations of influencing factors, including demographic character, government response, medical service, and resource. The resilience framework introduced in this paper integrated the above combinations, which provides an important insight for recognising the COVID recovery in different countries, and to what extent each combination give rise to COVID recovery and social resilience. According to our research finding, while demographic character has the most notable impact over of a region's COVID recovery, the resilience attributes including medical service and resources are also significant for pandemic recovery and resilience building at the country level. It is also revealed in our research that government response has the least influence in most cases. Therefore, whether the local governments are still taking strict pandemic control policies or chose to "live with virus", it is indisputable that increasing access and quality of effective medical service and reserve necessary resources are important. In general, it can be concluded that pandemic recovery and resilience building depend on combination of factors, rather than a single factor. Among such, the pre-disaster preparation including medical service and resources is a major approach to build a COVID resilient society.

At the moment, the world is far from full recovery of COVID-19 and continuous regional outbreaks are still expected. While the virus is still ranging, countries like South Korea and Singapore have already demonstrated strong ability to recover with high economic and vaccination rate. While the findings have been developed on 107 countries, it is believed that the resilience model we introduced in this paper is applicable for assessing recovery outcome and resilience of other countries and regions. However, future research may have to focus on applicability of single indicators. Take vaccination as an example, in areas where mandatory immunization is implemented, the vaccination rate will be artificially inflated. Still, this points out an important point for further research, that when applying social resilience frameworks on different countries and regions, whether existing frameworks reflect the priorities of community members in disaster recovery, and which indicators should be selected exactly before applying to examine local community.

Appendix 1 Contributing Factors and Sources of Data

Indicator	Category	Source	Time
Life expectancy at birth, total (years)	XA-Demographic Character	World Bank	2019
Mortality rate		WHO	latest
Population ages 15–64 (% of total population)		World Bank	2019
c1_school_closing	XB-Government Response	OxCGRT	2021.6.30
c2_workplace_closing		OxCGRT	2021.6.30
c3_cancel_public_events		OxCGRT	2021.6.30
c4_restrictions_on_gatherings		OxCGRT	2021.6.30
c5_close_public_transport		OxCGRT	2021.6.30
c6_stay_at_home_requirements		OxCGRT	2021.6.30
c7_movementrestrictions		OxCGRT	2021.6.30
c8_internationaltravel		OxCGRT	2021.6.30
e1_income_support		OxCGRT	2021.6.30
e2_debtrelief		OxCGRT	2021.6.30
h1_public_information_campaigns		OxCGRT	2021.6.30
h2_testing_policy		OxCGRT	2021.6.30
h3_contact_tracing		OxCGRT	2021.6.30
h6_facial_coverings		OxCGRT	2021.6.30
h7_vaccination_policy		OxCGRT	2021.6.30
h8_protection_of_elderly_people		OxCGRT	2021.6.30
Current Health Expenditure (CHE) per Capita in US$	XC-Medical service	World Bank	2019
Hospital beds (per 10 000 population)		WHO	latest
Medical doctors (per10000)		WHO	latest
Food production index	XD-Resource	World Bank	2018
GDP		World Bank	2019
Energy use (kg of oil equivalent per capita)		World Bank	2018
TOTAL_VACCINATIONS_PER100	Y-Recovery	WHO	latest
2019/2021 GDP change		IMF	latest

References

Ali, J., Khan, R., Ahmad, N., & Maqsood, I. (2012). Random forests and decision trees. *International Journal of Computer Science, 9*(5), 1694–1814.

Alizadeh, H., & Sharifi, A. (2021). Analysis of the state of social resilience among different socio-demographic groups during the COVID- 19 pandemic. *International Journal of Disaster Risk Reduction, 64*, 102514. https://doi.org/10.1016/j.ijdrr.2021.102514

Bruneau, M., Chang, S. E., Eguchi, R. T., Lee, G. C., O'Rourke, T. D., Reinhorn, A. M., Shinozuka, M., Tierney, K., Wallace, W. A., & von Winterfeldt, D. (2003). A framework to quantitatively assess and enhance the seismic resilience of communities. *Earthquake Spectra, 19*(4), 733–752. https://doi.org/10.1193/1.1623497

Busic, A., & Schubert, R. (2021). Social resilience indicators for pandemic crises. *SSRN Electronic Journal*. https://doi.org/10.2139/ssrn.3938198

Chung, C. Y., Wong, W. H., Fung, J. L., Kong, R. D. H., & Chung, B. H. (2020). Impact of COVID-19 pandemic on patients with rare disease in Hong Kong. *European Journal of Medical Genetics, 63*(12), 104062. https://doi.org/10.1016/j.ejmg.2020.104062

Cutter, S. L., Barnes, L., Berry, M., Burton, C., Evans, E., Tate, E., & Webb, J. (2008). A place-based model for understanding community resilience to natural disasters. *Global Environmental Change, 18*(4), 598–606. https://doi.org/10.1016/j.gloenvcha.2008.07.013

Dowd, J. B., Andriano, L., Brazel, D. M., Rotondi, V., Block, P., Ding, X., Liu, Y., & Mills, M. C. (2020). Demographic science aids in understanding the spread and fatality rates of COVID-19. *Proceedings of the National Academy of Sciences, 117*(18), 9696–9698. https://doi.org/10.1073/pnas.2004911117

Hynes, W., Trump, B., Love, P., & Linkov, I. (2020). Bouncing forward: A resilience approach to dealing with COVID-19 and future systemic shocks. Environment Systems and Decisions, 40(2), 174–184. https://doi.org/10.1007/s10669-020-09776-x

Katayama, Y., Kiyohara, K., Kitamura, T., Hayashida, S., & Shimazu, T. (2020). Influence of the COVID-19 pandemic on an emergency medical service system: A population-based, descriptive study in Osaka, Japan. *Acute Medicine Surgery, 7*, e534. https://doi.org/10.1002/ams2.534

Keck, M., & Sakdapolrak, P. (2013). What is social resilience? Lessons learned and ways forward. *Erdkunde, 67*(1), 5–19. http://www.jstor.org/stable/23595352

Khoja, L., Schubert, R., & Joerin, J. (2020). *Social resilience indicators for disaster related contexts: Literature review*. Singapore-ETH Centre (SEC), *Future Resilient Systems* (FRS) (Working Paper 2).

Obrist, B., Pfeiffer, C., & Henley, R. (2010). Multi-layered social resilience: A new approach in mitigation research. *Progress in Development Studies, 10*(4), 283–293. https://doi.org/10.1177/146499340901000402

Paton, D., & Johnston, D. (2006). *Disaster resilience: An integrated approach*. Charles C Thomas.

Rai, S. S., Rai, S., & Singh, N. K. (2021). Organizational resilience and social-economic sustainability: COVID-19 perspective. *Environment, Development and Sustainability, 23*(8), 12006–12023. https://doi.org/10.1007/s10668-020-01154-6

Saja, A. M. A., Goonetilleke, A., Teo, M., & Ziyath, A. M. (2019). A critical review of social resilience assessment frameworks in disaster management. *International Journal of Disaster Risk Reduction, 35*, 101096. https://doi.org/10.1016/j.ijdrr.2019.101096

Sallam, M. (2021). COVID-19 vaccine hesitancy worldwide: A concise systematic review of vaccine acceptance rates. *Vaccines, 9*(2), 160. https://doi.org/10.3390/vaccines9020160

Sarkis, J. (2021). Supply chain sustainability: Learning from the COVID-19 pandemic. *International Journal of Operations & Production Management, 41*(1), 63–73. https://doi.org/10.1107/IJOPM-08-2020-0568

Stanton-Geddes, Z., Jha, A. K., & Miner, T. W. (2013). *Building urban resilience: Principles, tools, and practice*. World Bank Publications.

Vasu, N. (2020). *Pandemic: The idea of social resilience* (No. 112; RSIS Commentaries). Nanyang Technological University. https://dr.ntu.edu.sg/bitstream/10356/143732/2/CO20112.pdf

Yi, F., Woo, J. J., & Zhang, Q. (2023). Community resilience and COVID-19: A fuzzy-set qualitative comparative analysis of resilience attributes in 16 countries. *International Journal of Environmental Research and Public Health, 20*(1), 474. https://doi.org/10.3390/ijerph20010474

Fangxin Yi works as a research fellow at the School of Social Development and Public Policy of Beijing Normal University. She earned a PhD from the National University of Singapore, specializing in urban resilience and public policy. She has published in the Journal of Cleaner Production, Land Use Policy, Urban Geography, Sustainability, etc. She was the recipient of the research funding from Ministry of Education Tier 2 research project of Singapore on Disaster Governance in Asia and received Peking University-Lincoln Institution Fellowship. She has years of urban planning consulting experience at the China Academy of Urban Planning and Design in China.

Yangtian Feng is a postgraduate at the School of Accounting of Southwestern University of Finance and Economics. She received her bachelor's degree from Zhejiang University of Technology. She has published in Sustainability. Her areas of interest include urban resilience and public budget.

Tao Li is a senior at BNU-HKBU United International College's Faculty of Humanities and Social Sciences, where he is pursuing a degree in Media and Communication Studies. His areas of interest include media psychology and journalistic professionalism.

"Surprisingly Real": Regressive Structural Transformations in the Wake of the Corona Crisis

Markus Heidingsfelder and Lihua Chen

1 A Functionally Differentiated World

Politics can seal off the economy—it cannot rule in it.
Freely adapted from Friedrich Schiller

This text takes its starting point from the basic assumption that the order of functional differentiation has become established worldwide. [1] Functional differentiation is understood in the following as the realisation of a kind of division of labour, a "sorting procedure" (Rudolf Stichweh), which has its origin in the concept of

[1] It is important to note that while we start from this sociological 'tenet', we do not confuse it with social reality, which we consider inaccessible. Whether this reality is de facto so constituted is of concern to us only with regard to the amount of truth to be gained thereby. In this we follow Luhmann's specification of a minimal ontology that presupposes existence but does not assert it (1984a, p. 30). Of course, this is not to say that we give up proving ourselves against reality. This text refers to the real world, to a real pandemic. What is impossible is merely a reality check beyond concrete measurement. Just like the epidemiologist, the systems theorist is bound to his own perspective—it orders reality but does not copy it.

All translations in the text are the authors' own unless otherwise indicated. We are indebted to Dirk Baecker, Peter Fuchs and Lars Clausen who provided essential advice.

M. Heidingsfelder (✉) · L. Chen
Beijing Normal University-Hong Kong Baptist University,
United International College, Zhuhai, China
e-mail: mgtheidingsfelder@uic.edu.cn

© The Author(s), under exclusive license to Springer Nature Singapore Pte Ltd. 2023

S. X. B. Zhao et al. (eds.), *Comparative Studies on Pandemic Control Policies and the Resilience of Society*, https://doi.org/10.1007/978-981-19-9993-2_5

vocation.[2] It says that within the comprehensive social system of society, individual subsystems emerge, each fulfilling a specific function for the system as a whole: so-called functional systems (such as economy, law, politics, art, health, education, etc.), but also organisations (banks, courts, galleries, hospitals, schools etc.) and networks.

The assumption is that all these systems operate autonomously in the sense that they produce their specific operations—communications—exclusively on their own, i.e., each has a primacy of reproduction.[3] Since this is true for all systems, no system can claim outstanding primacy. Instead, the picture that emerges is one of a parallelism of primacies. "[T]o limit one's interest in functional differentiation primarily to politics and the economy and to have only a residual idea of other functional areas" (Stichweh, 2013, p. 59) is therefore not advisable, as it absorbs the complexity of this social order. It may, however, be useful in terms of ideological or normative assumptions: "The theoretical decision then becomes the source of misdirected dissatisfaction, critique and protest" (Luhmann, 1997b, p. 70).

The pattern of structure formation is evident from the outset on a global scale, as the communications realised in the systems effortlessly transcend local boundaries. The reference context for any function system in contemporary society is thus defined by the system of world society as the most extensive societal context (Luhmann, 1997a, pp. 145ff., Stichweh, 2013, p. 62). There are no exceptions to this structure: no exclusion from global inclusion. A total function of the economy can no longer be determined in any country in the world, just as it can no longer be determined for politics (even though the corresponding ideologies of liberalism and socialism may provide for it) or any other of the world's systems. The collapse of the Soviet empire has shown the enormous support that functional differentiation has in world society. A regional boycott of this circumstance cannot be realized even with the strongest use of political and organizational means (Luhmann, 1997a, p. 161).

The irreversibility of world society is stabilised by the patterns of division of labour, which not only means differentiation of forms of communication, which are then organised in schools, hospitals, banks, courts, galleries. Division of labour also

---

[2] It points—in its strong Protestant and German versions—to professional specializations within a restricted professional sphere (Stichweh, 2013, p. 50). The concept was first explicitly formulated in the writings of Adam Smith (2003) and Ferguson (2012) as a theory of professionalisation, who, however, do not yet conceive of differentiation as a macrosocietal property, but rather as "collective bodies" (Ferguson, 2012, p. 47), for which the loss of collective agency is typical. Only the expansion and transfer of the concept of individuality by Schleiermacher (1988) and—following him—Simmel (1989), which is now no longer limited to natural persons but appears in the form of functional meaning domains as higher-level individuality, makes a theory of differentiation possible. For a detailed history of the development of the concept, see Stichweh (2013).

[3] Following Maturana and Varela (1980), Luhmann calls the process of self-production autopoiesis. Their neologism, which describes the capacity of living cells to reproduce and organise themselves, is transferred by him to social contexts that reproduce themselves not through life but through communication. Sociology is interested in the connection between this temporal reproduction, understood as dynamic because it progresses from moment to moment, whose only goal is self-preservation, and factual ('static') differentiation (Baecker, 2021, pp. 23f.). The fact that development in time simultaneously solves and exacerbates the problem of differentiation becomes particularly clear in the context of crises such as the pandemic.

concerns the environmental projections of these areas. To stay with the example: what for schools are hospitals, courts, and galleries; what for banks are schools, hospitals, etc. (Luhmann, 2018, p. 833). These are not ontological spheres of being that one only has to add up to get the whole of society, but rather spheres that can only relate to the whole from their own respective perspectives and thus produce a multiplicity of wholes.

This means that there is no representation of the unity of world society and its systems in world society. The attempt to reconstruct all or even a certain number of these functional contexts from the perspective of politics—or any other system—encounters a heterogeneity that cannot be reconciled (Baecker, 2021, pp. 37f.). The common social denominator can therefore only be formulated negatively: it lies in the non-congruence of all these different perspectives and in the socio-structural relevance of this non-identity (Luhmann, 2018, p. 835). The term that has come to be used for this circumstance is called 'polycontexturality', which we conceive as the inaddressability of world society.[4] From an observational point of view, it is not uniform, but manifoldly constituted; as a disparateness that excludes that there is a place from which another than an operative concept of unity could be gained. Such a concept means nothing more than the recourse to communications that produce communications in a worldwide network of communications. In this network, no above/below can be determined, no central organ that would be able to coordinate the different functional logics of the individual areas and guarantee the unity of this order. "Modern society has neither a center nor a top" (Luhmann, 1983, p. 31).

The pandemic has not changed this fact. Instead, the world's functional systems all responded to the challenge with their own on-board resources: politics made collectively binding decisions, though admittedly "more consequential" ones; science was not only able to identify and classify the virus and reconstruct the dynamics of its spread, but also to develop a vaccine in record time. The mass media had an exciting sequel, with the ever-changing variants—from Alpha to Beta to Delta to Omicron—enabling ever new seasons, including an episode addressing the 'scandal' that the WHO had skipped two letters (Ny and Xi) when naming the virus; and the business community was also able to adapt to the challenges: some made high turnover with 'mask deals', and in China, even the most run-down quarantine hotels charged high fees for accommodating their 'guests'. In short, the functionally differentiated society did what it always does—each of the systemic competences involved followed its own agenda. In medical terminology: they all formed different 'antibodies'. Corona—as the general name of a virus along with its variants—was and is something different for politics than for the economy, religion or the mass media. Neither the health system—as the field concerned with the deviation from normality of physical or psychological states of mind that can be identified with the help of science—nor science itself, to which this text ascribes itself, can claim a special role here.[5] It merely adopts a virtual perspective that observes the whole of society

[4] The term was introduced into scientific discourse by Gotthard Günther (1979).

[5] The Anglo-American science community may ask strange questions like "How Scientific are the Social Sciences" (Myrdal, 1972) and distinguish real or 'hard' science (natural science) from 'soft'

with the help of the code true/false. This does not prevent science from doing it anyway—and, like all other subsystems, from claiming a universal relevance of its own system.[6]

Some observers nevertheless try to positively formulate the common social denominator and urge a concordance of perspectives, i.e., strive for a representation of the unity of society and its functional systems in society—for example, with the help of a moral canon that claims universal validity. But these pretensions aimed at the whole make the fact of non-conformity all the more obvious and immediately provoke corresponding counter-reactions. Even if it wanted to, society can no longer use only one distinction; but since it cannot 'want to', the problem does not arise at all. In the language of systems theory: unlike its subsystems, society itself does not have a code, i.e., no guiding dual that generates operations. The resulting problem is the question of how it could describe itself today.[7] With Lyotard, we can call this the 'postmodern condition'.

It corresponds to the loss of the métarécit (metanarrative)—a 'total account', for instance in the form of a grand theory, is no longer possible. Paradoxically, Lyotard's 'report' can itself be understood as such a metanarrative: as a grand theory that describes the unity of the social order by means of a concept of post-modernity—precisely by means of that a 'post' that assumes a rupture between modernity and postmodernity (Luhmann, 1997a, p. 1144).

In what follows, we agree on the one hand with the thesis of the end of the grand narratives (as open to interpretation as it may be), but on the other hand we dispute the existence of a rupture and assume the fundamental continuity of the structures of modern society, since key characteristics of modernity persist. Even in museums there is no sharp epochal boundary.[8] From our point of view, the more interesting question than whether society is in fact postmodern is the fact of this description, that is, why society describes itself as postmodern in the first place. Luhmann's suggestion

science (social science), but from a systems-theoretical point of view this makes little sense: there is neither a hard truth nor a soft untruth, but only an examination of knowledge which is either considered to be connectable (true) or not (false).

[6] For Parsons (1971), this combination of universalism and specification is an important characteristic of modernity.

[7] We assume with Peter Fuchs (n.d.) that the system of the Internet processes such a description of society in society, which—paradoxically enough—*presents* this non-representability of the unity of society in an etymologically precise sense. The system copies the non-unity (or non-accessibility) of society into society by way of re-entry. This is in contrast to sociology's big brother, the mass media system, which allows competing self-descriptions of society and suggests by way of compression (foreshortening) that there is significance, global relevance, global urgency, and global noteworthiness. The internet eludes this—any!—pointing possibility. It stages a vaporising of every concatenation; it is, if you will, a perpetual atomising of every pretension of unity, and it is therefore in form: anti-fundamentalist, no matter how many fundamentalisms are put into the first-order document level.

[8] Luhmann is not so sure about architecture, the area which made Lyotard's diagnosis plausible. We regard the postmodern style simply as a particularly pronounced eclecticism, a historically reflected second order programme of architecture in which forms from different eras are all available simultaneously, and not as a threshold that would allow us to designate a before and after.

was that postmodern theory responds to the problem of undercomplex theorising: one has underestimated the dynamics of modern society, has offered overly static attempts at description, such as those found in connection with the concept of human rights or democracy (Luhmann, 1995b). Such a fixation must then necessarily lead to the discovery that the theory does not correspond to reality, but this can hardly be blamed on reality (or China!). The solution to this problem would be a shift from the semantic level represented by the postmodern self-description of society to its operational and structural realities.

We remain with the assumption that there can be no legal (also no: legalisable) central observer of society, thus also no universally recognised directive donor who could point out the paths that society and its functional systems would have to follow. Lyotard’s question: “Where, after the metanarratives, can legitimacy reside?” (Lyotard, 1984, p. xxv) we answer with: legitimacy is dispersed in the respective functional areas of the systems. Therefore, it makes no sense to ask ‘society’ for changes.[9] Only the instances of the functional systems that represent them can be addressed—schools, hospitals or banks, for example, and these cannot agree on common goals. Not even when the survival of humanity is at stake, as the attempts to set a minimum target for climate policy demonstrate.

2 Frequency Reduction or: Too Bad for the Facts

If our assumptions are correct so far, then the tendencies that became visible during the pandemic require an explanation, because they are in clear contrast to them. The assertion that ‘globalization’ has come to an end, that it is, as it were, going into reverse (‘deglobalization’), stands in contrast to the irreversibility of world society that we asserted at the beginning.[10] The temporary dominance of politics—in conjunction with the health care system—stands in contrast the thesis that no functional area is above the other, that there is no longer any central authority in modern society. And the fact that some observers see Corona as the lost point of unity, as if the narrative function had recovered its great hero, contrasts with Lyotard’s account of

[9] The attempts of the self-proclaimed ‘lost generation’ to call the world society to action against climate breakdown, i.e., ‘collective suicide’, as honorable as they may be, are therefore misguided from the start: the world society is just as unaddressable as Santa Claus. Instead, it is the legal system that feels addressed when activists stick themselves to famous paintings or goalposts.

[10] We write globalization in quotation marks to draw attention to the fact that we understand the global system decidedly as a society, with Parsons: as system of modern societies (1971), not as a ‘world-system’ characterized by the interaction of different regional societies (Wallerstein, 1989). One of the advantages is that one can move away from designing a segmentally differentiated system according to nation-states in favor of a system differentiated according to different functional systems. For a detailed discussion of the two concepts, see Wittmann (2014). As she correctly notes: “Both of them do not have a consensus definition at their disposal” (2014, p. 199). But as unclear as the concept of world society may be, “(T)he concept of the international system is even more unclear” (Luhmann 1997a, p. 160)—because how can an ‘inter’ be a system?

disbelief in metanarratives and our assessment that the common social denominator can no longer be positively formulated.

First, on the question of deglobalization. It indeed seems as if Corona has empirically verified Zhao's (2020) diagnosis: that we live in a 'failed world' in which the individual rationality of the individual nation states still guides action and thinking—even though world society no longer allows them to be considered as entities in themselves. The territorial principle, so important for segmentary societies, i.e., a definition of one's own unit that is primarily based on space, gained renewed significance. Once again, national borders functioned as a common integrating factor. It is therefore no wonder that space theorists in particular lay claim to an adequate description of the situation (see for instance Löw et al., 2021).

Second, the assessment that Corona challenges the structure of functional differentiation can't be dismissed out of hand either. The temporary dominance of politics—in conjunction with the health care system—seems to contradict the thesis that no functional area is above the other, that there is no longer any central authority in modern society. Politics became the most important 'systemic individual', brusquely rejecting the claims of the other functional areas. The state, i.e., the segmentary, territorial order of power control, emerged as organizational unit with a claim to a functional monopoly in society, and at the same time eliminated the problems of legitimation—as well as motivation—associated with such a claim. Very few citizens disputed its claim to this monopoly, most were prepared to let him get away with such a temporary primary differentiation. Seen from this perspective, one crisis has ended another (cf. Luhmann, 2018, p. 250). Interestingly, while the centralization of the political function gained credibility, its former counterpart, the church, did not benefit from the crisis—man's calamity is no longer God's opportunity, it seems.[11]

Dirk Baecker (2020) has used the metaphor of frequency for this circumstance, to clarify that the crisis has not ended functional autonomy, but merely throttled the pulse—the frequency of communicative events—of the other systems. If the frequency of religious events, for example, had initially decreased during functional differentiation, at the same time an increase in industrially and urbanly coded events could be observed. At present, we observe an increased frequency of events declaring gender difference obsolete, which can be traced back to the emancipation of women. During the pandemic, politics had considerably slowed down the pulse of the other systems and "implemented measures deemed appropriate by virologists and epidemiologists through the means of recommendation, order, and threat" (Baecker, 2020, p. 73). The frequency of family events had - by necessity - increased just as much. The state, one might say, went to great lengths to fulfill the central role assigned to it

[11] Exceptions confirm the rule, but mainly contribute to the amusement of the global audience. A good example is the sermon of the preacher Kenneth Copeland: "In the name of Jesus, I execute judgement on you, Covid-19! I execute judgement on you, satan, you destroyer, you killer [...] get off this nation, I demand judgement, I demand, I demand, I demand!" That this incantation would make an excellent metal lyric has been well recognised by Portuguese Youtube star André Antunes (Antunes, 2021). Subscriber Jeff Gordon states: "Kenneth Copeland is now my favourite metal lead vocalist. Great talent. Lyrics straight from hell." And user TheVlads frontman ironically states artistic progress: "Man, Christian Metal has gotten a lot better."

by some sociologists (Giddens, 1985, 1990). It almost seemed like it were interested in transforming the epistemic blockage of social theory into a blockage of modern society itself.

At the same time, the principle of stratification experienced a revival. A renewed recognition of rank differences came into sight, with the help of which the decision-making burdens created by the pandemic were managed. Observers noted a re-feudalisation brought about by Corona (the global South being marginalised), an increasing orientation towards wealth and class boundaries that distinguished between those who could benefit from health services and those who could not, those who could work in a home office and those who had to sit at the supermarket checkout and expose themselves to the risk of infection. This was accompanied by a digitally conditioned difference in rank, as the possibility to dispose of the data processing acted as a bottleneck: it excluded those who did not dispose of it from the next society in the present one (Baecker, 2020).

Confronted with these old forms acquiring new validity, one could thus come up with the same thought that gripped Musil's 'man without qualities' when looking at the imperial Hofburg: that on the one hand we are dealing here merely with "a world that has not been cleared away", but which on the other hand makes us feel a strong, strange quality: it proves to be "surprisingly real" (Musil, 2003, p. 86).

Just as real as the fact that some observers used the pandemic to reassert the unity of society as a principle: after the Risk Society (Beck, 1992) and the Society of Fear (Bude, 2014), the Corona Society (Volkmer & Werner, 2020) was being proclaimed.[12] Thanks to an infectious organic structure, we would once again have a common denominator at our disposal that would make it possible to lead society, which is fragmented into a multitude of different contexts, back to a common model of reality. Corona would have given us back the grand récit: as the pivot around which world society gravitates, the lost perpetrator who pulls the strings, some of which can even be 'traced'. Not systems theory, but network theory would then be the appropriate programme of reflection—and with it the difference between connection/contagion: Who is connected to whom and what flows across ties? Since the pandemic was able to add up widespread experience, the appeal of this proposal was quite understandable.

It would not be without irony if modern society would owe the new métarécit to a toxic slime (the original meaning of virus), of all things. In the sense of: Every society the narrative it deserves. But it would make perfect sense, because the great narratives of modernity have always been narratives of crisis management and overcoming (Nassehi, 2021). The narrative of reason defeats unreason, the narrative of the nation defeats disorder after revolution or reconciles order and progress, the narrative of freedom defeats precisely this freedom by telling about the necessity of submission to the reasonable. And the Corona narrative, should there be one?

[12] Not all of them were as clear-cut as Volkmer and Werner—most attempts at determination used Corona to characterize society as fundamentally sick and deficient. We will come back to this in detail in the last section.

Most of the candidates who laid claim to this title referred to the points we have mentioned above. One strand of the Corona narrative concerned the end of the "post national condition" (Habermas), and another one the constraints on systems, especially those that hit the economy. But since one of the most visible conflicts during the Corona crisis was the state's curtailment of basic individual ("human") rights in favor of health protection and social control, the third strand was unsurprisingly about freedom—and even less surprisingly, this pathos formula was used mainly in the discourses of the West. For example, in the form of "the promise of modernity" (Nassehi) that one can counter challenges such as Corona with something, namely actions that are both free and necessary, which thus profess a freedom that knows how to limit itself, i.e., is not entirely free without therefore being forced from above.[13]

The focus was on the responsibility of the individual actors in order to establish—quite in the sense of Foucault (2020a, 2020b)—an "I want what I should" in them: Keeping one's distance, wearing a mask, getting vaccinated or isolated, etc., i.e. practising consideration for others.[14] The opponents of vaccination and Corona deniers were thus accused of a lack of historical consciousness, since they summarily equated freedom with mere egoism, failing to appreciate the decades-long struggle of democracies to reconcile self-determination and social expectations, subjective rights and social order.[15]

The reason China did not initially have to deal with the problem of intransigent citizens is the fundamentally different cultural context.[16] While Western democracies place the emphasis on the rights of the individual vis-à-vis the state and society: on the autonomy based idea of subjectivity of the individual human being who positions himself (generic masculine) against political and religious powers (historically: empire and church), in China the duties of the individual towards society are paramount and the exercise of fundamental rights and freedoms must not interfere with the interests of the state and society (Gan, 2013), in this case: the strict "control and maximum suppression" policy of the government known as 'zero-Covid'. Whereas the German citizens considered the measures an "imposition to democracy" (as their top representative, Angela Merkel, put it in 2020), the Chinese citizens were

---

[13] In fact, even the supposed self-empowerment of the subject is not, as Nassehi (2021) notes, but an external empowerment or one imposed from above.

[14] Foucault's example from *The History of Sexuality*: 'I should and I want to marry, i.e. contain my sexuality in the covenant of matrimony'.

[15] This is what Nassehi (2021) reproached the protesters with: to develop a historical consciousness and to prove oneself worthy of the acquired rights. Whether tutoring may be the right way to go here is, in our view, rather questionable. Moreover, one could argue with Charles Taylor that this "dark side of individualism" has always been carried along in the promise of modernity, as "a centering on the self, which both flattens and narrows our lives, makes them poorer in meaning, and less concerned with others or society" (Taylor, 1991, p. 10).

[16] It should be clear that the distinction made here is a strong simplification that neglects the great differences existing within cultures (see, for example, Marsella et al., 2000, who point out the similarities in behaviour across cultural boundaries). We accept the associated disadvantages in favour of a didactically motivated plasticity. The distinction finds its counterpart in disciplinary juxtapositions, whose limitations cannot be overcome by interdisciplinary research efforts, but can be made productive.

initially more or less willing to put up with them, or more cautiously: these measures restricting their freedom by the state concerned only with their well-being (and we do not mean this ironically) were more familiar to them. More familiar, at any rate, than the conception of conflict that emphasises the freedom or the rights of the (infected or at risk of infection) individual.

It is only recently that China has seen increased resistance to this strategy, which at times locked 22 million people in their homes.[17] Due to the highly infectious omicron variant, the continuing adherence to the strict zero-Covid policy is no longer considered appropriate from an epidemiological perspective, and even in China itself is growing resistance to the role that politics has usurped. One could say that the functional subsectors assert their rights here—for example, in the form of representatives of academia such as Tong Zhiwei (P. Zhang, 2022) or Xu Zidong (Lu, 2022).[18]

However, these cultural differences, which explain different responses and narrative preferences, do not yet answer the question of how to explain deviations from the social norm when global functional differentiation knows no exceptions and cannot be reversed. The aphorism attributed to Hegel, "If the facts do not agree with the theory—too bad for the facts" we do not want to adopt in the following. This arrogance may be due to a philosopher, sociologists cannot afford it.[19]

A first clue is provided precisely by the fact that both the global order and the restriction of degrees of freedom—the definition of integration (Anderson, 1960; Luhmann, 1997a, pp. 601ff.)—have been discussed so hotly at all. This suggests that we are dealing with an anomaly, an irregularity which of all the candidates currently

[17] The fact that the rights guaranteed in the constitution cannot be enforced in court in China is well known and is not disputed by us. The Chinese courts cannot apply constitutional rights (Guo, 2009, p. 168). The compatibility of laws with the constitution is not examined, and constitutional complaints are not provided for. In accordance with Confucian tradition, China prioritises economic and social rights over civil rights. Cheng Boqing (2009) has argued with Scheler that for many in China there is a sense of lack of control over one's own destiny due to the lack of legal certainty, leading to a state of 'resentment' that is repeatedly discharged in flashes (shitstorms, etc.). This finding can be supported by systems theory: If legal procedures can no longer assume the function of absorbing disappointment, then individuals have no choice but either to resign or to rebel (Luhmann, 1983, pp. 47f.). Leaving the absorption of disappointment to the individual is quite risky because the bill is presented to politics. In the West, the opposite problem can be observed: that of entitlement inflation, which has devalued the legal mechanism (Luhmann, 1983, p. 48).

[18] Legal expert Tong points out the social function of law: "Local governments and officials need to act strictly according to the constitution and laws and cannot destroy the rule of law for convenience" (P. Zhang, 2022). Historian Xu compares the drastic nature of the measures taken by the pandemic control personnel—who vandalize private homes in the name of questionable disinfection rituals—to the devastation of private homes by the Red Guards during the Cultural Revolution.

[19] In fact, the phrase originated in France in the early nineteenth century ("Tant pis pour les faits") and is found nowhere in Hegel's work.

vying for the position of social self-descriptor probably has the most promising chances, not least because of its negative content: the crisis.[20]

3 Crises or: The Superproblem of Inacceptable System Overruns

For several decades now, crises have been talked about everywhere and all the time. Anyone who watches a news programme today will inevitably encounter the term. Luhmann (1984b) therefore diagnoses a 'crisis fashion': oil crisis, Corona crisis, Ukraine crisis, climate crisis—the list is long. In Germany and Europe, respectively, people are currently talking about a gas crisis, in the USA about a water crisis. Sociologists are not sparing in their use of the term crisis either. Habermas (1973) heaped rationality crisis, economic crisis, legitimation crisis and motivation crisis on top of each other, Luhmann (1984a) himself starts his reinvention of sociology with the diagnosis of a theory crisis, more recent entries in the crisis register deal with a competence crisis of politics (Nassehi, 2016), a crisis of liberalism (Reckwitz, 2019) or a crisis of symbiotic relationships (Folkers & Opitz, 2020).

If the inflationary use of words is an indication of the social, crisis-ridden reality, then "we would have to live in an all-encompassing crisis" (Koselleck, 2010, p. 203). Indeed, some observers hold this view. They understand the crisis as the 'fate' of modern society, as a permanent state. Not Corona but the crisis would be the metanarrative of modern society. But drawing conclusions from the inflationary use of words to social reality contributes little to the diagnosis of the situation (Koselleck, 2010, p. 203). In fact, the word is predominantly used as a metaphor, otherwise society could not simply move from one crisis to the next without further ado.

What is certain is that crises are frequently observed. The associated 'swellllling' of communication is at least an indication of the reality of a crisis, because one of the functions of society is error correction and the coordination of behaviour in such situations. One could speak of a social immune response of society, an emergency management that organises different communicative connections; connections that - in the Case of Corona - seek a relationship to the virus (Luhmann, 1984a, p. 549). Crisis communication can thus be understood as a kind of social alarm system that suggests related action—as a call to arms (*à l'arme*).[21] In this sense, a crisis suggests urgency and speed: "We have not much time, approaching an either/or situation" (Luhmann, 1984b, p. 59). This is precisely why crises can also justify hasty, precipitate intervention, as can the drastic nature of the measures taken. Favourably, it coincides with what the population expects from politics in times of crisis—to make

[20] "The negative has, by virtue of its own indeterminateness, important strategic advances" (Luhmann, 1984b, p. 59).

[21] It should only be noted that crisis communication is part of the daily bread of the mass media. The reality they produce is the reality of the mass media (Luhmann, 1995a). It is therefore necessary to relate this reality to the realities produced by other functional areas.

decisions that hold out the prospect of mastering such circumstances (Baecker, 2016; Heidingsfelder, 2020, p. 132); that it acts as a loss compensator, security broker and provider of resources. The role of freedom limiter is then gladly conceded to it. Not least because the crisis not only legitimizes such unusual measures, but also marks them as temporary.

Unlike Habermas (1973) who assumes a legitimation crisis of society, we do not understand social crises as functional crises of overarching overall institutions, as there are no indispensable total functions in modern society (Luhmann, 2018, p. 858). Our concept of crisis sees them as systemic or functional crises. They occur when a problem cannot be solved at the individual system level. The solution is then expected of other systems. In the case of such a co-fulfilment, Luhmann speaks of the 'superproblem of system overruns' (Luhmann, 2018, pp. 839, 861f.). Such overruns can be understood quite literally, as a shifting of the problem burden.[22] A problem is pushed to where it can be solved most rationally. Such problem overruns are not 'structural faults', nor are they wrongs. They are tailored to the specialized competencies of other systems and therefore not, in principle, intended to trigger structural changes. However, they come in acceptable and non-acceptable forms. They are non-acceptable when they create crises, which can in turn initiate structural transformations in a regressive direction—the very return to earlier social forms that occupied us in the previous section. The fact that they require restructuring in all the subsystems concerned and thus cancel the normal system/environment perspectives explains the diffuseness of crisis behaviour. The crisis acts as a compulsion to reactivate the unity of society at the operational level (Luhmann, 2018, p. 860). In this, the concept of crisis shares some common features with the concept of morality. Both subvert the distinctions of individual social areas or systems.[23]

By contrast, problem overruns are acceptable if they do not in principle trigger structural changes, i.e., if the functionally differentiated society can, as it were, continue to go about its business. In the case of Corona, we are dealing with such acceptable regressions. However, irreversible regressive transformations are something we always have to reckon with from an evolutionary theoretical perspective (Luhmann, 2018, p. 863). As mentioned above, the problem is not so much one of overriding or crisis, but to ensure that problems find "the best place of their solvability or their most innocuous persistence" (Luhmann, 2018, p. 839). The double strategy that society has come up with relies on simply letting the process run all by itself and creating problem-specifically conceived special organisations for residual problems that have not been solved.

The relevance of the Corona crisis for society resulted on the one hand from the fact that the problem-specifically conceived special organisations were 'overtaxed' or insufficiently equipped with the solution, and on the other hand that it threatened

---

[22] Luhmann takes the term from finance, which uses tax shifting to describe the permitted shifting of tax burdens from one economic subject to another.

[23] On the role of morality in monitoring the coronavirus pandemic in Croatia, see Žažar (2022). On the damage that mass outbreaks of moral controversy can do to the scientific system, see, regarding the Ukraine crisis, Gordin (2022).

to exceed the normal adaptation possibilities of all other subsystems. Even if that did not happen, the concern was more than justified. It would hardly have been possible to intercept it in terms of policy measures if infected patients could not have been treated, since the hospitals cannot turn away anyone who has been included by the virus. No state would be able to cope with the loss of services in its own health system. So simply letting the process 'run its course' was out of the question.[24]

The costs were primarily borne by the environment (Luhmann, 2018, p. 839)—which, since Corona itself represents a diffuse environmental factor, initially seems quite justified. Except that the measures hardly impressed the virus, unlike the people who share this environment. All of a sudden, "we, the people" became aware of the fragility and the wealth of preconditions of our own everyday actions based on the social order of functional differentiation—those routines that we otherwise use as a matter of course. Families paid a high price for the fact that they suddenly no longer had access to the familiar(!) repertoire of changing contexts of their members in the workplace, in educational and care facilities.[25] In Shanghai, the consequence of the long lockdown was a large number of divorces and suicides.[26] Apparently, the number of mental illnesses due to social isolation has also increased significantly. While such mental 'overload' symptoms are certainly treatable, the situation looks bleaker in the case of treating the immensely increased climate problems. Conversely, the more diffuse environmental factors act on a system—and the virus, i.e., the people who become infected with it, are just as much a diffuse environmental factor as climate change—the more likely crises become.

In other words, functional differentiation is crisis-prone. It is this very order that is the condition of the possibility of crises. Therefore the 'functional consciousness' of systems is always simultaneously performance consciousness and crisis consciousness (Luhmann, 2018, p. 862). The fact that they are realised on a global scale makes the whole world vulnerable to crises. All it takes is for one of its highly specialised sub-sectors to have problems fulfilling its function and 'hand it off to the outside', i.e., force other systems to co-perform, and they can occur. The different time requirements of the individual functional areas add to this. Politics, for example,

---

[24] In this context, it might be worthwhile to look again more closely at the question of the significance of the health system for our notions of a state. The assumption that it is so vehemently protected by the state because the latter owes it a large part of its legitimacy—just like the education system, which not coincidentally differentiated at the same time.

[25] However, families in the West are still much more dependent on exit possibilities than other families of the world, even though families in Asia, for example, are currently undergoing major changes in the context of world society. Ting and Chiu have pointed out that leaving home in China is by no means "a simple logic of individualism propelled by modernization" (Ting & Chiu, 2002, p. 614). The authors deny that 'leaving home' has gained in popularity; rather, this exit is still closely related to the marriage transition and practical considerations. Their analysis highlights the complex relationship between tradition and modernization, which will concern us later with regard to the question of regression: functional differentiation does not mean that the older forms of social order die out.

[26] G. Zhang (2022) recalls the psychological stresses to which the citizens trapped in Wuhan were subjected in an excellent visual essay dedicated to not letting their experiences be forgotten too quickly.

runs faster than the law, mass media faster than politics, and law so slowly that one can sometimes have the impression that it is standing still (Luhmann, 2018, p. 846). These pulse divergences create synchronisation problems even in 'normal' times. The Corona crisis exacerbated these problems, which are most noticeable in the form of time limits and uncertainties. They distort and massively burden factual preferences.

If the changes associated with crisis behaviour are increasingly established and solidified, the effects triggered by them can become permanent. This is precisely what the opponents of the Corona measures have been warning about: the irreversibility of the structural changes. Is the epidemic possibly the prelude to a new, abnormal normality? Will we live in a constant, seasonal alternation between the normal frequency of society and a pulse that is greatly damped in some areas (Baecker, 2020)? Is the new Western form of government the 'pandemocracy' (the beautiful neologism of our colleague Matthias C. Kettemann)?

There are already many indications that the short-lived 'total integration' of politics and the healthcare system has not changed the way society operates. To vary an iconic Thai phrase: *Different different—but same.* Because if the competencies of the systems can be regulated up, they can of course also be regulated down again, and since it was only the crisis that made it necessary and possible to increase the frequencies of politics and the health care system alike, there is no reason to maintain their high pulses after its end. Not least because a downward adjustment of the economic, educational, artistic, etc. pulses that lasts too long would in turn have a critical impact and provoke a new crisis.[27]

However, it is not quite that simple, because after all, the primacy of a function-oriented world society made the pandemic possible in the first place. And since it cannot be reversed, the 'next catastrophe' is already just around the corner. This is precisely what virologists and epidemiologists assume. They see the current crisis as an exercise and assume that the coping practices we are currently using will be used more often in the future, which is why pandemics must be 'better managed': "New diseases are inevitable", warns the editorial board of the *Economist*. "Ensuing global calamities are not" (EBE, 2020, p. 7). After all, the working conditions for viruses have never been more favourable than they are today. In this sense, Corona would not be an end, not a conclusion, but the beginning of a fundamental change not only in the economy, politics, and law, but also in our everyday behaviour. Then an epoch of world history would indeed start from here and now.[28] Then the prophecy of the Chinese philosopher and politician Kang Yuwei about the age of the 'Great Community' would come true:

---

[27] The West is currently at great pains to point out to China the high costs of a continued zero-Covid policy. See as representative example *The Financial Times* (EBFT, 2022), but *BBC, CNN, Time* or *The New York Times* also join the chorus.

[28] Thus, Goethe's famous sentence, allegedly addressed to the soldiers of the anti-French coalition on the evening of the Valmy cannonade: "From here and today, an epoch of world history emanates, and you can say that you were there" (cf. Conrady, 1994, p. 565). In contrast to the Corona crisis, however, this artillery duel was more of an insignificant event to which only Goethe was able to literally ascribe world-historical status—no one but he has handed down the beautiful sentence.

> Throughout the world, health policy will be paramount; nutrition, clothing, house construction and sanitation will be constantly monitored by doctors. All community institutions will also be controlled by doctors. Everyone is examined daily. Doctors also control the hygiene of the home [...] Medical skills must be highly developed, because doctors bear the brunt of responsibility. Only the best will then be allowed to work in the health sector. On the other hand, the physicians will then also have the greatest influence in the community [...] The age of the 'Great Community' will be a world of health care. (Heberer, 2020, p. 222)

4 Epoch Overlap

We do not have to fear either a health care dictatorship nor a return to earlier stages of differentiation, not even a decline or worldwide decrease of interconnectedness in favor of regional conditions. The numerous changes triggered by the virus did not cancel out the primacy of the structural principle of functional differentiation in favour of older principles—even though all the above-mentioned tendencies were increasingly observed in the course of the pandemic. How can this contradiction be explained?

By a simple circumstance: that we are in fact dealing with "a world that has not been cleared away"—or more precisely: with many worlds not cleared away. The epochs have not disappeared, they overlap.

For example, the global society still proves to be clearly differentiated regionally, especially regarding cultural traditions and levels of development. It therefore makes sense that the effects of the "COVID-19 shock" (Sforza & Steininger, 2020) are being felt above all in those countries that are strongly integrated into the functionally differentiated global production network; at the same time, it shows the above-mentioned heterogeneity of this network, the density of which varies according to sector, region and country and also depends on the factor of the geographic distribution of industries in each region and country (Sforza & Steininger, 2020). Global society is characterised by a highly uneven spatial context of networking and dependency, in which disconnected regions and exclusions emerge as the other side of networking nodes and global cities (cf. Harvey, 2006; Löw et al., 2021, p. 10; Sassen, 2001). This unequal distribution has to do with the fact that, unlike China, global society does not govern itself through targets or directives, the regional observance of which would then be examined and possibly corrected. Instead, the regions themselves must respond to the fluctuations generated by the financial markets (Luhmann, 1997a, p. 808). Another reason why regional interests continue to be increasingly brought to bear is the continued existence of nation states. We can illustrate this with the German proverb 'Wenn zwei sich streiten, freut sich der Dritte' (When two quarrel, the third is happy)—when two states compete on the international stage for capital for regional investment purposes, the region is the profiteer (Luhmann, 1997a, p. 808).

These blatant differences of the individual regions of the earth are the reason why some authors still cling to a regional concept of society. But one is under a misapprehension here, because if one makes a historical comparison instead of a regional comparison, then several consistent trends towards world society clearly emerge (Luhmann, 1997a, p. 161). Even Corona has not been able to stop the trend towards the dissolution of diversity with the spatial differentiation of the world or the fact that, in the context of world society, differences are no longer secured by spatial segregation. They emerge in the form of cultures of communication, which, even if they are only small cultures, can certainly include communicative addresses worldwide and in this way are able to reach critical sizes (Stichweh, 2009, 2013). This strengthening of site-specific references takes place within a global framework.[29] The global inclusion realised by world society does not allow for exceptions. This is shown by the high costs that such attempts entail for all participants, not only for the countries of the 'global South'. Corresponding political and mass media demands, for example, for the detachment or independence of "questionable" or "controversial" states such as China (because of Taiwan, Uyghurs, autocracy) and Russia (because of Ukraine, autocracy) come up against global reality, which we want to understand with Luhmann as resistance of operations against operations. A less integrated world leads to a loss of income and prosperity. Especially export-oriented economies like Germany can hardly afford to turn back on themselves. The German Ifo Institute has calculated the various scenarios and comes to the—unsurprising—conclusion that a German attempt to turn back globalisation and realise renationalisation, i.e. to bring back production to Germany on a larger scale, for example, is not a current solution to the crisis (Flach et al., 2021): Reshoring (the extensive shifting of production back to the home country) would reduce the gross domestic product to a level almost 10% lower; nearshoring (the soft variant, a shift to neighbouring, 'non-controversial' countries) would lead to a loss of value added of 4.2%; and unilaterally turning away only from China would supposedly be bearable, but would cause costs almost four times as high as Britain's EU exit, not taking into account the possible retaliatory measures by China. Deglobalisation therefore does not mean less global society, but simply less prosperity. The researchers recommend functional differentiation: in other words, leaving it up to the economy itself to weigh up between cost efficiency and ideology. One could also say it is a recommendation to German politics not to authoritatively interfere in the decisions of companies. Put more cautiously: Not to turn investment decisions into political decisions.

According to Bergsten (2022), what the Ifo Institute has found for Germany also applies to the USA: the interconnectedness of China's with other economies is too great to rely on a strategy of containment, not least because the country is the most important trading partner for the countries of the world (far ahead of the USA). We would agree with Bergsten that the different 'values' of the two countries need not be an obstacle to economic cooperation, let alone a common environmental strategy; implicit agreement and informal cooperation are quite sufficient. Bergsten is cautiously optimistic for the time after American dominance—on condition that Europeans and Japanese succeed in cooperating with China. Moreover, hardly any economist would dispute that the IMF's rules should have been changed long ago due to China's growing global economic importance. Finally, Bergsten is pleased to

[29] Volkmer (2014) has used this insight to redefine Habermas's concept of the public sphere, which she sees as fragmented into a multitude of 'micro spheres'.

point out that the Americans have not always kept to the rules either (which they accuse the Chinese of, key words: subsidies for state-owned enterprises, protection of patents). What could persuade them to take a rational stance towards China is that the economic costs of containment have a political impact (exactly what the West is trying to convince China of with regard to zero-Covid).

To what extent these counter-movements to globalisation are a new phenomenon is another question. It seems that Corona has only made more visible what has been happening all along anyway. Many regressive tendencies have nothing to do with the Corona crisis, which is why social researchers have already been looking at some of the opposing tensions. They can only be inadequately described with globalisation versus nationalization or progression vs.regression. Moreover, they cannot be observed in all areas, and their social causes would have to be investigated separately. (One possible consideration concerning regression is that the ageing of the functional differentiation type could be related to it, cf. Heidingsfelder, 2020, pp. 298f; Luhmann, 1984a, p. 516). The return of the extra-functional element of religion to politics in the form of civil religion and wokeism (which some consider a new civil religion in itself), the attacks on organisation and institutions we are witnessing (keywords: 'deep state' and 'fake news') that seem to dock onto older patterns of differentiation in order to subvert functional differentiation, or the 'democratic regression' observed by Schäfer and Zürn (2022), all this took place long before the Corona crisis. The moment of transnationalisation, i.e., country-specific patterns of networks and exchange relations (cf. Mau, 2007; Schmidt-Wellenburg & Bernhard, 2020), should not be neglected either.

What sets them apart is the attempt to take back the rigorous repudiation of the past and declare it binding once more, and thus automatically the achievements of modern society—the modern state—non-binding. Instead of distancing oneself from ancient forms of social organisation, for example, from one's own person in terms of role behaviour, from family forms of cooperation, or from old, outdated economic strategies such as protectionism, there is an attempt to reintroduce ('re-enter') what has been excluded. One simulates continuity, the selective relationship no longer takes place regarding a historical past, but with regard to one's own present. The future is to be opened to possibilities that belong to yesterday—the prefix 're' points to this in all clarity. One could speak with Simon Reynolds (2011) of a "retromania", a comprehensive addiction to its own past that has gripped world society.

But economic protectionism and the rise of left- as well as right-wing nationalist populists with decidedly anti-globalist attitudes, challenging commentators to new interpretations, from the crisis of globalisation (Diamond, 2019) to that of the terrestrial age (Latour, 2018), are found in the same society that remains globally oriented (which is already evident from the fact that the political reactions to Corona are similar worldwide). What they have in common with Corona is their high risk of contagion—which is why some observers describe fundamentalism itself as a disease that has the property of infecting its opponents (Luhmann, 1997a, p. 155, footnote 211, see also Strong, 1990). These 'breakthroughs' do not endanger the global social order, but they do give indications of unsolved problems in world society which are looking for their own solutions. In other words, they too are indications of crises. This also applies to the above-mentioned phenomenon of populism, which we understand as a solution to the problem of motivation ('voter fatigue').

The epoch overlap is particularly evident below the primary level of function systems. Here we find numerous (Stichweh: "thousands") of subsystems which are internally differentiated along complex mixes of hierarchical, segmentary, and functional lines (Stichweh, 2013, p. 59). The Corona crisis makes us realise that the theorem of functional differentiation does not mean that we have left the segmentary or stratified forms completely behind, but only that they no longer apply without further ado and can no longer lay claim to general validity. In other words, the order of functional differentiation has not ended stratification, it has merely taken away its function as a primary principle of differentiation.[30] The primary subsystem formation, however, still takes place following functions such as collectively binding decision-making (politics), gaining knowledge (science), ensuring future satisfaction of needs (economy) or curing diseases (health, i.e., disease system) (Luhmann, 1983, p. 30).

In other words, the strength of the correlation between inclusion and stratification, for example, is strongly 'dimmed down', but not eliminated. Stratification still plays a role, for example, when it comes to acquiring property or the possibility of attaining higher education. This weakening occurs in several ways, which we do not want to explore in detail here (but see Luhmann, 2018, p. 852). Nevertheless, stratification-like differentiation defects are reproduced continuously—for example, via differences in interests or differences in starting conditions (think for instance of how much talent is negated as a factor in the education system because it calls inclusion equality into question). The same applies to the currently hotly debated obsolescence of gender difference. From the perspective of functional differentiation, this obsolescence makes sense: there is no primary gender coding, modern society is indifferent to gender differentiation. And yet it continues to make itself insistently felt—not least in the attempt to overcome it, which is then in turn responded to by emphasising masculinity and femininity.[31]

The systems exploit these differences 'icily', as it were. The global transfer of education and training processes to schools, of all things, allows regional differences to have a reinforcing effect on inequality. The individual regions of the global society

[30] Luhmann observes the Freemasons as an auxiliary mechanism that continues to facilitate intraclass exchange among the upper classes in a functionally differentiating society as a contact network (Luhmann, 2018, p. 855). An equivalent to this in China is the parasitic and network like "contact structure" (Luhmann) called 'guanxi', which operates across system boundaries and is therefore repeatedly the cause of accusations of corruption (as an abuse for a different system logic). The ubiquitous presence of 'guanxi' in China enables the government to weed out potential opponents at will. However, since motives are not accessible to us, we prefer to focus on the legitimisation of these 'selections'. Here we can firstly state: The government acts in conformity with the social structure, i.e. in the sense of functional differentiation. Second, the ubiquity of corruption (not only in China) is a clear indication that purely functional differentiationat the organisational levelcannot be sustained. The reason is, as it were, architectural: organisations are not built along the coding of functional systems.

[31] Nassehi (2021) brings perceptibility into play here as a stabilizer of gender differentiation, which of course has its limits.

participate to very different degrees in the advantages and disadvantages of functional differentiation, "to the extent that disadvantages prevail, the already differentiated functional systems [...] seem to mutually impede each other" (Luhmann, 1997a, p. 163). But this does not change the unity of the context that realises functional and segmentary differentiation in a world-unifying way. The current type of society thus remains with us (which many a sociologist eager for variety may regret). It is therefore no coincidence that states come to the fore in this situation and gain a new visibility. For just like families, these sub-areas have always been engaged in such reconstruction.

Corona or not, deviations from the principle of equality, unlike in stratificatory society, are perceived as disturbing in modern society; they demand their elimination, so to speak. The same applies to deviations from the principle of world society. As has been shown, the overall political monopoly was accompanied by a minimization of social output, which is based precisely on a disaggregation of the various levels. The scope of these spheres, which had previously been guaranteed by a unified world social system, experienced and continues to experience a considerable restriction. To limit the virus is to limit the systems, is to limit the people. But the reality of globalisation does not contradict a renationalisation of observations and conflict resolution—otherwise it would not be an issue at all. "The normality of movements around the globe becomes evident precisely in the interruption" (Löw et al., 2021, p. 9). On the contrary, this renationalisation is a global phenomenon. It is precisely when we reject the reality of interconnectedness and invoke our 'own' that it asserts itself all the more forcefully.

5 Social Cohesionism

The world and the concept of the world have not only coincided since Corona. But the influence that every shock has on all other states, as already noted by Kant,who cites the interconnectedness of all parts of the world as the reason (2020), has never been so visible. The virus acts like one of those dyes that neurologists introduce into the brain to watch it at work. It makes the structures of world society, which normally carry events quietly in the background, stand out clearly. No climate catastrophe has been able to make this realisation of the global, worldwide scope of social relations so vivid and comprehensible. It is a warning—not of the consequences of climate change or a lung disease, but of the implications of networked, inclusive globality. "We have created a world," is the contemporary variant of Kant's dictum, "where a shock anywhere can become a catastrophe everywhere."

We also knew that a virus—'disease X'—could benefit from the structural conditions for this union (Daszak, 2020; Menon, 2020). The Global Preparedness Monitoring Board (GPMB), a joint arm of the WHO and the World Bank, and as such itself an effect of world society, pointed out several times the risks threatening this structure from a pandemic: "This report provided a snapshot of the world's ability to prevent and contain a serious global health threat. It also called for seven urgent priority actions leaders must take to prepare across five areas: leadership, building multisectoral country systems, research and development, financing, and robust international coordination" (GPMB, 2019). As early as 2015, Bill Gates had warned of "the next outbreak" in retrospect of the Ebola virus, which was contained—"thanks to some very good luck"—and repeated these warnings in vain over several years. And Andrew Lakoff (2017) had emphasised that a pandemic would hit world society unprepared despite all the highly developed scenarios. These were not least and in convincing, drastic form played out many times long before Corona made them come true by Hollywood, for example Steven Soderbergh's *Contagion* (US, 2011). The suspicion fuelled by some US Republicans that China could have maliciously bred the virus and used it as a biological weapon follows the script of *Outbreak* (US, 1995)—with the difference that a U.S. officer was the villain then.

But our world leaders, one could paraphrase Kant, initially had no money left for what was best for the world, i.e. for pandemic protection measures. The threat was not acute and therefore of little interest for the present-oriented reproduction of the systems; a presentness that is directly related to the decentralised form of synchronisation of social perspectives. The high degree of division of labour—in addition to the strong interconnectedness of goods, information and people that represent the biggest problem as 'virus carriers'—explains not only why this order is so susceptible to crises, but also why the warnings have not been heard. It is therefore not surprising that the researchers' findings (and the filmmakers' visions) did not initially pass the systemic relevance filters.

The virus has also drawn our attention to the close connection between tourism and the new technologies of transport. Without the airplane that makes almost every point on earth accessible and has also compressed the time required for this to a period of a few hours, the virus would not only have spread much more slowly, but probably not have reached certain regions of the world in the first place.[32] Now, thanks to the global infrastructure, it can bridge great distances in a few steps, creating physical connections in remote places. It also benefits from the human monopoly. For the enormous reduction in the diversity of animal and plant species on earth—a reduction in the diversity of forms of life that seems to be an inevitable consequence of global society (Stichweh, 2009, 2013, p. 26)—has not affected the intrinsic world of microscopic life forms. They could continue to exist and evolve relatively unaffected by world society—only to precisely affect this world society all the more strongly.

If it were a marketing campaign, the product would have to be certified as highly efficient. In his book *The Tipping Point* (2000), Malcolm Gladwell worked out "how little things can make a big difference"—except that the "fantastically large outcome" (Timothy Noah) achieved by the little thing called the Corona virus is now apparently viewed critically by society: not so fantastic. Its amazing career has been made possible by the structure of world society—and not, as Gladwell envisaged with the typical American's view of heroic actors, by the three types of connectors, mavens

[32] However, in this context, it is also worth noting another problem that contributed to Corona's likelihood: The fact that people are using more and more land in wildlife areas and intensifying livestock farming.

and salesmen. So does a 'viral' marketing campaign, a financial crisis follow the same patterns of contagion that we are currently observing at the biological level? In fact, the virus has nothing in common with communication because it is not something that one of the two processors involved in it does. It can only be realised as a co-production of these two processors, which is why the still popular transport metaphor does not do justice to what is happening: a virus is 'sent' and 'received', but no communication participant sends or receives a communication. This is the reason why the same reactions to a crisis do not occur, even if they all show a strong emotional engagement and a tendency to simplification and in this sense—as mentioned—seem to have an epidemic character themselves (Strong, 1990). The emotional engagement may be contagious, but only in the sense that one connects to it oneself, not in the sense that one is simply connected without one's doing, as it happens in the case of an infection. In other words, no communication passively penetrates us. But in both cases, a highly networked world is the prerequisite. The virus exploits the worldwide spaces of possibility, it is, as it were, world-conscious, even 'worldly'—but in contrast to Kant, the 'citizen of the world' who never left the greater Königsberg area, it is necessary for it to 'get around spatially'.

True, the virus is not a world event in the original sense like 9/11 or the World's Fairs. Corona did not bring together all the participants considered important in the world in a single place for a closed period of a few days. But world events are not only characterised by their temporal limitations. They can also take place as spatially distributed events.

As important as the techniques of transporting people and goods may have been for the formation of world society, it does not depend on the transport of viruses for its self-description. We learn more about modern society in the crisis than about the 'toxic slime' called Corona. The virus stages the world, stages world society. In the language of Hegel: it brings the world event to itself, because it realises a global configuration of participants that is no longer dependent on Wuhan—and those who are not infected can participate thanks to the mass media. It blocks other world events (G8 summits, World Economic Forum, Olympics, climate conferences, Aids conferences, etc.), it can be used for opposition and aggression against the structural effects of world society, and just like 9/11, it proves its world-reorganising power.

The preliminary conclusion of the GPMB chairs Brundtland and Sy confirmed Lakoff's fears and turned them into a social critique:

> The COVID-19 pandemic has revealed a collective failure to take pandemic prevention, preparedness and response seriously and prioritise it accordingly. It has demonstrated the fragility of highly interconnected economies and social systems, and the fragility of trust. It has exploited and exacerbated the fissures within societies and among nations. It has exploited inequalities, reminding us in no uncertain terms that there is no health security without social security. COVID-19 has taken advantage of a world in disorder. (GPMB, 2020, p. 2)

From the GMBP's point of view, the problem is thus not the virus, nor the lack of preventive measures, but 'the world'.

It seems that the organization has eagerly seized the opportunity to follow up on Kang Yuwei's vision and mark the function of the health system as universally relevant. The production of health is no longer to take place with the help of the usual, isolable medical techniques, because that is not enough: the whole world is to be 'healed', because it is 'sick' (i.e., in disorder), and what is aimed at is no longer only the production of health, but the unrestricted well-being of the world. As presumptuous as the claim is, it certainly follows the system's own logic, because since its differentiation, health understands itself as the complete comprehensive physical, mental and social well-being, from which it follows that all are sick and in need of treatment (Luhmann, 1983, p. 35). Thanks to Corona, it can loudly claim that social primacy which, according to its own assessment, it has been entitled to since its differentiation. While this hypostasizing of its own function is nothing special as we have seen, all systems make this claim to universal relevance, the pandemic that has lent apparent legitimacy to this projection. It is to the system's credit that health is an undisputed supreme value in modernity—"indeed, arguably the only supreme value that stands outside all ideological controversy" (Luhmann, 1983, p. 42). According to the WHO, to meet this value is only possible if the world as a whole is put back in order. Vaccination, lockdown, masks: don't cut it. (The special role that the health system ascribes to itself is confirmed by observers who see it as the most important motor for the current economic and social development. The so-called theory of the long waves or Kondratieff cycles, for erxample, sees the health system as the trigger and carrier of the current, sixth wave and justifies it, for example, with the number of new jobs created in the health sector or the innovations in the field of psychosocial health and biotechnology.Whatever one thinks of this cycle theory, hardly any of its critics will deny that the pandemic has provided a decisive further boost to innovation and thus confirmed the importance of the health system to some extent).

One might reject the demand to set the world right, even to cure it, in the same way that Edward Chamberlayne was rejected in T. S. Eliot's play *The Cocktail Party*. When he asked to be admitted to a sanatorium, because he had ceased to believe in his personality, the 'psychiatrist' Sir Henry Harcourt-Reilly—in fact an angelic being—informs him: "You are no case for my sanatorium. You are much too ill" (Eliot, 1950, p. 104, see also Luhmann, 1983, p. 36).

Whether we like it or not, we have to get along with the society we have—there is no other one hiding behind it, a better, fairer one that is unfairly withheld from us, by whomever. Such a doubling of the levels of attribution may have a relieving function, but it does not do justice to the reality of society. If at all, it is this insight that makes the crises disappear—crises that are only crises if they are understood as undesirable developments that can be corrected, as something that can be solved with a lot of good political will (Luhmann, 1997a, p. 1117). If we look at them instead as the result of evolution, a completely different picture emerges. Certainly not one that continues to hold out the prospect of a cure. Society is 'much too ill'. The hopes associated with the global civic society for an end to the disturbed equilibrium, the inequality of peoples and parts of the world were once again disappointed. We are still far from the general cosmopolitan state that Kant (2020) had in mind—as the womb "wherein all the original faculties of the human species" would be developed. "Obviously, society cannot live up to its own promises" (Luhmann, 1997b, p. 69).

However, there is no question of disorder. We are dealing with a highly complex, diverse order, a polycontextural world that, in contrast to the previous one, cannot bring everything into a hierarchical scheme. It instead follows a principle of juxtaposition and simultaneity. This order integrates not only nation-states, but also "fissures" and inequalities—and not least the overestimation of formative action that is also evident in the statements of the GPMB: the idea of being able to solve the 'social question' by granting equal rights, and so on.[33]

From our point of view, social stability can only ever be dynamic—as a temporary solidity that has to be reproduced, that proves itself to be 'solid' solely thanks to its continual reproduction. The "general big society" (Carl Gottlieb Suarez) thus rests on certain structural pre-decisions that are not necessarily the way they are, but only possible and therefore changeable. Stability is precarious, not in the sense of problematic, but in the sense of 'revocable, uncertain'. In fact, it is not the Corona crisis but the current world social order that is the exception. Whether and how it will continue is not only at stake since Corona. We would therefore agree with Dirk Baecker's assumption: "Perhaps fragility is its most important cement" (2021, p. 22).

The reason for complaints about a world in disorder, a 'collective failure' which sees crises as aberrations, is a notion of world society as a kind of political community of solidarity. If one starts from such a premise, it is indeed obvious to see the "malaise of modernity" (Taylor, 1991) in the loss of social cohesion. Apart from the fact that the appeals to this cohesion, which demand a belonging with a collective purpose, joint efforts to improve the world, are not without danger, modern society simply does not permit to act as one. These demands come up again and again against those limits which world society itself with its systems has realized as a precondition of each of our actions, of each intervention we undertake. Society is a super-medium into which all that is socially possible can be inscribed, including a pandemic, i.e., the way people deal with a high risk of infection. This medium conditions what is socially possible. The possibility of short-circuiting all agents, which is so impressively realized in the mass choreographies of Chinese state television (especially the annual New Year's Eve show), cannot be realized in China either.

This already applies to the projection of the future. Since all subsystems of society prefer different designs for the future, this divergence makes not only a 'collective coping' extremely unlikely, but also the cultivation of *one* future horizon. Instead, we are dealing with a multiplicity of 'futures'. What they have in common is merely the fact that it is no longer the old semantics of decay (i.e., cultural decline) that serves to mobilize against the present, but a semantics of catastrophe (cf. Horn, 2014).

Some Western sociologists were enthusiastic about the possibilities for political action that were briefly unlocked by the crisis, albeit with a guilty conscience: because

---

[33] Paradoxically, this notion of feasibility is a direct consequence of functional differentiation, which precisely opposes it. The fact that society adjusts its structure to regulate communications and can only be a differentiated society in this way, has then also created, so to speak, pre-missive illusions, from which arguments are then made. "These include, in temporal terms, the illusion that everything can be done, provided only that the partner does not resist; in factual terms, the illusion that there is no lack, provided only that maldistributions are eliminated by society itself; in social terms, the illusion of perfect humanity" (Luhmann, 2018, p. 603).

how can one begin to rave about an alternative future that appears here in the face of this enormous human tragedy? "We can change the world", Hartmut Rosa (2020) showed himself crisis-drunk in a newspaper article. While very much aware of the state of emergency that, if continued, would lead to a systemic collapse, he argued that there is no need to go back to the old mode, to re-start the "acceleration machine" again. One could simply invent a new one! Armin Nassehi (2021) has pointed out that this reinvention is not derived from the possibility of total governing, from the exceptional situation of the lockdown, but takes its starting point from the need for change. In doing so, Rosa overlooks what we have already pointed out: no lockdown changes anything in the way society operates. No crisis does either.

The attempt to politically combine all forces towards a common goal—whether it is to fight a virus or climate change—must necessarily fail, even if we sometimes had a different impression in the course of the pandemic. The very simultaneity of reproductions stands in the way:

> One cannot tell science to stop research if it is to be feared that the results could be misused; for knowledge of dangers is also knowledge. One cannot tell educators to stop trying to educate if the resulting demands on life chances could not be satisfied after all; for education can also be used for that. All the more, one cannot ask a doctor to do a little less for health so that the financing possibilities of the entire system would not be blown up. All appeals of this kind (and appeals to politicians are part of this) are misguided, they start in a way that is contrary to the structure and are futile. (Luhmann, 1983, p. 30)

Unless, of course, such restrictions are enforced against the reality of functional differentiation. Corona has shown how difficult this is even in times of crisis. One only has to look at the recent attempts of Chinese scientists to conform to the government's narrative of pandemic origin outside the country and to offer alternative explanations (from frozen food imported from abroad to a laboratory of the U.S. military to 'foreign', i.e., bats from Southeast Asia). Not surprisingly, they have been unsuccessful so far; the scientific community rejects them all as rather improbable (Cohen, 2022). To paraphrase Luhmann: One cannot tell science to stop saying the virus originated from China if it is to be feared that this result could be misused; for knowledge of misuse is also knowledge. The advice to Chinese scientists would therefore be to focus on this misuse instead of questioning the worldwide scientific consensus—that the virus naturally jumped from bats to an "intermediate" species and then to humans—and ultimately sabotaging their own credibility for future statements (Heidingsfelder, 2020, pp. 235f.).[34]

Functional differentiation also limits what the institution of the family is capable of today, as the most basic form of group membership and the most 'natural' form of the imposition of solidarity (with Hegel: the most moral). Corona has brought this home to us once again: when modern individuals are forced to endure a kind of 'total closeness' or total inclusion, when the "together mode" (Lu et al., 2020, p. 17) of

[34] The fact that the frozen food route thesis has at least made it into the WHO final report of 2021—despite the doubts of renowned scientists—can be explained by the ambivalent status of the organization, which neither strictly follows the code of the health system nor that of science or that of politics and is therefore susceptible to 'corruption' (see footnote 31, see also Heidingsfelder, 2012, pp. 423f.).

family is absolutized, without the possibility of temporarily outsourcing its members, the result here is also overload. To transfer the function of education completely to the family exceeds its capabilities. The family system is just as dependent on the other functional systems as they are on the family. Inclusion in society therefore only exists in the plural: as inclusions. To make use of medical care, to participate in monetary transactions, to have the ability to practice law and religion—all these possibilities of inclusion are conditional on and need each other. The institutions of marriage and family only function because one can also leave these 'private protective spaces'. They serve precisely as a possible retreat—and not as a necessary, enforced one. It is the reciprocity that constitutes the social context, not the superiority of any social sphere.

After all the negativity, we would like to end this text on a positive note—or, to be more precise, on two, even if our attempt to characterize society conclusively in positive terms again resorts to negations: (1) No crisis endangers social cohesion; (2) In the Corona crisis, it is not society that is at stake.

1. Crises have no impact on 'social cohesion', because what holds society together—at its 'core'—is not cohesion (i.e., the bond between the members of a group, a sense of community, *esprit de corps*, team spirit, connectedness, we-feeling). This idea of cohesion is built all too close to notions of interactional socialization, to the idea of "mankind as one community" (Locke, 1953). The experience of good, friendly interpersonal interaction, of a fair balance of interests, as we know it from our circle of friends, is extrapolated to society, which is understood as a social association or large group, and the 'stubbornness' and momentum of the functional areas are negligently underestimated.[35] From this perspective, the Corona crisis appears as a disturbance of a community that is actually oriented towards balance, solidarity and justice, as something that can be solved with a lot of good political will.

What holds society together would not be, as equally the German Federal Ministry of the Interior, for Building and Home Affairs and the editors of Corona Society

[35] More than that, it could be that this very solidarity stands in the way of a peaceful world of cooperation. Smith and Szathmary (1999) hold the different culturally acquired beliefs responsible for the disunity of the world, which are secured by rituals and narratives: those fictions, which are also based on the idea of the nation, with which the different sub-groups—Muslims, Serbs, or Croats, Shia or Sunni, Tutsi or Hutu, Jews or Arabs, Protestant or Catholics, Democrats or 'Autocrats'—identify themselves to instill group loyalty. In other words, it is not "the corruption and viciousness of degenerate men", which, according to John Locke, opposes the "great and natural community" and are the reason that men associate into "lesser combinations" (1953, p. 181)—but it is the narratives of these supposedly corrupt, degenerate others. Smith/Szathmary, therefore, clearly reject the optimistic ideas of cooperation on which Zhao Tingyang's philosophy is based: Co-operation is induced by myth and ritual, not by reason, and the capacity to be influenced by them is innate. Neither rationality—in the form of laws or social contracts—nor equality are therefore equal opponents; the latter not because modern society requires the independent co-operation of independent and competing entities (in our language: the division of labour). A quick look at our own work situation should make this clear: our superiors expect us to cooperate, but at the same time the organization fuels ambition, be it for better pay or more reputable positions on the other. "One effect is the de-solidarisation of competitors" (Fuchs & Heidingsfelder, 2019, p. 82). This is exactly where the aforementioned concept of the family comes into play to counteract this circumstance: another narrative.

(Volkmer & Werner, 2020) assume: a basic consensus of common values; social rules of togetherness; attachment to the community; trust in institutions; a cultural consensus of values, without which modern society would not function. That these values are factually right cannot be verified, only assumed. For values are not provable, "thus not capable of truth" (Luhmann, 2021). This is precisely why it makes little sense to take them as the basis of factual behavior (Luhmann, 1984a, 1984b, p. 176). Social order does not rest on a consensus of values, on an agreement about these values. An operative consensus in the form of a fictitious basis for action is quite sufficient. All that is necessary for the continuation of society is time or the connection to—contingent—selections.[36]

Our considerations thus amount to an unplanned, evolutionary transformation of society—and as we know, evolution cannot be controlled. It has no plan, on the contrary, the goals of the planners only serve as material for variations and selections—including the beautiful proposal by Armin Nassehi (2021) to save the climate by changing the logic of consumption. To recognise this—instead of denouncing this state of affairs as unsatisfactory and to demonstrate against it—and "to come to terms, once and for all, with a society without human happiness, and of course, without taste, without solidarity, without similarity of living conditions" (Luhmann, 1997b, p. 69) can in turn be understood as an excessive demand ('Über-Forderung'). But the alternative is not very tempting from our point of view either: to cling to the leftover vocabularies of tradition, to insist on aspirations of community, to expose oneself to a continuous disappointment of expectations. So why not choose the cognitive option: and learn?

2. Corona does not endanger society. Because contagion and mortality affect the human body—that is, the outside of society. Even if they represent an extremely important environmental sphere for this reproduction, which shows not least the existence of a specific social subsystem formed in relation to this body (Luhmann, 1983, p. 41). As long as society reproduces itself—and as we already formulated at the beginning, if it has a purpose, then it is its self-preservation—it is doing well. And it does so without much trouble even when a virus partially decimates the population. Society is relatively insensitive to an increase or decrease in the number of people. From the 'cold' (some say cynical) perspective of systems theory, too many or too few people are a problem of organisation and supply—especially when, as in the case of the Corona crisis, rapid change has to be accommodated. So, when wars, tsunamis, tornadoes and pandemics decimate people, when the basic consensus on shared values crumbles, democratic basic orders are exchanged for totalitarian ones and trust in institutions dwindles, it is not society that is endangered. At most, it is certain social orders and states of development that people consider 'good'—people

[36] A brief look at the actions of politicians and businessmen who refer to values, but continually contradict them in social practice (in the maintenance of the respective system), should make this plausible. Thus, instead of determining what happens when values are to be realized, systems theory pays attention to what strategies are available to sustain a system.

who belong to the Western hemisphere, as a rule the political order of democracy. And most of all: certain levels of prosperity.[37]

This does not mean that society will necessarily continue. No value, no sociologists appealing to cohesion, no God can guarantee a duration beyond this stability. It only means that its coherence is exclusively endogenously determined (Baecker, 2021, p. 20). There may well be such a thing as urgent final decisions. The decisions to prevent our self-destruction and ensure our survival on this planet would qualify for this rank (Heidingsfelder, 2020, pp. 41ff.). Accessing the concept of crisis as a final decision makes the most sense here. In our view, the Corona crisis is not one of them. The pandemic does not hold out the possibility of a complete decimation of humanity, as a nuclear war could do (the probability of which has recently increased again, see Mearsheimer, 2022).

It follows that we need not worry about society. All problems can be delegated to its subsystems. Because of these step functions, the current social order—because of its fragility!—can be considered ultra-stable (Ashby, 1981, p. 48). Political difficulties are not automatically economic difficulties, nor do scientific theories had to be changed because of Corona. When the sub-systems cannot solve a problem at the individual system level, they pass on functional burdens and 'overtax' other systems, which can cause crises and thus a temporary integration. Corona was such a crisis, it has disruptively interrupted, even paralyzed society, or with Dirk Baecker: ensured a down-regulation of the frequencies of certain functional areas. But it has not changed its routines, or its nature if you will.

We therefore do not go as far as Nassehi (2021) in elevating society's permanent crisis mode to its great strength. This reinterpretation of the crisis, as honourable as it may be from a therapeutic point of view, because it wants to take the crisis out of the crisis, misses the point in our view, even if the basic diagnosis is correct: society creates these crises for itself. But as history shows, crises do indeed pass. The Corona crisis is currently ending—even in the Philippines the schools have opened again. This is not to say that there is a non-crisis state in modernity. That is also not to say that world society can solve the problem of its susceptibility to crises. To reiterate: A complete mutual coordination that puts an end to crises once and for all is not possible. Society is 'too functional'. This is the price that the current social order pays for its central feature: the autonomy of the individual areas, from whose assumed existence our considerations took their starting point. It only means that crises alternate. Unlike social orders, such anomalies cannot be made permanent. Otherwise, one could get used to them, and then it would be superfluous to find a designation for these conditions that marks them as critical.

[37] The mass media continue to focus on the fact that Corona has suddenly disrupted our daily lives, the security of the world's wealthy niche, but studiously ignore the fact that the poorer regions of the world are used to insecurity: "We are prepared to scandalise border closures or flight bans when they limit our tourism, but not when they endanger the lives of refugees. Challenges of a fundamental rights nature are these for the refugees, not the tourists" (Paul, 2021).

References

Anderson, R. (1960). Reduction of variants as a measure of cultural integration. In G. E. Dole & R. L. Carneiro (Eds.), *Essays on the science of culture in honor of Leslie A. White* (pp. 450–462). Crowell.

Antunes, A. (2021, November 27). *Judgement on COVID-19 goes heavy metal [Kenneth Copeland Remix]* [Video]. Retrieved August 21, 2022, from https://www.youtube.com/watch?v=0JPRvxTjfOk

Ashby, R. W. (1981). The set theory of mechanism and homeostasis. In R. Conant (Ed.), *Mechanisms of intelligence: Ross Ashby's writings on cybernetics* (pp. 21–49). Intersystems.

Baecker, D. (2016, November 29). Die blanke Waffe [The naked weapon]. *Neue Zürcher Zeitung.* Retrieved August 21, 2022, from http://nzz.ch/feuilleton/zeitgeschehen/trumps-erfolgsrezept-die-blanke-waffe-ld

Baecker, D. (2020). Corona und die pulsierende Gesellschaft [Corona and the pulsating society]. In M. Heidingsfelder & M. Lehmann (Eds.), *Corona—Weltgesellschaft im Ausnahmezustand? [Corona—World society in a state of emergency?]* (pp. 71–96). Velbrück.

Baecker, D. (2021). *Katjekte* [Catjects]. Merve.

Beck, U. (1992). *Risk society.* Sage.

Bergsten, F. (2022). *The United States vs. China: The quest for global economic leadership.* Polity.

Bude, H. (2014). *Gesellschaft der Angst* [The Society of Fear]. Hamburger Edition.

Cheng, B. (2009). Resentment and recognition—A sociological exploration. *Journal of Jiangsu Administration Institute, 5,* 59–65.

Cohen, J. (2022, August 18). Anywhere but here. *Science.* Retrieved August 21, 2022, from https://www.science.org/content/article/pandemic-start-anywhere-but-here-argue-papers-chinese-scientists-echoing-party-line

Conrady, K. O. (1994). *Goethe: Leben und Werk* [Goethe: Life & work]. Artemis & Winkler.

Daszak, P. (2020, February 27). We knew disease X was coming: It's here now. *The New York Times.* Retrieved August 21, 2022, from https://www.nytimes.com/2020/02/27/opinion/coronavirus-pandemics.html

Diamond, P. (Ed.). (2019). *The crisis of globalization: Democracy, capitalism and inequality in the twenty-first century.* I. B. Tauris.

EBE (Editorial Board Economist). (2020, June 27). The next catastrophe (and how to survive it). *The Economist.* Retrieved August 21, 2022, from https://www.economist.com/weeklyedition/2020-06-27

EBFT (Editorial Board Financial Times). (2022, July 19). China's zero-Covid policy risks economic damage spiralling. *Financial Times.* Retrieved August 21, 2022, from https://www.ft.com/content/7dee81e2-9be1-430e-9b35-ae3651455996

Eliot, T. S. (1950). *The cocktail party.* Harcourt Brace.

Ferguson, A. (2012). *An essay on the history of civil society.* In F. Oz-Salzberger (Ed.). Cambridge University Press. (Original work published 1767)

Flach, L., Gröschl, J. K., Steiniger, M., Teti, F., & Baur, A. (2021). Internationale Wertschöpfungsketten—Reformbedarf und Möglichkeite [International value chains—Need for reform and opportunities]. *ifo-Studie,* Konrad Adenauer-Stiftung.

Folkers, A., & Opitz, S. (2020). Symbiosozialität—Zwischen Leben und Gesellschaft [Symbiosociality—Between life and society]. *Zeitschrift Für Theoretische Soziologie, 2,* 238–259.

Foucault, M. (2020a). *Discipline and punish: The birth of the prison.* Penguin.

Foucault, M. (2020b). *The history of sexuality volume 1: The will to knowledge.* Penguin.

Fuchs, P. (n.d.). Das WorldWideWeb—ohne Technik [The WorldWideWeb—without technology]. Retrieved August 21, 2022, from https://www.fen.ch/texte/gast_fuchs_www.htm

Fuchs, P., & Heidingsfelder, M. (2019). Psyches and organizations. *Journal of European Studies, 35*(3), 65–85.

Gan, S. (2013). Menschenrechte in China [Human rights in China]. In P. Brunozzi, S. Dhoubib, & W. Pfannkuche (Eds.), *Transkulturalität der Menschenrechte: Arabische, chinesische und europäische Perspektiven [Transculturality of human rights: Arab, Chinese and European perspectives]* (pp. 243–254). Karl Alber.

Gates, B. (2015, April 4). *The next outbreak? We're not ready* [Video]. Retrieved August 21, 2022, from https://www.youtube.com/watch?v=6Af6b_wyiwI

Giddens, A. (1985). *The nation-state and violence*. Polity Press.

Giddens, A. (1990). *The consequences of modernity*. Stanford University Press.

Gladwell, M. (2000). *The tipping point: How little things can make a big difference*. Little & Brown.

Gordin, M. D. (2022). A century of science boycotts. *Nature, 606*(7912), 27–29.

GPMB (Global Preparedness Monitoring Board). (2019). *A world at risk*. World Health Organization.

GPMB (Global Preparedness Monitoring Board). (2020). *A world in disorder*. World Health Organization.

Günther, G. (1979). Life as poly-contexturality. In *Beitrage zur Grundlegung einer operationsfähigen Dialektik* [Contributions to the foundation of an operational dialectic] (Vol. 2, 283–306). Felix Meiner.

Guo, S. (2009). Implementation of human rights treaties by Chinese courts: Problems and prospects. *Chinese Journal of International Law, 8*(1), 161–179.

Habermas, J. (1973). *Legitimationsprobleme im Spätkapitalismus* [Problems of legitimacy in late capitalism]. Suhrkamp.

Harvey, D. (2006). *Spaces of global capitalism: Towards a theory of uneven global development*. Verso.

Heberer, T. (2020). Chinas Kampf gegen Corona. Historische Erfahrungen, innen- und außenpolitische Implikationen [China's struggle against Corona. historical experience, domestic and foreign policy implications]. In M. Heidingsfelder & M. Lehmann (Eds.), *Corona—Weltgesellschaft im Ausnahmezustand?* (pp. 213–241). Velbrück.

Heidingsfelder, M. (2012). *System pop*. Kadmos.

Heidingsfelder, M. (2020). *Trump—beobachtet. Eine Struktursuche* [Observing Trump. A structural analysis]. Springer.

Horn, E. (2014). *Zukunft als Katastrophe [The future as a catastrophe]*. Fischer.

Kant, I. (2020). Idee zu einer allgemeinen Geschichte der Menschheit in weltbürgerlicher Absicht [Idea for a general history of mankind with a cosmopolitan intention]. Kindle Edition. (Original work published 1784)

Koselleck, R. (2010). Einige Fragen an die Begriffsgeschichte von "Krise" [Some questions to the conceptual history of crisis]. In *Begriffsgeschichten. Studien zur Semantik und Pragmatik der politischen und sozialen Sprache* [Conceptual histories. Studies in the semantics and pragmatics of political and social language] (pp. 203–217). Suhrkamp.

Lakoff, A. (2017). *Unprepared: Global health in a time of emergency*. University of California Press.

Latour, B. (2018). *Das terrestrische Manifest* [The terrestrial manifesto]. Suhrkamp.

Locke, J. (1953). *Two treatises of civil government*. Everyman's Library. (Original work published 1689)

Löw, M., Sayman, V., Schwerer, J., & Wolf, H. (Eds.). (2021). *Am Ende der Globalisierung. Über die Refiguration von Räumen*. [At the end of globalization. On the refiguration of spaces]. Transcript.

Lu, C. Y., Liu, D., & Liu, Q. (2020). Letters against separation. In M. Heidingsfelder & M. Lehmann (Eds.), *Corona—Weltgesellschaft im Ausnahmezustand?* (pp. 17–39). Velbrück.

Lu, F. (2022, 13 May). Zero Zweifel. *Zeit Online*, Retrieved August 21, 2022, from https://www.zeit.de/kultur/2022-05/corona-china-lockdown-shanghai-null-covid?wt_zmc=sm.int.zonaudev.twitter.ref.zeitde.redpost.link.x&utm_medium=sm&utm_source=twitter_zonaudev_int&utm_campaign=ref&utm_content=zeitde_redpost_link_x

Luhmann, N. (1983). Anspruchsinflation im Krankheitssystem. Eine Stellungnahme aus gesellschaftstheoretischer Sicht [Claim inflation in the health care system: A statement from a

socio-theoretical perspective]. In A. P. Herder-Dorneich, A. Schuller (Eds.), *Die Ansspruchsspirale—Schicksal oder Systemdefekt* [The spiral of claims—fate or systemic defect] (pp. 29–49). Kohlhammer.
Luhmann, N. (1984a). *Soziale Systeme* [Social systems]. Suhrkamp.
Luhmann, N. (1984b). The self-description of society: Crisis fashion and sociological theory. *International Journal of Sociology, 25*(12), 59–71.
Luhmann, N. (1995a). *The reality of the mass media.* Stanford University Press.
Luhmann, N. (1995b). Why does society describe itself as postmodern? *Cultural Critique, 30*, 171–186.
Luhmann, N. (1997a). *Die Gesellschaft der Gesellschaft* [The Society of the Society]. Suhrkamp.
Luhmann, N. (1997b). Globalization or world society: How to conceive of modern society? *International Review of Sociology, 7*(1), 67–79.
Luhmann, N. (2018). *Die Systemtheorie der Gesellschaft* [Systems theory of society]. Suhrkamp.
Luhmann, N. (2021). *Die Grenzen der Verwaltung* [Limits of administration]. Suhrkamp.
Lyotard, J.-F. (1984). *The postmodern condition: A report on knowledge.* University of Minnesota Press.
Marsella, A. J., Dubanoski, J., Hamada, W. C., & Morse, H. (2000). The measurement of personality across cultures: Historical, conceptual, and methodological issues and considerations. *American Behavioral Scientist, 44*(1), 2000.
Maturana, H. R., & Varela, F. (1980). *Autopoiesis and cognition: The realization of the living.* Reidel 1980.
Mau, S. (2007). *Transnationale Vergesellschaftung: Die Entgrenzung sozialer Lebenswelten* [Transnational socialisation: The dissolution of boundaries in social life worlds]. Campus.
Mearsheimer, J. J. (2022, August 17). Playing with fire in Ukraine: The underappreciated risks of Catastrophic Escalation. *Foreign Affairs.* Retrieved August 21, 2022, from http://foreignaffairs.com/ukraiune/playing-fiore.-ukrainde
Menon, A. (2020). Global disease surveillance systems and cooperation. In M. Heidingsfelder & M. Lehmann (Eds.), *Corona—Weltgesellschaft im Ausnahmezustand?* (pp. 117–126). Velbrück.
Merkel, A. (2020, May 23). Bundeskanzlerin Merkel zum Tag des Grundgesetzes: Corona-Virus ist Zumutung für Demokratie [Chancellor Merkel on the day of the basic law: Corona virus is an imposition on democracy]. Pressemitteilung [Press release]. *Bundesregierung.de.* Retrieved August 21, 2022, from https://www.bundesregierung.de/breg-de/suche/tag-des-grundgesetzes-1754788
Musil, R. (2003). *Der Mann ohne Eigenschaften [The man without qualities].* Rowohlt.
Myrdal, G. (1972). How scientific are the social sciences. *Journal of Social Issues, 28*(4), 151–170.
Nassehi, A. (2016). *Die Wiedergewinnung des Politischen. Eine Auseinandersetzung mit Wahlverweigerung und kompromisslosem Protest* [Recovering the political. An examination of electoral refusal and uncompromising protest]. Konrad-Adenauer-Stiftung.
Nassehi, A. (2021). *Unbehagen: Theorie der überforderten Gesellschaft* [Discomfort: Theory of the Overstrained Society]. C. H. Beck.
Parsons, T. (1971). *The system of modern societies.* Prentice Hall.
Paul, S. (2021, August 15). Beschreiben, um zu verstehen [Describing in order to understand]. *ZU Daily.* Retrieved August 21, 2022, from https://www.zu-daily.de/daily/schulterblick/2021/02-15_lehmann-heidingsfelderbeschreiben-um-zu-verstehen.php
Reckwitz, A. (2019). *Das Ende der Illusionen: Politik, Ökonomie und Kultur in der Spätmoderne* [The end of illusions: Politics, economy and culture in late modernity]. Suhrkamp.
Reynolds, S. (2011). *Retromania: Pop culture's addiction to its own past.* Faber & Faber.
Rosa, H., (2020, May 4). Wir können die Welt verändern [We can change the world]. *Die Zeit.* Retrieved August 21, 2022, from https://www.zeit.de/2020/19/corona-pandemie-oekonomie-zukunft-klimakrise?utm
Sassen, S. (2001). *The global city: New York, London.* Princeton University Press.
Schäfer, A., & Zürn, Michael (2022). *Die demokratische Regression* [The democratic regression]. Suhrkamp.

Schleiermacher, F. (1988). Monologen. In G. Meckenstock (Ed.), *Kritische Gesamtausgabe* (KGA), (Vol. 3, pp. 38–44). De Gruyter. (Original work published 1800)

Schmidt-Wellenburg, C., & Bernhard, S. (Eds.). (2020) *Charting transnational fields: Methodology for a political sociology of knowledge*. Routledge.

Sforza, A., & Steininger, M. (2020). *Globalization in the time of COVID-19* (Working Papers 8184). Cesifo.

Simmel, G. (1989). Über sociale Differenzierung: Sociologische und psychologische Untersuchungen ([On social differentiation: Sociological and psychological studies]. In H.-J. Dahme & O. Rammstedt (Eds.), *Georg Simmel Gesamtausgabe* (Vol. 2, pp. 109–295). Suhrkamp. (Original work published 1890)

Smith, A. (2003). *The wealth of nations*. Bantam. (Original work published 1776)

Smith, J. M., & Szathmary, E. (1999). *The major transitions in evolution*. Kindle Edition. Oxford University Press.

Stichweh, R. (2009, January 9). *Das Konzept der Weltgesellschaft—Genese und Strukturbildung eines globalen Gesellschaftssystems* [The concept of world society—genesis and structural formation of a global social system] (working paper). Sociological Seminar of the University of Lucerne.

Stichweh, R. (2013). The history and systematics of functional differentiation in sociology. In M. Albert, B. Buzan, & M. Zürn (Eds.), *Bringing sociology to international relations: World politics as differentiation theory* (pp. 50–70). Cambridge University Press.

Strong, P. (1990). Epidemic psychology. *Sociology of Health & Illness, 12*(3), 249–259.

Taylor, C. (1991). *The malaise of modernity*. House of Anansi Press.

Ting, K.-F., & Chiu, S. W. K. (2002). Leaving the parental home: Chinese culture in an urban context. *Journal of Marriage and Family, 64*(3), 614–626.

Volkmer, I. (2014). *The global public sphere: Public communication in the age of reflective interdependence*. Polity.

Volkmer, M., & Werner, K. (Eds.). (2020). *Die Corona-Gesellschaft: Analysen zur Lage und Perspektiven für die Zukunft* [The Corona society: Analyses of the situation and prospects for the future]. Transcript.

Wallerstein, I. (1989). *The modern world-system, volume III: The second era of great expansion of the capitalist world-economy, 1730–1840s*. University of California Press.

Wittmann, V. (2014). World society and globalisation. *Journal for Multicultural Education, 8*(3), 194–206.

Žažar, K. (2022). Fighting the virus, "hunting the witches": Moralizing in public discourses during the coronavirus pandemic in Croatia. *Kybernetes, 51*(5), 1833–1848.

Zhang, G. (2022). Short videos as affective contagion: (Un)locked WeChat chatlogs on viral videos. *Visual Studies*, 1–18. Retrieved August 21, 2022, from https://doi.org/10.1080/1472586X.2022a

Zhang, P. (2022, 9 May), Stop Shanghai's Covid-prevention excesses, academics urge in online appeal. *South China Morning Post*, Retrieved August 21, 2022, from https://www.scmp.com/news/china/politics/article/3177050/stop-shanghais-covid-prevention-excesses-academics-urge-online

Zhao, T. (2020). *Alles unter dem Himmel: Vergangenheit und Zukunft der Weltordnung* [Everything under the sky: The past and future of world order]. Suhrkamp.

Markus Heidingsfelder is Professor for Communication and Media Studies at United International College in Zhuhai, China. His areas of interest include media theory, sociological theory, culture theory, film theory, education research, and popular music studies. Among his recent books are Trump—beobachtet. Eine Struktursuche (Observing Trump. A Structural Analysis) (2020), Corona: Weltgesellschaft im Ausnahmezustand? (Corona: World Society in a State of Emergency?) (W. M. Lehmann, 2020), George Spencer-Brown's Design with the NOR. With Related Essays (W. S. Roth, L. Clausen, K. B. Larsen, 2020).

Lihua Chen is a PhD student in BNU-HKBU United International College and Hong Kong Baptist University. His primary research interests include fandom/anti-fandom, the culture industry, tourism, and discourse analysis, from both socio-cultural and critical perspectives. He received his master’s degree from the University of Liverpool.

International Relations in the Post-Pandemic World: Globalization, Global Governance and Leadership

How the Pandemic Affects Global Leadership: A Comparative Analysis of the G2 (United States and China)

Selçuk Çolakoğlu, Chaofan Chen, and Jiawei Chen

1 Introduction

The coronavirus has hit the world severely: not only the health sector is facing an unprecedented challenge but also is the whole world system, which involves global power distribution, global governance mechanisms, global leadership roles, and so on. In this chapter, we will focus on chances in global leadership in the post-pandemic world: How does the pandemic affect global leadership capacity and roles through a comparative analysis of the G2: the United States and China.

The United States and China are the two most influential and powerful countries in today's world. The discussion regarding their relations, interactions and competitions has become more popular soon after the coronavirus outbreak (Zhao et al., 2021). Charles P. Kindleberger argued that the disastrous decade of the 1930s was a result of the U.S. failure to provide global public goods after the United States had replaced the United Kingdom (UK) as the leading power (Kindleberger, 1986). After the end of World War II, the Bretton Woods System, the dollar-centric international monetary system, and the United Nations-centric global governance system were built under the leadership of the United States. Before the rise of China, the United States has played its role well in establishing world order and continually providing global leadership and public goods matching its status of the leadership role in the global governance system.

China used to be the leader of hierarchical tribute system in East Asia for centuries. From fourteenth century to nineteenth century, the tribute system had evolved into a set of rules and institutions with China obviously the hegemon in East Asia. China

S. Çolakoğlu (✉) · C. Chen · J. Chen
Beijing Normal University-Hong Kong Baptist University, United International College, Zhuhai, China
e-mail: selcukcolakoglu@uic.edu.cn

© The Author(s), under exclusive license to Springer Nature Singapore Pte Ltd. 2023

S. X. B. Zhao et al. (eds.), *Comparative Studies on Pandemic Control Policies and the Resilience of Society*, https://doi.org/10.1007/978-981-19-9993-2_6

had fixed the rules of international relations in Asia and other East Asian states benefitted from the system (Kang, 2012).

China, however, is a relatively new member of the modern world system since its Reform and Opening Up in 1978. China joined the U.S.-oriented international system, including the World Trade Organisation (WTO) and the World Bank and became one of the most active members in this system. China's international status has been improving from a silent participant into an ambitious organiser thanks to its developmental strategy. Beijing's international outreach has deepened after President Xi Jinping has initiated more global engagement for China through foundations of the Asian Infrastructure Investment Bank (AIIB), the Belt and Road Initiative (BRI) and the New Development Bank (NDB). China has begun to develop some financial alternative or the "Chinese solution" to U.S.-led international institutions, which have numerous problems.

In this chapter, we will first examine the United States and China within the framework of the G2 to analyse the global leadership and the pandemic's impacts. Then, we examine differences between the G2 and the G20 and the roles of the G2 in global leadership. We will also focus on the complex behaviours the G2 countries have done since it falls into conflict with the concept of the "Kindleberger Trap," referring a global leadership vacuum (Nye, 2017).

Finally, we will discuss the G2 countries' roles and positions to increase their global leadership capacities/capabilities in the post-pandemic era, what has been changed in this process, and what we could expect in the future.

This chapter adopts the methodology of literature review and data analysis. We examine the previous works on the G2, the G20, global leadership, and global governance. In addition, we refer to the different ideas from the Chinese and Western scholars on the issues. This chapter reads the national strength, public goods provision, capabilities of health sector of the G2 to analyse its roles in global leadership during the pandemic.

2 The G2: Is a New Designation for Global Leadership?

The developments throughout in the 1990s right after the end of the Cold War were indicative of the arrival of the expiration date of the G7, which comprised of the seven leading industrialised countries (Germany, UK, France, Italy, United States, Canada, and Japan) in the West. Even the G7's expansion to include Russia in 1997 - thus changing the group to the G8 until Russia's annexation of Crimea from Ukraine in 2014 - was insufficient in eliminating the austere crisis of confidence that loomed large over the entire globe.

In this regard, the successive financial crises in Asia, Mexico, and Russia provoked inflated fears, as they resonated far beyond national borders. At this time, the launch of the G20 in 1999 was an important attempt to incorporate the increasing dynamism of developing countries into the developed world so that the governance of global affairs could be led in a more inclusive and representative manner (Çolakoğlu et al., 2016).

In addition to the G8 leaders, top representatives from 11 influential developing countries (namely Argentina, Australia, Brazil, China, India, Indonesia, Mexico, Saudi Arabia, South Africa, South Korea, and Turkey) alongside those from the European Union (EU), the World Bank, and the International Monetary Fund (IMF) were invited to take part in the new group. The main purpose of the G20 is to provide opportunities for dialog on national policies, international cooperation, and international financial institutions, with the goal of strengthening the global financial architecture, supporting global growth, and minimizing the chances for future crises (Pempel, 2012).

The term of G2 (United States and China) has developed within the evolving process of the G20. There are different perspectives, descriptions, and expectations for the G2. The main consensus is that the United States and China are the most influential and dominant players within the G20 in the post-2008 global financial crisis (GFC).

Garrett (2010) claims that a de facto "G2" has emerged almost by default, even though neither China nor the United States will give their relationship this title. China and the United States have also decided that their G2 should be firmly nested in a multilateral G20 that will sit on top of but not replace the existing institutions of global economic governance. According to Garret, China and the United States support this ambitious agenda that puts the G20 at the centre of global economic governance. For China, the G20 represents the leading edge of worldwide recognition of its status as a "global power," draped in the less threatening cloth of a broader balancing between the old powers of the twentieth century and the rising powers of the twenty-first century. The G20 is also big enough for the spotlight not to shine too brightly on China, allowing it to grow slowly into a global leadership role it remains uncomfortable about.

Both China and the United States know that they will have de facto vetoes in the G20 without ever having to appear heavy-handed. The G20 will not replace existing institutions, but it steers them. The G20 also allows the United States to encourage China to become a "responsible stakeholder." However, Garret considers that there is a problem at the core of this nascent G2 in G20 architecture. Sino-American relations are poised to turn increasingly conflictual after the 2008 GFC. Sino–American economic frictions will be more intense and more important than ever before (Garrett, 2010).

Park (2017) underlines that China became one of most influential members of international trade after its joining to the WTO in 2001. When the 2008 GFC was at its peak, China has assumed the leading position of the G20 for discussing global governance with the United States. The discussion of the G2 symbolizes China's global economic and political power, which is comparable to that of the United States. According to Park, assessments of China's global behaviour are divided. For many Western scholars, China has shown renewed leadership ambitions in Asia and has assumed a position of global power comparable to that of the United States. China therefore deserves its position in the G2 within the G20. The G20 thus became a venue for the United States and China to form strong "global leadership cooperation" for the governance of global issues.

However, many Chinese scholars appear hesitant to assume the role of one pole of the global bipolar system. China is not yet ready to take responsibility for forming a new "global order." For many Chinese academics, China appears to have already emerged as a "regional superpower" but is not prepared the role of "global power" comparable to the United States (Park, 2017).

Lin (2012), reflecting a Chinese point of view, argues that the G2 has and will continue to play an important role within the G20 mechanism. As the representative of emerging economies and established power respectively, China and the United States wield great influence over the other G20 members. The better relations are between China and the United States within the framework of the G20, the more effective role will the G20 can play in world affairs. Therefore, Lin says that China and the United States have the responsibility to search out common ground for global governance and to help build the G20 into a broad platform that can help tackle any topic in the name of global peace and global prosperity (Lin, 2012).

Pempel (2012) agrees with Park that the most pervasive economic story of the 2000s has been the "rise of China." Staggering double-digit growth rates in China's annual GDP have captured the global attention, as has the fact that in 2010 China's GDP surpassed that of Japan, jumping next only to the United States. Beijing has yet to show much willingness to surrender any significant measure of national control to be the generous provider of public goods that would be intrinsic to its taking a leadership role as part of the G2 (Pempel, 2012).

As an example of reluctance of other countries, Cooper and Mo (2013) argue that the "middle powers" in the G20 are not happy to hear of the G2 concept, because any move towards that scenario will decrease everyone else's room for manoeuvrability including that of middle powers (Cooper & Mo, 2013). Mexico, Indonesia, South Korea, Turkey, and Australia formed a middle power grouping, namely MIKTA in 2013 to increase their roles within the G20 against the "big powers" including the United States and China (Çolakoğlu, 2018). Furthermore, it is not surprising that the other G7 and Brazil, Russia, India, China, and South Africa (BRICS) members will not be happy to see a strong G2 within or outside of the G20.

Walter (2011) considers the G2 as a fragile and problematic foundation for global economic governance that it is neither necessary nor sufficient to resolve important global order challenges. Nor is the G2 a necessary component of an effective G20. On the contrary, the increasingly bilateral focus of many important global issues, from climate change to global imbalances, has more often been an obstacle to their effective solution (Walter, 2011).

In conclusion, the unofficial term of the G2 has come to surface just after the 2008 GFC, as the G20 has begun to play dominant roles for global economic governance. In this regard, the G2 discussions have referred that the United States and China as the two biggest economies of the world would play more roles for global governance. Nevertheless, the G2 concept has not become dominant in the global governance literature. Unlike the debate of Sino-U.S. global competition and the rise of the "new Cold War," the G2 concept refers more cooperation rather than competition between the two biggest economies. The G2 likely urges China and the United States to cooperate for their "global leadership" roles as a coordinated effort to solve the global problems that the international community face.

The pandemic underlines the importance of joint global leadership to tackle with very severe problems from the health issues to re-opening economies in the global scale. As the two most capable countries, the United States and China, could play more roles and the G2 is likely to help to define a possible joint global leadership.

3 The Kindleberger Trap and Global Leadership in the Age of Power Transition

Joseph Nye (2017) published an article called "The Kindleberger Trap," recalling Charles P. Kindleberger's argument that the disastrous decade of the 1930s was caused by the failure of the United States to take up the responsibility of the UK. Nye's article is a counterargument to Graham Allision (2017) who raised the "Thucydides Trap," which considers China is too strong rather than too weak and warns the world that when a rising power threatens an established power, war is the most likely result. Nye claims China is too weak rather than too strong in providing global leadership and public goods to the world during its challenging process towards the United States. Nye considers China will keep being a free rider rather than shouldering its responsibility while the United States is more and more incapable or unwilling to realise its duty, resulting in the world into the Kindleberger Trap in the end and repeating the tragedy of the 1930s (Nye, 2017).

Nye warns the world that given that the current Sino-U.S. situation is worsening and complex, people should rather than focus on the Thucydides Trap but pay more attention to the Kindleberger Trap, the hegemon fight might cause the failure that happened in history: a global leadership vacuum (Nye, 2017).

Soon after Nye's article in 2017, global opinion leaders paid attention to the Kindleberger Trap. Chinese Foreign Minister Wang Yi considered that China and the United States would not fall into the Thucydides Trap or the Kindleberger Trap by enhanced dialogue and coordination with each other (Wang, 2017).

Two dimensions are most vital in assessing the Kindleberger Trap: the capacity and willingness of provision. Carla Freeman (2020) focuses on China from capacity of willingness, benefit, and support from home and abroad and considers that China has made its commitment to preventing the Kindleberger Trap. However, China's contributions to global leadership are "selective and incremental," which might cause unintended side effects. Nevertheless, Freeman raises her concerns about the interactions of the Kindleberger Trap and the Thucydides Trap, the Sino-U.S. rivalry would increasingly cause the tension to "propel both states into conflict."

Jin Jiyong (2020) focuses on the hegemonic system crisis with the growing nationalism, populism, xenophobia hampering global efforts at health governance. The broken global health governance system makes the world approach the edge of the Kindleberger Trap. Although China has made its promise in promoting global cooperation and improving global health governance, challenges remain when the United States has been losing its appetite for global leadership by its isolationism and populist policy under the Trump administration (2017–2021). Jing raises two main challenges: the rising geopolitical tensions between China and the United States and the resurgence of populism worldwide, which pose serious challenges and dilemmas to improving the global health governance system. Jing considers that only the world overcome those challenges brought by politics could the Kindleberger Trap be avoided.

He Kai and Feng Huiyun (2020) present that the existing international institutions are playing a significant role in stabilising and coordinating the tensions within great powers, especially between the United States and China, an established power and a rising power. He and Feng claim that the decline of the United States would not push the world into the Kindleberger Trap, but the reputation of the international institutions that those hegemons provide leadership might lose confidence from international communities. Therefore, He and Feng suggest that the United States and China should re-embrace multilateralism, and the dynamic institutional balancing and changes in the U.S.-China competition context should not deprive their provision of global leadership and public goods to international society (He & Feng, 2020).

Riham Bahi (2021) expresses his concerns about exacerbated great power competition worsened by the pandemic and the global order experiencing fragmentation and decomposition. Bahi reviews the Kindleberger Trap in a G2 context: world order could not be functional by one or two countries, both the United States (an established power enjoys priority) and China (a rising power eager to play a greater role) should take its responsibilities. The G20 could be a good platform embodying China's great power status without asking China to play a global role it is not yet ready to embrace. In addition, the G20 could play as a "steering committee" of all global governance institutions, including embodying the G2 into the *de jure* G20.

The Kindleberger Trap, a concern re-emerged in the Sino-U.S. rivalry scenarios in the Trump Administration, has raised discussion around the world. The core issues of this trap are the global leadership and public goods provision problems: the established power (United States) is more likely to reject taking up its responsibilities in global governance while the rising power (China) is not strong enough to replace the established one. However, once the involved parties could embrace multilateralism to make more coordination, communication, and cooperation within the international governance institutions like the G20 and/or the G2, they might play a coordinating role to provide global leadership to stabilise the world and prevent the world from falling into the Kindleberger Trap.

4 Evaluation of the Material Capabilities of the G2 During the Pandemic

The pandemic has left dramatic effects on countries' material capabilities. Full economic lockdowns, rocket-high increase of health expenditure and particular total collapse of some sectors like hospitality and transportation industry have damaged many countries' economies (Monaco, 2021). Unlike other health crisis and disasters, there is no country that has remained unaffected from the COVID-19 pandemic. There is a huge global economic plunge in the first year of the pandemic in 2020. In the case of the G2, the United States and China have also been deeply affected by the pandemic.

The total GDP of the U.S. and China has increased in the past four years (2018–2021). However, it is worth noting that the U.S. economy declined under the severe impact of the pandemic in 2020, even lower than the total GDP of the previous year (2019). On the other hand, China showed some economic resilience in the pandemic, with an incremental decrease but no recession. Both countries recovered quickly from the pandemic in 2021 (see Table 1).

The global GDP growth declined from 2.8% in 2019 to −3.1% in 2020, signaling a dramatic downsizing of the world economy because of a global health crisis. The U.S. economic growth levels are almost parallel to those of the world average. While the U.S. economy had 2.3% growth rate in 2019, it was −3.4% in 2020. However, China was distinct. Although the Chinese GDP growth declined from 6% in 2019 to 2.3% in 2020, China was one of the few countries still having growth rate. The Chinese GDP achieved a higher growth rate than the global rate with 8% in 2021, while the U.S. GDP growth in 2021 is on par with global level with 6% in 2021 (see Table 2 and Chart 1).

Table 1 GDP of US and China from 2018 to 2021 (current billion US$)

Countries\year	2018	2019	2020	2021
U. S	20,530	21,370	20,890	22,940
China	13,840	14,340	14,870	16,860
China GDP/U. S. GDP (%)	67.41	67.10	71.18	73.50

Source IMF, https://www.imf.org/external/datamapper/NGDPD@WEO/WEOWORLD/USA/CHN

Table 2 Nominal GDP growth of US, China, and World average (annual %)

Countries\year	2018	2019	2020	2021
US	2.9	2.3	− 3.4	6
China	6.8	6	2.3	8
World Average	3.6	2.8	− 3.1	5.9

Source IMF, https://www.imf.org/external/datamapper/NGDP_RPCH@WEO/WEOWORLD/USA/CHN

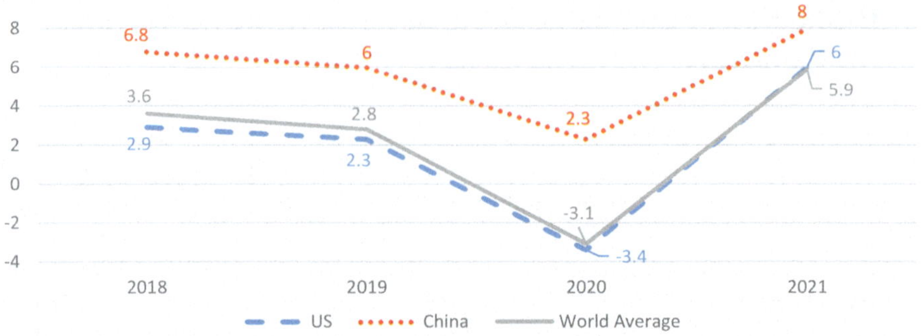

Chart 1 Nominal GDP growth of U.S., China, and World Average from 2018 to 2021 (annual %) (*Source* IMF, https://www.imf.org/external/datamapper/NGDP_RPCH@WEO/WEOWORLD/USA/CHN, Chart created by authors)

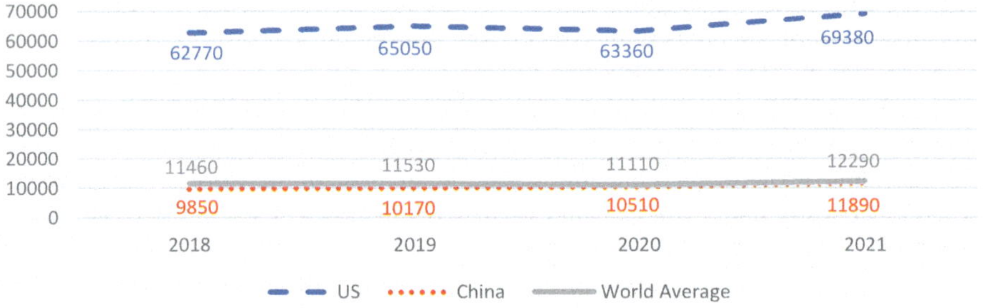

Chart 2 GDP per capita of U.S., China, and the World Average from 2018 to 2021 (current US$) (*Source* IMF, https://www.imf.org/external/datamapper/NGDPDPC@WEO/OEMDC/ADVEC/WEOWORLD, Chart created by authors)

Table 3 Human Development Index (HDI)

Countries\year	2018 (HDR 2019)	2019 (HDR 2020)
United States	0.925	0.926
China	0.755	0.761
World Average	0.734	0.737

Source United Nations Development Progress Human Development Index (HDI), https://hdr.undp.org/sites/default/files/hdr2020.pdf

Except China, the GDP per capita of both the U.S. and the world average declined slightly under the impact of the pandemic in 2020, but the overall trend shows that all three are growing. The U.S. GDP per capita is well above the world average (5.5 times). In the past four years (2018–2021), China's GDP per capita has consistently increased and reached the world average level (see Chart 2). The Chinese government declared that China's GDP per capita had exceeded the world average in 2021, reaching 12,550 US dollars (National Bureau of Statistics of China, 2022) (see Table 3).

Total foreign trade—imports and exports (goods and services) of both U.S. and China have increased rapidly since the pandemic outbreak. Their combined share of the world's total imports and exports has increased in 2019, a 0.16% up from 2018. The total exports decreased by 0.1% in 2020. The 2020 and 2021 average share of imports increased by 0.63%; exports increased by 0.51%. U.S. imports have always been higher than China's (2018–2021), and the gap between the two countries fluctuates. It is difficult to see whether the gap is widening or narrowing. In contrast, the export volume gap between the two countries has always (2018–2021) been higher than that of the U.S., and this gap is widening rapidly - the export value gap from 2018 to 2021 is: 339.3, 382.7, 757.8 and 1157.3 billion U.S. dollars, respectively (see Chart 3).

The whole figure implies that the U.S. economy has kept its leading position despite some challenges and there is no material capability loss for the United States because of the pandemic. On the other hand, China has better economic performance than the world average and Chinese material capability is on the rise. The G2 has become stronger in the post-pandemic period to play global leadership role.

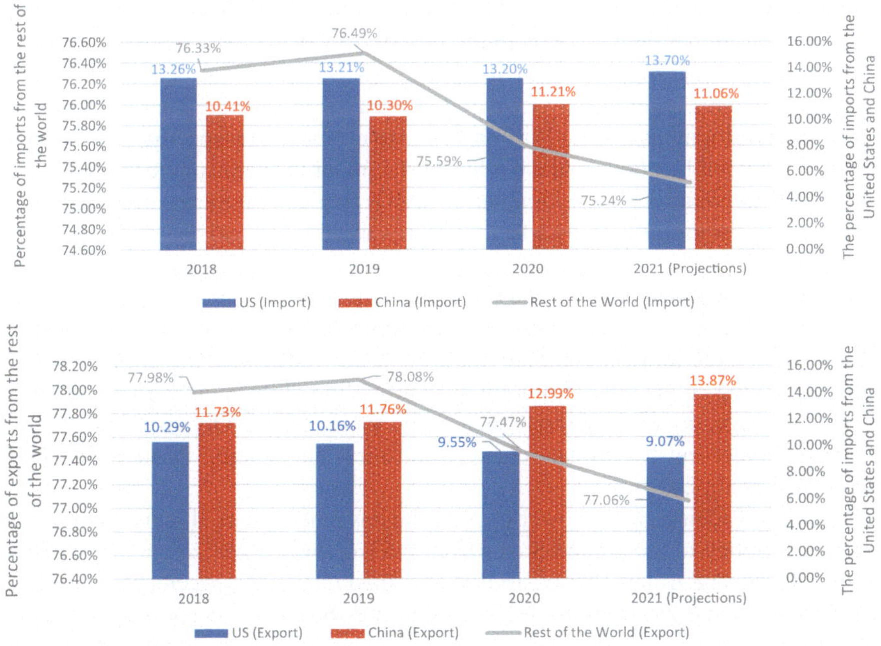

Chart 3 Total foreign trade—import/export—share of world imports of U.S., China, and World total from 2018 to 2021 (in goods and services, %) (*Source* OECD Data, https://data.oecd.org/trade/trade-in-goods-and-services-forecast.htm#indicator-chart, Chart created by authors)

5 G2's Policies on Global Public Goods

Global public goods are those whose benefits affect all citizens of the world and mostly related to environment, culture, technology, and public health (Chin, 2021). The importance of global public goods in everyday lives becomes more salient with each new crisis—COVID-19 pandemic has increased demand for global public health, migration crises for balanced world development, and climate change for sustaining the global environment. Despite having no certain consensus on how to define global public goods, we tend to adopt a broader scope to measure global leadership capability. For example, official development assistances (ODAs), foreign direct investments (FDIs), and financial aids and loans to other states and international organizations are considered helpful tools to promote global development, poverty reduction and environmental protection. The failures of governments that underprovide public goods are amplified when it comes to global public goods. If poor countries have become economically more capable with the help of ODAs, FDIs and financial aids and loans, it is assumed that they will be able to contribute more to global public goods. Vaccines, medicines, and health equipment donations are considered as global public goods in the struggles against the pandemic.

The United States and China have set a series of frameworks in institutions and policies in the global provision of public goods. The United States act as the main contributor of global public goods in the post-World War II era. The United States is also acting as the biggest donor to most of the international organizations and the most influential role in the UN-based system to affect the global political and economic development course. Washington also established a specific institution to provide global public goods to the world in 1961 under the name of the United States Agency for International Development (USAID), which is the executive agency to implement assistant programmes from diseases eliminations to economic recovery around the world.

The proactive policy image of the U.S. as an active provider of global public goods has changed during Donald J. Trump's presidency from 2017 to 2021. The Trump Administration put more attention to domestic needs rather than directing resources overseas. Trump cut over $4 billion funds in providing global public goods, decided to quit the international institutions such as the World Health Organization (WHO), the Paris Agreement for climate change, and the Trans-Pacific Partnership (TPP) (Bresnahan et al., 2019).

When Joe Biden came to power in 2021, he has made a partial revision of disengagements policy of the United States from international organizations (The White House, 2020). The Biden Administration re-joined the Paris Agreement, ceased the disengagement process from the WHO, and established a new economic forum (the Indo-Pacific Economic Framework, or IPEF) in May 2022 instead of the TPP (Iqbal, 2022). In the 2022 U.S. government budgets, even though the foreign aid and international budget have been raised compared with the Trump Administration, most of the funds are prepared for domestic needs to create jobs, boost the economy, and raise military budget (Garver, 2021). The Biden Administration has still showed low interest in international cooperation and provision of global public goods.

With the rise of China, the Chinese policy frameworks on global public goods contribution were raised particularly in the post-2008 GFC. Under Chinese President Xi Jinping, the term of global public goods (国际公共产品) has become dominant in official documents (Xi, 2017). China's policies in providing public goods are rooted in two considerations. One is the rising responsibility in morals and strength as a rising power to provide global public goods in accordance with its international status. Second, Chinese policies in provision of global public goods represent the Chinese diplomatic thinking—"all under heaven idea" (天下观) (Zhao, 2009). Beijing now values the traditional Confucianism thinking in its diplomatic policy-making process. The guiding ideology "a Community of Shared Future for Mankind" (人类命运共同体) became the inner logic of Chinese policies in global public goods provision.

In 2021, Beijing released the policy white book on Chinese international cooperation and foreign aid (The State Council Information Office, 2021). Four global public goods are listed in the white book. First is ODAs. China is raising its provision of ODA size, technologic aid, humanitarian aid, and material aid to the Global South. Second, the BRI is regarded as the flagship of China's global public goods. Third, China provides public goods and services to achieve the UN Sustainable Development Agenda 2030 by reducing poverty, providing aid and technologies in agriculture, education, gender equality, infrastructure, and medical fields. The AIIB is the latest institutional product for China to supplement the provision of public goods that the existing World Bank and Asian Development Bank (ADB) might lack. The final part is the global pandemic control.

6 G2's Contributions to Global Public Goods During the Pandemic

Although there was a dramatic need of global cooperation against fallout of the pandemic, many countries have resorted to nationalistic agenda in the first place. Contribution to global public goods witnessed a declining trend for many countries during the pandemic. There was also a dramatic decline in FDIs and ODAs globally. However, humanitarian aids, health equipment, medicine and vaccine donations were on the rise.

According to the public data collected over four years (2017–2020), China tends to surpass the United States in terms of foreign investment inflows. The outflow of FDI reflects a country's comprehensive national strength and influence on a certain extent. Under the Trump Administration, the U.S. drastically reduced the amount of OFDI; even an amount of $128.32 billion of OFDI funds was withdrew in 2018. It has since rebounded, but that inevitably took some of the weight off the U.S. global influence. China's outward FDI remained stable from 2017 to 2020 (see Chart 4).

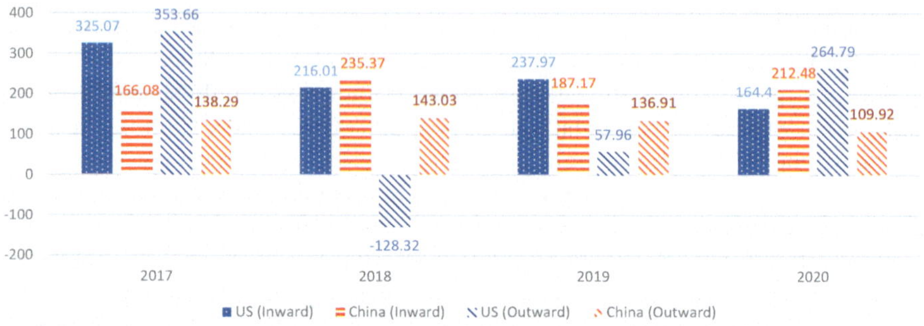

Chart 4 FDI of U.S. and China (in billion US$) from 2018 to 2021 (*Source* OECD Data, https://data.oecd.org/fdi/fdi-flows.htm, Chart created by authors)

Table 4 ODA or "ODA-like" offered by U.S. and China to the world in million US$

Countries\year	2018	2019	2020
United States	34152[a]	32906[a]	34456[a]
China	5900[b]	5900[b]	5400[b] (projection)

[a]OECD, https://www.oecd-ilibrary.org/sites/45472e20-en/index.html?itemId=/content/component/45472e20-en
[b]Donor Tracker, https://donortracker.org/insights/new-era-trends-chinas-financing-international-development-cooperation

Through searching and screening the ODA data of China and the United States from 2018 to 2020, it can be found that China and the United States differ significantly on ODA, with the United States about 5.8 times that of China (see Table 4). It is worth noting that the ODA data of the United States can be found in official documents over the years, while China rarely has detailed and specific data (The statistic of China is conducted by a Japanese Institution—*JICA Ogata Research Institute*). China likely uses multiple financial sources to cover its ODA, but it is difficult to collect all these data precisely. What's more, different from the definition of the Organisation for Economic Co-operation and Development (OECD) donor countries and the direct free donation of other developed countries, China has its own interpretation for ODA. China prefers Other Official Flow (OOF) in foreign aid, namely, concessional loans with commercial considerations—seeking "mutual win." At the same time, Beijing feels inaccurate and dissatisfied with accusations that China has led recipient countries into debt crisis when other countries' institutions, for example, AidData from the U.S., measuring China's ODA and other foreign aid by the mainstream measurement method in the world and believes that this is slandering China's aid achievements.

When the COVID-19 vaccines were invented, some countries received a portion of the donated vaccine and then purchased another portion of vaccines. It is therefore difficult to determine whether the source of vaccines in a country/region is only by donation from other countries, so donations and purchases fall into the same category.

Table 5 Anti-COVID-19 vaccine donations/distribution of the U.S. and China

Countries	Vaccine donations already donated
US	453 million doses[a]
China	184 million doses[b]

[a]U.S. Department of State, https://www.state.gov/covid-19-recovery/vaccine-deliveries/
[b]Bridge Consulting, https://bridgebeijing.com/our-publications/our-publications-1/china-covid-19-vaccines-tracker/

Unlike the U.S. government (www.state.gov), the Chinese government does not have an official real-time dashboard to show the number and status of vaccine donations, so the authors used *Bridge Consulting*, a Sino-US joint Consulting firm, as their data source. Its COVID-19 Vaccine Tracker collects detailed information about vaccine distribution from China to the world from Chinese official government documents (see Tables 5 and 6).

The COVID-19 Vaccine Global Access (COVAX) initiative was created and led by some international organizations in the vaccine field. COVAX's role can be summarized as the sharing of research and development risks, the acceleration of production, the lowering of vaccine prices, and the equitable distribution of vaccines globally (see Table 7).

The gap between China and the U.S. in the proportion of financial support and investment in international organizations is noticeable. The United States' financial support for the well-established international organization is nearly twice China's amount. Also, China was absent from participating in some organizations, such as the OECD's financial support. However, with the continuous enhancement of China's comprehensive strength, in seeking more significant international influence, China's share of financial support in the WTO is close to the level of the United States. At the same time, China led to establish the AIIB to strengthen cooperation between China and other countries in region (Khanna, 2016) (see Table 8).

The United States has maintained its leadership role to contribute global public goods and there was no significant change for that during the pandemic. In addition, China has increasingly contributed to global public goods depending on its enlarging material capability. The pandemic also revealed that China has an increasing role for further global responsibilities. Together with the United States, China was the leader of vaccine donations and deliveries to other countries. It seems that the pandemic has pushed the G2 to take more responsibilities at the international level as a reflection of their rising global leadership position.

Table 6 Anti-COVID vaccine's destinations of distribution from the U.S. and China

China vaccines donated and sold[a]	U.S. vaccines donated and sold[b]
Receiver countries/regions:	Receiver countries/regions:
Africa:	**Africa:**
Morocco, Tunisia, Algeria, Libya, Mauritania, Senegal, Guinea-Bissau, Guinea, Sierra Leone, Côte d'Ivoire, Burkina Faso, Benin, Toga, Niger, Nigeria, Chad, Cameroon, Equatorial Guinea, Gabon, Congo, Central African Republic, Arab Republic of Egypt, Sudan, Djibouti, Ethiopia, Somalia, Kenya, Uganda, Rwanda, Burundi, Tanzania, Democratic Republic of Congo, Angola, Zambia, Mozambique, Comoros, Madagascar, Zimbabwe, Botswana, Namibia, Lesotho	Angola, Benin, Botswana, Burkina Faso, Cabo Verde, Cameroon, Central African Republic, Chad, Côte D'Ivoire, Democratic Republic of the Congo, Djibouti, Eswatini, Ethiopia, Gabon, The Gambia, Ghana, Guinea, Guinea-Bissau, Kenya, Lesotho, Liberia, Madagascar, Malawi, Mali, Mauritania, Mauritius, Mozambique, Namibia, Niger, Nigeria, Republic of the Congo, Rwanda, Senegal, Seychelles, Sierra Leone, Somalia, South Africa, South Sudan, Sudan, Tanzania, Togo, Uganda, Zambia, Algeria, Egypt, Libya, Morocco, Tunisia
Americas:	**Americas**:
Mexico, El Salvador, Nicaragua, Cuba, Jamaica, Dominican Republic, Dominica, Trinidad and Tobago, Barbados, Antigua, Barbuda, Venezuela, Colombia, Ecuador, Guyana, Peru, Brazil, Bolivia, Chine, Argentina, Uruguay	Antigua and Barbuda, Argentina, Barbados, Belize, Bolivia, Brazil, Canada, Colombia, Costa Rica, Dominica, Ecuador, El Salvador, Grenada, Guatemala, Guyana, Haiti, Honduras, Jamaica, Mexico, Panama, Paraguay, Peru, Saint Kitts and Nevis, Saint Lucia, Saint Vincent and the Grenadines, Suriname, Uruguay
Asia:	**Asia:**
Lebanon, Palestine, Jordan, Syria, Iraq, Georgia, Armenia, Azerbaijan, Iran, Uzbekistan, Afghanistan, Kazakhstan, Kyrgyzstan, Tajikistan, Pakistan, United Arab Emirates, Oman, Nepal, Bhutan, Bangladesh, Sri Lanka, Myanmar, Vietnam, Laos, Thailand, Cambodia, Malaysia, Brunei, Indonesia, East Timor, Philippines, Mongolia, Singapore, Kiribati, Papua New Guinea, Solomon Islands, Vanuatu, Maldives, Turkmenistan	Cambodia, Fiji, Indonesia, Laos, Malaysia, Mongolia, Papua New Guinea, Philippines, Solomon Islands, South Korea, Taiwan, Thailand, Timor-Leste, Vietnam, Iraq, Jordan, Lebanon, Palestinian Territories (West Bank and Gaza Strip), Tunisia, Yemen, Afghanistan, Bangladesh, Bhutan, Kyrgyzstan, Maldives, Nepal, Pakistan, Sri Lanka, Tajikistan, Uzbekistan
Europe:	**Europe:**
Belarus, Ukraine, Moldova, Hungary, Serbia, Bosnia and Herzegovina, Montenegro, Albania, North Macedonia, Turkey	Georgia, Kosovo, Moldova, Ukraine
Total number: 115	**Total number: 110**

[a]U.S. Department of State, https://www.state.gov/covid-19-recovery/vaccine-deliveries/
[b]Bridge Consulting, https://bridgebeijing.com/our-publications/our-publications-1/china-covid-19-vaccines-tracker/

Table 7 Donation to COVAX of the U.S. and China

Countries	Proceeds to GAVI from donor contributions and pledges (2021–2025) until 30 June 2021 (Amount in US$) in total
US	4390 million to Gavi COVAX AMC (Gavi COVAX Advance Market Commitment)
China	20 million to Gavi COVAX AMC

Source Gavi, https://www.gavi.org/investing-gavi/funding/donor-profiles/

7 Conclusion

All these figures and data analyses have helped us to attest our arguments. First, the pandemic has no or little effect on global leadership capacities of the G2. The U.S. economy has kept its leading position despite some challenges. That means there is no material capacity loss for the United States because of the pandemic. With the help of vaccine developing capacity, the United States has become the most influential country on the struggle against the pandemic. In addition, China has better economic performance than the world average and Chinese material capacity is on the rise as it was before the pandemic. In fact, the G2 has become stronger in the post-pandemic period to play global leadership role.

Second, there is no significant change for willingness to contribute to global public goods by the G2 during the pandemic. It is widely claimed that the United States was unwilling to lead in the provision of that global public goods after the 2008 GFC. This trend can clearly be seen during the Trump Administration (2017–2021). However, the Biden Administration has turned to conventional U.S. policies to contribute to global public goods. China's capacity to contribute to global public goods is on the rise before and after the pandemic. Nevertheless, China's rising role is still complementary to the U.S. one rather than being a supplementary global leadership. It can be considered within the framework of the increasing willingness and capacity to play for global leadership role within the G2.

Furthermore, the United States has a resilient leadership role to contribute to global public goods and there is no significant change for that during the pandemic. The 2022 Russia-Ukraine war has unexpectedly caused to revive the Western/NATO alliance as it was in the early 1990s. The United States has re-emerged as the leader of the West and got a chance to show up its strong global leadership role in world affairs. In addition, China has increasingly contributed to global public goods depending on its enlarging material capability. The pandemic also revealed that China has an increasing role for further global responsibilities. Together with the United States, China was the leader of vaccine donations/deliveries to other countries. It seems that the pandemic has pushed the G2 to take more responsibilities at the international level as a reflection of their rising global leadership position.

The Kindleberger Trap argues that the disastrous decade of the 1930s was caused by the failure of the United States to take up the responsibility of the UK when it replaced it. Within this framework, the Kindleberger Trap claims that China is not

Table 8 Financial contribution to international organizations of the United States and China

Countries\capacity	**United Nations:** regular budget share 2019–2021(%)[a]	**World Health Organization:** share of net contribution payable for 2021(%)[b]	**U.N. Peacekeeping Operations:** financial contributors (Approximate Rate of Assessment, 2020–2021, %)[c]	**World Trade Organization:** members' contributions to the consolidated budget 2021, (contribution %)[d]	**Organization for Economic Co-operation and Development:** member countries' percentage shares of part I budget contributions for 2019, (%)[e]	**Asian Infrastructure Investment Bank:** members' total share, (contribution %)[f]
United States	22	24	27.9	11.7	20.5	Non-member
China	12	12	15.2	10.4	Non-member	30.8

[a]UN, https://undocs.org/en/a/res/73/271
[b]WHO, https://www.who.int/about/finances-accountability/funding/2020-21_AC_Summary.pdf?ua=1
[c]UN Peacekeeping, https://undocs.org/a/73/350
[d]WTO, https://www.wto.org/english/thewto_e/secre_e/budget_e/budget2021_member_contribution_e.pdf
[e]OCED, https://www.oecd.org/about/budget/member-countries-budget-contributions.htm
[f]AIIB, https://www.aiib.org/en/about-aiib/financial-statements/.content/index/pdf/AIIB-_Annual-Financial-Statements-20201231-signed.pdf

willing or does not want to take responsibility to take global leadership role to prevent a chaotic international system.

However, our research shows that China is eager to take more global responsibility before and after the pandemic for international stability. China is the first or second donor/contributor to global public goods in the pandemic too. Beijing wants to see a coordination rather than competition for the global leadership with the help of international organizations and other major powers. Beijing also wants to see a joint global leadership rather than a dominated by a single great power. Within this framework, the G2 within the G20 may reflect an ideal platform to build a joint global leadership by the two most powerful economies, the United States and China.

Nye (2017) claims China is too weak rather than too strong in providing global leadership and will keep being a free rider rather than shouldering its responsibility while the United States is more and more incapable or unwilling to realise its duty, resulting in the world into the Kindleberger Trap in the end and repeating the tragedy of the 1930s. However, our research indicates that China is neither too weak nor too strong in providing global leadership. China's contribution to the global public goods makes a significant effect, but Beijing cannot fill the global leadership gab for its own. It is true that Beijing does not want to play a global leadership role singlehandedly, but China does not refrain itself to take global responsibility too. What China indicates during the pandemic is that multilateral solutions are must to solve global challenges. In this point, we argue that only the G2 can play a leadership role in the post-pandemic period with the coordination of other influential global players like in the G20.

Another assessment of our research is that China is on the rise, but the United States is not in decline. In fact, the U.S. material capabilities have been remaining strong in the post-pandemic period. The United States is not a declining dominant power or hegemon, as the UK was in the 1930s and will likely to stay the main dominant power for global leadership role in next decades. Again, the Kindleberger Trap is not sufficiently valid in the 2020s world. This supports our argument that the United States and China can play a joint global leadership role under the framework of the G2 to handle the global challenges.

References

Allison, G. (2017). Destined for war: Can America and China escape Thucydides's trap? s. l. Houghton Mifflin Harcourt.

Bahi, R. (2021). The geopolitics of COVID-19: US-China rivalry and the imminent Kindleberger trap. *Review of Economics and Political Science: REPS, 6*(1), 76–94.

Bresnahan, J., Scholtes, J., & Levine, M. (2019). Trump kills plan to cut billions in foreign aid. https://www.politico.com/story/2019/08/22/white-house-backs-off-foreign-aid-cuts-1472130. Accessed 10 March 2022.

Chin, M. (2021, December). What are global public goods? *Finance & Development*.

Çolakoğlu, S. (2018). MIKTA in global governance as a middle power grouping: A Turkish perspective. In R. Marsudi (Ed.), *MIKTA: Current situation and the way forward* (pp. 51–72). Indonesian Ministry of Foreign Affairs. ISBN: 978-602–51358-0-4.

Çolakoğlu, S., Hecan, M., & Sakaoğlu, E. T. (2016, April). *The G20 and global governance: An analysis on the vision and capacity of the group* (USAK Rep. No. 44).

Cooper, A., & Mo, J. (2013). Middle power leadership and the evolution of theG20. *Global Summitry Journal, 1*(1), 1–14. https://doi.org/10.7871/2291-4110.1001.

David, K. (2012). East Asia when China was at the centre: The tribute system in early modern East Asia. In M. Beeson & R. Stubbs (Eds.), *Routledge handbook of Asian regionalism*. Routlege.

Freeman, C. P. (2020). Reading Kindleberger in Beijing: Xi Jinping's China as a provider of global public goods. *The British Journal of Politics and International Relations, 23*(2), 297–318.

Garrett, G. (2010, January). G2 in G20: China, the United States and the world after the Global Financial Crisis. *Global Policy, 1*(1), 29–39.

Garver, R. (2021). Biden budget substantially boosts foreign aid, diplomacy, but raises defense by 1.7%. https://www.voanews.com/a/usa_us-politics_biden-budget-substantially-boosts-foreign-aid-diplomacy-raises-defense-17/6206352.html. Accessed 10 March 2022.

He, K., & Feng, H. (2020). International institutions, institutional balancing, and peaceful order transition. *Ethics & International Affairs, 34*(4), 487–501.

Iqbal, B. A. (2022, May 31). The Indo-Pacific economic framework: Issues and challenges. *The Turkish Center for Asia Pacific Studies*. http://www.asianpacificcenter.org/indo-pacific-economic-framework.html. Accessed 10 June 2022.

Jin, J. (2020). Beware the Kindleberger trap in global health governance. *China Quarterly of International Strategic Studies, 6*(3), 333–353.

Khanna, P. (2016, January). China's infrastructure alliances. In *Geo-economics with Chinese characteristics: How China's economic might is reshaping world politics*. The World Economic Forum. https://www3.weforum.org/docs/WEF_Geoeconomics_with_Chinese_Characteristics.pdf. Accessed 10 March 2022.

Kindleberger, C. P. (1986). *The world in depression, 1929–1939*. University of California Press.

Lin, H. (2012, March/April). The G20 and the G2. In *Contemporary International Relations (CIR)* (Vol. 22, No. 2, pp. 49–55).

Monaco, E. (2021). Introduction. In S. X. B. Zhao, J. H. C. Wong,·C. Lowe, E. Monaco & J. Corbett (Eds.), *COVID-19 pandemic, crisis responses and the changing world*. Springer.

National Bureau of Statistics of China. (2022). *盛来运: 逆境中促发展 变局中开新局——《2021年国民经济和社会发展统计公报》评读* (*Sheng Laiyun: Promoting development in adversity and create a new situation in changing circumstances—Review of statistical communiqué of the people's Republic of China on the 2021 national economic and social development*). National Bureau of Statistics of China. http://www.stats.gov.cn/tjsj/sjjd/202202/t20220227_1827958.html. Accessed 14 March 2022.

Nye, J. S. (2017). *The Kindleberger trap*. https://www.belfercenter.org/publication/kindleberger-trap. Accessed 9 February 2022.

Park, S. (2017). China's G20 membership and quest for leadership in East Asian regional architecture. In C. Schlager, D. Chen, A. Rosenplanter & H. Zhang (Eds.), *China and the group 20: The interplay between a rising power and an emerging institution* (pp. 169–192). World Century.

Pempel, T. J. (2012). Global financial regulation: G2 or G20? In J. Park, T. J. Pempel & G. Xiao (Eds.), *Asian responses to the global financial crisis: The impact of regionalism and the role of the G20* (pp. 158–169). Monograph. https://doi.org/10.4337/9781781003916.00016.

The State Council Information Office, o. t. P. R. o. C. (2021). *China's international development cooperation in the new era*. The State Council Information Office of the People's Republic of China.

The White House, W. (2020). *The Biden-Harris administration immediate priorities*. https://www.whitehouse.gov/priorities/. Accessed 10 March 2022.

Walter, A. (2011). *Global economic governance after the crisis: The G2, the G20, and global imbalances*. https://personal.lse.ac.uk/wyattwal/images/globaleconomicgovernancea.

Wang, Y. (2017). 王毅: 中美不会走入"修昔底德陷阱" (*China and the US will not walk into Thucydides trap*). https://language.chinadaily.com.cn/a/201703/21/WS5b2c9e74a3103349141ddda1.html. Accessed 9 February 2022.

Xi, J. (2017). *共同构建人类命运共同体 (Jointly build a community with a shared future for mankind)*. http://www.qstheory.cn/dukan/qs/2021-01/01/c_1126935865.htm. Accessed 10 March 2022.
Zhao, S., Zhao, S. X. B., & Zhang, Z. (2021). US-China rivalry and its implications for the post-pandemic world. In S. X. B. Zhao, J. H. C. Wong, C. Lowe, E. Monaco, & J. Corbett (Eds.), *COVID-19 pandemic, crisis responses and the changing world*. Springer.
Zhao, T. (2009). A political world philosophy in terms of all-under-heaven (Tian-xia). *Diogenes, 56*(1), 5–18.

Selçuk Çolakoğlu Associate Professor at Faculty of Humanities and Social Sciences, BNU-HKBU United International College. Dr. Çolakoğlu received his Ph.D. in the Department of International Relations at Ankara University. His research interests include Foreign Policy, International Security, Globalization, Global Governance, and the G20. He has published monographs, 40 articles, and book chapters. His books are: *Turkey and China: Political, Economic, and Strategic Aspects of the Relationship* (London: World Scientific, 2021); *Northeast Asia in International Relations* (Ankara: USAK Press, 2009); *Korean Society, Culture, Politics* (with Bengü Çolakoğlu, Ankara: Orion Press, 2008); *China in World Order and Turkey-China Relations* (Ankara: USAK Press 2012). He has held positions in universities in South Korea, Turkey, the United States, and China.

Chaofan Chen studies Globalisation and Development at Beijing Normal University-Hong Kong Baptist University United International College. Chen's main research interests include International Relations, Diplomacy, Chinese Politics, Chinese Foreign Policy and Comparative Politics. Chen has founded the Sustainable Development Research Group at UIC, an organisation that promotes student research on UN SDGs and global governance. Chen is also a member of World Ocean Day Youth Advisory Council since September 2022.

Jiawei Chen studies globalization and development at Beijing Normal University-Hong Kong Baptist University United International College. Chen's main research interests focus on geopolitics and global governance issues. Chen's collective research project on the successful development experience of small countries was supported by the college faculty. Chen also actively participates in government research; In 2022, he partook in the investigation of public infrastructure resources in Foshan City.

Political Considerations and Rationales for Different Control Policies in COVID-19 Pandemic: An Analysis from the State Theory Perspective

Chaofan Chen, Jiawei Chen, and Bo Yan

1 Introduction

The concept of State has been one of the oldest and most important ideas in modern political philosophy. Since the Westphalian System was established in the seventeenth century after the Thirty Years' War, State has become a precise term to describe "an independent political society" with its territorial and political sovereignty (Simmons, 2006). The government of a sovereign country has the supreme authority to decide the policy and governmentality path of its domestic and international behaviours. Due to the different historical, cultural, economic, and political backgrounds, different governments of states act variously from many perspectives.

In 2020, the coronavirus (used interchangeably with COVID and COVID-19 hereafter) sparked a worldwide crisis in multiple dimensions, from the health sector to global trade and even the landscape of international relations. In the face of the pandemic, the globalisation process that lasted for decades has also experienced significant severe setbacks. National actors become the primary role in dealing with this problem. As the most basic unit in the international community, the state becomes responsible for coping with the pandemic with its power and policies. Different states generate various pandemic control policies with different considerations and rationales.

Overall, there developed two major pandemic control policies: "zero COVID" strategy and "live with COVID" strategy. To understand and analyse the differences between two pandemic control policies from the aspect of government decision-making considerations and rationales, this chapter employs the theory of the state, an essential theory in the political sciences and state construction.

C. Chen (✉) · J. Chen · B. Yan
Beijing Normal University-Hong Kong Baptist University, United International College, Zhuhai, China
e-mail: p930032001@mail.uic.edu.hk

© The Author(s), under exclusive license to Springer Nature Singapore Pte Ltd. 2023
S. X. B. Zhao et al. (eds.), *Comparative Studies on Pandemic Control Policies and the Resilience of Society*, https://doi.org/10.1007/978-981-19-9993-2_7

The chapter will firstly review the core concepts of the theories of the state and two major types will be detailedly discussed: the Neo-Weberian approach and the Liberal approach. Then, it elaborates on their formation processes, legitimacy, and the decision-making processes of the state power and government. The third part of the chapter will examine the effectiveness of the pandemic control policies conducted by the two different approaches of state theory as well as the political considerations and rationales behind them. The chapter then goes to analyse the consequential cost and implications under different pandemic control policies with comparative studies of China and the United Kingdom. Finally, the chapter offers a brief examination of the possible challenges from the emergence of new variations in COVID-19 and concludes the chapter.

2 The Theories of State

In the sixteenth and seventeenth centuries, the political philosophy of the West experienced significant changes, and the establishment of the Westphalian System remarked the development of early modern political philosophy. From then on, scholars from different schools express different thinking on the theories of the state. Matthias vom Hau (2016) summarises the state theories into four approaches: class-analytic approach, liberal approach, neo-Weberian approach, and culturalist approach. As will be analysed below, this chapter will mainly refer to two approaches: the Liberal approach and neo-Weberian approach.

For the liberal approach of the state theory, Hau (2016) considers its rooted thinking to be the natural freedom and social contractarian idea. The thoughts of distrust to government, and containment or dispersion of its power are main considerations of this approach. The state is seen as an arena of strategic action among individuals and groups. The liberal approach of state theory considers that state capacity depends on the policy choice of the ruler and how the state ensures consent. Liberals see the state legitimacy as based on a "quasi-voluntary compliance among citizens," though no penalty for the non-compliance (Levi, 1988).

The Neo-Weberian approach sees the state as a set of administrative and coercive organisations which claim the legitimate monopoly on national territory and the binding decision (Anter, 2014; Evans et al., 1985; Weber, 1978). Hau (2016) introduces two theoretical questions: one focuses on to what extent the state makes policy and action independently and the other one is on how to execute the decision the state made in its territory. To answer the questions above, the Weberian approach focuses more on state-society relations and state capacity. Weberian thinking considers the state as a potentially autonomous actor (Evans, 1995), which means certain relations between state and non-state actor (society) allow the state to do autonomous and well-informed policies. The Weberian approach also thinks that the state capacity is valued by the bureaucratic competence and capacity of territorial reach out by the view of this approach. Therefore, the state power consent comes from the output legitimacy, nationalist claims and nationalising activities and state ideological work (Gellner, 1983; Wimmer, 2002).

John Simmons (2006) emphasises two great "divides" in the discussion of the theories of the state, one is the divide between "political naturalism" and "political antinaturalism" and the other one is the transition from the complex and hierarchical political structure (with overlapped religious and contractual relations) to political society with "the form of modern, sovereign, territorial states." Due to the nature of culture, Hau pointed out that the culture provides the state capacity with the identity-based disciplining effect which establishes the scheme of moral control and popular mobilisation that could be used by the state. Therefore, Hau concludes the consent of the state power originates from the rituals, everyday practices, and world cultural models.

Simmons (2006) concludes that political naturalism as an idea implies that people are organised politically and obey political authority as national conditions. Political naturalism indicates that the government and subjection are parts of the natural order, some are born to rule while some are born to be subjected (Simmons, 2006). Aristotle's view of humankind as "by nature a political animal" supports this idea. (Aristotle, 1995) In the medieval era, the religious leaders practised political naturalism well, those who were born to rule (e.g., emperors, pope) were natural with the naturally granted power by God. This can be interpreted as the "rule by 'divine right'," a power that connects with subjects' obligations of passive obedience and rulers are accountable to God only (Simmons, 2006). Political nationalism provides an approach to explaining the relationship of rule and subjection by the meaning as a natural or determined fact.

Political antinaturalism represents a concept that is more closed to the modern liberal approach to political theories, people are born with freedom and are not controlled by political authority naturally. This idea considers that freedom and equality are natural conditions and deems that the established powers are not granted by God (Simmons, 2006). Rousseau expresses a similar idea that God has not ordained any political order for mankind in his *Social Contract*. Rousseau and most political antinaturalists are contractarians, they believe that legitimate political authority comes from the contract, pact, or covenant (Simmons, 2006). The parties of the contract are not individuals but are generally considered to be the existing organised people and rulers/governments at the founding moment of political society (state). The rulers/governments can obtain all rights of the people to maintain the necessary stability and peace of society and to fulfil the best interest of all. Otherwise, individuals keep the right to withdraw or resist the contract or the rule. It provides the modern idea of the government power and its authorisation granted by the people of the state.

Political antinaturalists are also keen on advocating the "state of nature" by defending the natural rights and freedoms as well as the contract or consent as the course of political power's legitimacy. Simmons (2006) concludes that "natural rights as individual moral prerogatives or protection for certain spheres of individual choice or pursuit of certain interests," this idea provides a moral justification for acting contrary to the commands of political authorities. Although the contract is seen as the active authorisation of the rulers or government by individuals, Hobbes raises his idea that the contract made by the means of conquest and intimidation might be also binding. This Hobbesian idea has often been concluded as "might makes right" (Simmons, 2006). These ideas reveal the boundaries and relations between the power of government and people, as well as the ways that a government is given authorisation.

Through reviewing the different theories of the state and its studies in the context of the ongoing COVID-19 pandemic, this chapter will tentatively argue that the different pandemic control policies in different political societies are based on different theories of the state and the political considerations and rationale.

3 Zero COVID Strategy and Neo-Weberian Approach

The zero COVID strategy is one major pandemic control policies that has been adopted by many countries, such as China, Japan, South Korea, Australia, and New Zealand at the preliminary stage of pandemic (Economist Intelligence Unit, 2021). A typical zero COVID strategy contains measures like lockdowns, mass COVID-19 testing, strict quarantine of certain groups of people and necessary compromise on socio-economic development (Sobhan, 2022). The zero COVID strategy requires a mass scale of mobilisation of the population and national resources. Also, most of the measures are compulsory and mandatory for citizens to follow. Many zero COVID pursuing governments have pleaded that the zero COVID strategy is the most effective way to control the spread of pandemic, save people's lives, and ease the public health burden (Silver, 2021). However, the zero COVID strategy has been criticised for its way of compromising too much individual rights and economic interests, which violates modern citizenship and is not cost-effective (*The Economist*, 2021). To be noted, most of these critics come from societies and media in the West, a concentrated area showing the Liberal theory of state. As per the literature review, from the nature, performance, and measure of the zero COVID strategy, we can see the strategy has theoretical underpinning from the Neo-Weberian approach.

The Neo-Weberian approach of the state theory treats the state as a set of administrative and coercive organisations that claim the monopoly of legitimate force over territories and make collectively binding decisions. It emphasizes an organisational logic of state actions. Hau (2016) concludes that this idea vests the state with the certain power and forms of autonomy, and the state is treated as a political force *sui generis*, which is different from the autonomy of non-state actor. Peter Evans (1995) called it the "embedded autonomy," suggesting the special relations with non-state actors authorises the state to make "autonomous yet well-informed" policy. Therefore, the state legitimacy is conferred as state performance, collective awareness, nationalisation, and the socialisation of citizens. The state capacity is valued by leveraging resources and popular mobilisation ability.

The Neo-Weberian approach is reflected in the implementation of the zero COVID strategy. Firstly, the government, as an independent organisation with autonomous power with monopoly status, made independent judgment about the pandemic and its impact. In the early stage of the pandemic, knowledge of the virus was limited. The states, including those decided to "live with COVID," resorted to varying degrees of travel restrictions or lockdowns to prevent the spread of the virus. The instant effectiveness of prevention and control has given legitimacy to the government's mandatory zero COVID strategy. The states leverage publicity channels to continuously exude the positive influence and benefits of zero COVID strategy for society and arouse a wide range of collective awareness. Promoting the sacrifice of personal freedom as some governments claimed, for example, could prevent more people from being caught up in COVID-19. This measures the rationality of the states' collective restriction policy. In this process, both social groups and citizens obey the states' policy decisions. National capacity is also reflected and tested in the implementation of the zero COVID policy, which included policy implementation scope, resource scheduling capacity, infrastructure services, financial situation, and mass mobilization capacity. The organisational logic contained in the zero COVID strategy, and the government performance provided by the policy further provide a reasonable explanation and political incentive for the government to continue this policy. At the same time, the government used the monopoly position to mobilize resources and mobilize the masses within the state, which in turn strengthened the state capacity.

One of the most typical showcases of the Neo-Weberian approach and the zero COVID strategy implementation is China. Using China as a representative case and analysing the effectiveness of the Chinese Zero COVID strategy, the chapter will present how the Neo-Weberian approach of state theory provides political rationale to support the implementation and continuation of the zero COVID strategy, and how an effective implementation of zero COVID strategy in turn validate the government's theoretical selection.

3.1 Case Study: China

In the case of China's zero COVID strategy, we might see the political considerations and rationale behind the state's leverage of its autonomy to set the policies and use the state power to reinforce the legitimacy of the state as a well organised institution.

First, the Neo-Weberian approach sees the state as a formal organisation with autonomy, it requires us to examine two basic abilities of the state. One is the capacity of making independent goals and decisions, another is the capability of executing the goal and policy in state territory (Hau, 2016). The COVID-19 pandemic firstly broke out in Wuhan, and China is the first country to deal with mass domestic infection in the world. The zero COVID strategy was established by the Chinese government in the context of limited understanding of the virus and a shortage of effective medication and vaccine to cure and prevent the virus. With the development of COVID-19 research and the emergence of new variants, the Chinese government switched from an absolute pursuit of zero COVID strategy to a dynamic "Zero COVID strategy" (*dongtai qingling* 动态清零) in August 2021 (Liang et al., 2022).

As for the implementation ability, Chinese state actors maintain a strong executive capability through its party-led governance. The execution of policies by the central government (when it comes to the pandemic control policy) is mostly finalized and led by the Communist Party of China forces and its members. The heads of the local command headquarters are mostly the local Party Secretary or the senior party members. Also, the party members are encouraged or half-mandated to go to the frontline in combating the pandemic (*People's Daily*, 2021). Following the outbreak and mass lockdown of Wuhan, Chinese governments (at all levels) have implemented numerous lockdowns. In addition, the mass COVID testing all over China has become day-to-day practice. A mass surveillance system on citizens' trajectory has been established and working well. A set of observation, quarantine, and cure system has also been set up. The Chinese state actor shows its strong executive capability in implementing the zero COVID strategy within its territory, which reflects the characteristics of a formal organisation of state in the Neo-Weberian approach in return.

Second, to achieve the goal set up by the state organisation, it requires the state capacity to function at its finest condition. The Neo-Weberian approach regards the governments exercising their control putting policy choice into practice by relational and organisational properties, including government resource disposal, and the popular mobilisation ability and the partnership with non-state actors (Hau, 2016). For the disposal of resources, the Chinese state actor leverages its advantage in coordinating the nation-wide interest. The zero COVID strategy demands mobilization at all levels of society. For this measure, the state needs to coordinate the resources on medical personnel, medical materials, laboratory capacity and financial strength. The long-standing execution of this measure is an evidence of state capacity undergirding the zero COVID strategy. China's inter-regional resource redistributions on medical resources, human resources, daily necessities, economic support take place consistently. The adequate state resource and strong distribution ability of the state provide the government with competent capability to implement zero COVID strategy, which is in accordance with Chinese state organisational logic. For the mobilisation and partnership with non-state actors, Chinese state actors resorted to both the combination of party organisms and market economy. One such example is the conduction of mass testing. Party organisms in the community are the executive units for testing. They use multiple incentives to drive the residents to engage in the testing. Incentives comprise positive encouragement, like rewarded shopping cards, free eggs and vegetables, and other possible material incentives. The Chinese state has also used its monopoly right to "force" people into the engagement, such as restricting access to social activities of those who do not conform to take COVID tests. Local resident organisations (like neighbourhood committee, 居委会), operating as semi-state social units, reinforces the ability of the state to control mobilization. Private enterprises also establish partnerships with the government. A certain amount of the PCR (polymerase chain reaction) testing is conducted by private laboratories in China to compensate for the inspection ability of public labs and hospitals (Meng, 2020).

Third, the legitimacy of the state and its policies are assessed by the state's performance in the Neo-Weberian approach. The Neo-Weberian approach of state theory considers that the governed would see the legitimacy of the state as shown by the organisational competence of service and the service reaching out through the territory. So, we may need to examine what objective of the zero COVID policy and the state is trying to achieve.

As the most prominent zero COVID pursuing country, China has targeted at eliminating the infected cases since the outbreak of the pandemic as of writing (August 2022). The dynamic zero COVID strategy saw China's rapid response and precise prevention and control of the virus. Since 2020, China established the goal in pandemic prevention and control as "preventing the virus from re-entering the territory and rebounding inside" (外防输入, 内防反弹). To accomplish this goal, China has perpetuated mass COVID testing within the territory which would be vital for the state performance assessment. The goal of a low-level infection rate can be regarded as the most essential indicator of state performance during pandemic, which provides the fundamental rationale and legitimacy of the state's zero COVID strategy.

It is noted that the "zero" goal comes at significant cost even for China, a state with big government capacity in policymaking and strengths in state resource mobilisation. At the State Council press conference on 11 December 2021, the head of the China National Health Commission's Leading Group, Liang Wannian, said: "Dynamic Zero COVID strategy does not mean zero infection. Now there is no ability to eliminate the risk of local cases (CCTV, 2021)." Indeed, although China has adopted strict pandemic control policies and practices, such as regular mass COVID-19 testing, strict quarantine rules for entrants, and the health code system used on daily basis, it can hardly prevent the regional outbreak from happening. As the Fig. 1 shows below, regional outbreaks in China have not ended, particularly when the Delta and Omicron have become the dominant variants.

High costs haunting China though, the vitality of the Neo-Weberian in China's selection of the zero COVID strategy has been confirmed by China's effectiveness in keeping both the number of infected cases and death cases at a very low level. China has 2,290,634 confirmed cases and 14,767 death cases in total as of August 2022, which is a very low infection and death rate compared with the population of China and one of the best records among the major economies. (JHU, 2022) The comparatively good policy performance provides a solid foundation to legitimatise the state policy of zero-COVID.

The effective of China's pandemic control policy out of the Neo-Weberian approach is also verified by China's relatively early resumption of business and commercial activities. By using the "Baidu Index" (百度指数, Baidu is China's largest online search engine), we can observe how Chinese see normal life and the zero COVID strategy and its effects. We have collected the trending search keywords on Baidu from 1 January 2020, when the large-scale pandemic was yet to break out, to 2 August 2022. We select keywords that are representative of social activities, which are: (1) Pandemic (疫情) (2) Restaurant (餐厅) (3) Transportation (交通) (4) Travel/Trip (旅游) (5) Shop (商店)—to observe the impact of lockdown measures under China's zero COVID strategy on social activity after regional outbreaks (Fig. 2).

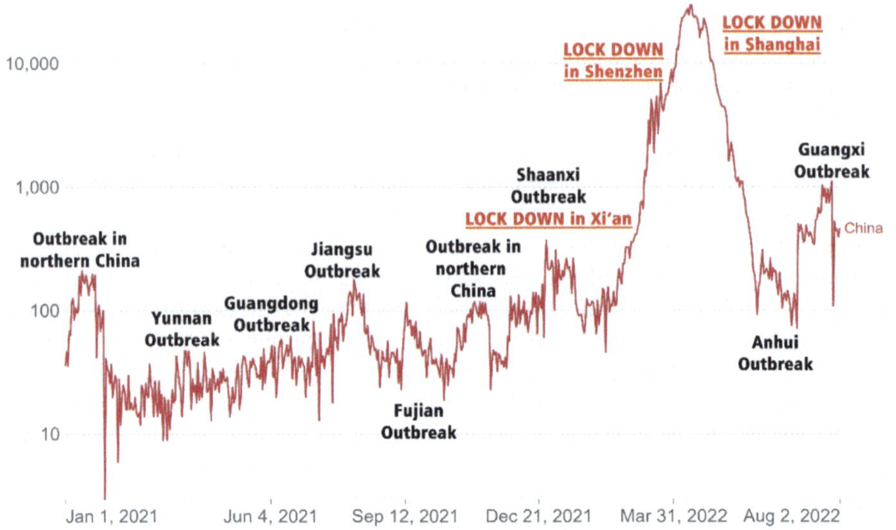

Fig. 1 Daily new confirmed COVID-19 cases and the major regional outbreak of mainland China by period (January 1, 2021–August 2, 2022) (*Source* https://ourworldindata.org/coronavirus)

The volumes of the keywords search show Chinese netizens' accommodation on the zero COVID strategy and the dynamic changes between pandemic control and normal life. This chapter would suggest these searches represent the mass acceptance of zero COVID policy among Chinese people. For example, when the Chinese Lunar New Year is approaching (point 1, 3, and 6 in the keyword "pandemic"), the attention on social activities (other keyword search volumes) has diminished. Searches of "restaurant" and "travel" decreases markedly. Self-selection of social activities during pandemic shows the China dynamic zero COVID strategy is well-received by its citizens, and citizens have accepted the life of dynamic pendulum swing between necessary pandemic control and resumption of normal life. Chinese citizens seem satisfied at the Chinese state's implementation of the zero COVID strategy. Edelman Trust Barometer 2022 indicates that the Chinese demonstrate a high-degree confidence in the Chinese state. In this report, Chinese government gained a mark of 91 out of 100 in the citizen's trust index, a nine-point up from 2021 (Edelman Holdings, 2022).

4 "Live with COVID" and Liberal Approach

The Liberal approach of the state theory embedded discussions on political naturalism and political individualism. Political naturalism believes that people are politically organised, government and subject are natural. The power or authority is mandated by some people. The political antinaturalism disagrees with that notion, considering

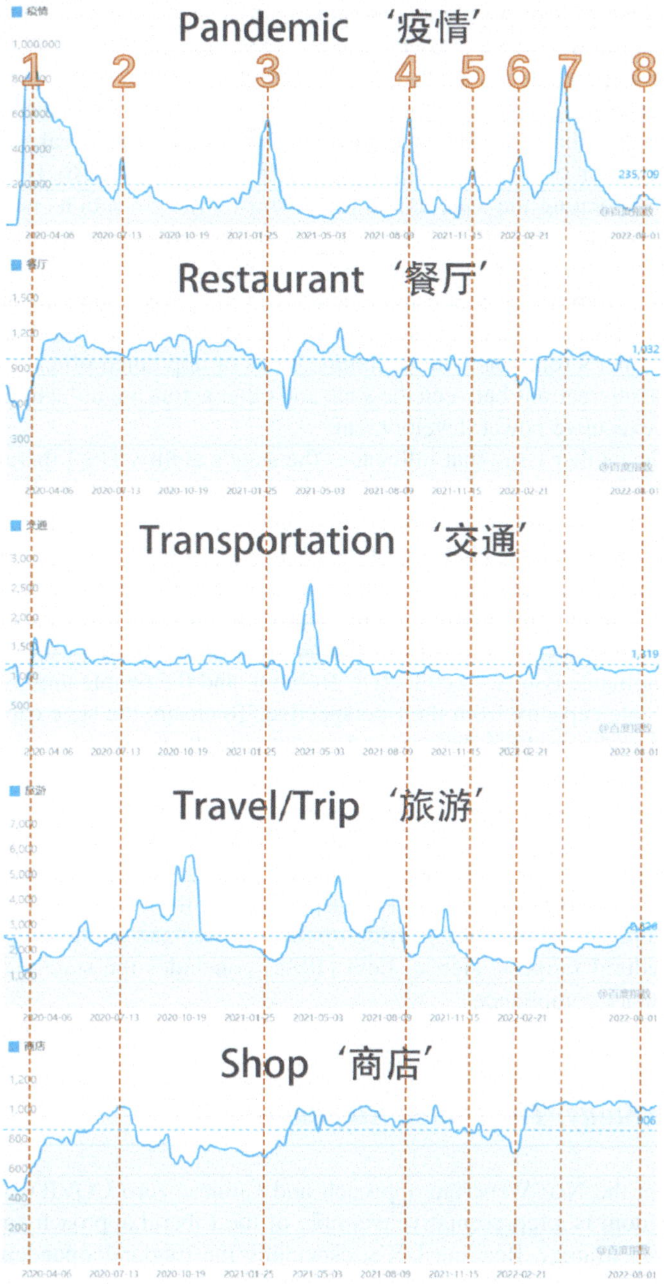

Fig. 2 Observation on the Chinese netizens' attentiveness and potential relations between selected words (*Source* Baidu Index, http://index.baidu.com)

the natural conditions of people as free and equal. The state power is based on the concept of natural freedom and social contract (Simmons, 2006).

The Liberal approach of state theory is rooted in the idea of social contract. State is place of strategic interaction among individuals and groups. The Liberal approach does not conceptualise the state as an independent source of political power, but to see it as a reflection of social development and changes (Hau, 2016). Hau considers the concept of treating the state as the platform of strategic actions has impactful influence on the state capacity and consent analysis. From this perspective, the state's competence in contacting society and implementing policies is closely connected to the choice of governors. Structure and efficiency of the governments are also defined by the relationship between the state and certain groups of actors. Hau raises the examples of ruler's relations with economic elites or appointed officials on the tax policy. These interactions between the state and other actors are influential on policy making and executive power development.

Consent is another issue that influences the state's ability. The Liberal approach of state theory sees the state power as based on the consent of the natural rights alienation from the individuals to the state. Individuals' authorisation on its rights to the state does not mean the abandon of the rights. Individuals have the power to not comply with the political order and the social contract idea provides a solid moral protection. So, another presentation of the state capacity is to secure the consent and authorisation from the constituents, or the people. Therefore, the Liberal approach sees the compliance issue as a collective problem, and the people might be the free rider to the state capacity from their perspective. To ensure the state capacity, state leaders must make credible threat to solve non-compliance problem to counteract the potential free rider.

However, when the state force of violence would not be workable on forced compliance for people, the legitimacy plays its role. As same as the core thinking of social contract, Liberal state theory sees the legitimacy of the state as originating from the respect to the state policy. The respect comes from two dimensions: one is the high acceptance of the state-provided public service and goods, and the second one is procedural fairness. Hence, Levi (1988) concludes the state legitimacy as "quasi-voluntary compliance."

4.1 Case Study: The United Kingdom

Varying from the Neo-Weberian approach and Chinese zero COVID strategy, the United Kingdom is a representative example of the Liberal approach and the "live with COVID" strategy. Reading UK's, especially the England's pandemic control policy from the state theory, we need to read its root concept, the state-society relations, state capacity and the consent to the state power.

As one of the oldest Western democracies with a parliamentary system, the state power of the UK is vested in the parliament. The political party that wins the majority of seats in the House of Commons has the opportunity to form government and exercise the power of the state. Citizens are conceptualized in this system as voters. Voters support Member of Parliament candidates through learning their manifesto and vision of their service in the House. The process by which voters support different candidates or parties through their ballots can also be understood as the signing of a contract between individuals and the state. Political leaders represented by the Prime Minister, after gaining the majority seats, are symbolically granted power by the Queen to form government (UK Government, 2022). By the collective individuals' choice of government, the government gets authorisation from elections to make policy decisions. In 2020, the UK Parliament passed the "Coronavirus Act 2020" to authorize the government to handle the COVID-19 pandemic with emergency powers. This is the source of the UK government's authority to implement policy against the pandemic. The bill gave the UK government the discretion to restrict and suspend public gatherings, detain individuals suspected of having COVID-19, intervene or relax a range of legal provisions to limit the spread of the disease and reduce the burden on the National Health Service (UK Parliament, 2020).

Chancellor of the Exchequer Rishi Sunak, Prime Minister Boris Johnson and Deputy Chief Medical Officer Jenny Harries delivering the press conference on COVID-19 at 10 Downing Street. An in-person press audience is present. The 'Stay at home, protect the NHS, save lives' slogan is displayed on their podiums. (Shotted on: 20 March 2020; Photo Source: Prime Minister's Office, 10 Downing Street)

The relationship between the state and society is regarded by Liberals as an area of strategic actions. The Liberal approach thinks that the ability to reach out to society and implement government goals is closely related to the policy choices of government leaders. The relationship between government policy makers and government policy-related actors will affect government administrative capacity and efficiency. The government's strategic COVID-19 prevention and control policies have won broad support from the public (Halliday et al., 2022), which was evidenced by the high vaccination rates and resurgence of social activities mentioned later.

The first case of COVID-19 in the UK was officially reported in January 2020, and the government has imposed entry controls on travellers from places including Hubei, China, and Italy since February (Department for Transport & Public Health England, 2020). The UK Government's Chief Medical Officer has identified four stages of containment in the UK: Contain, Delay, Research, and Mitigate. The government devised the slogan "Catch it, bin it, kill it." In a speech in early February 2020, Prime Minister Boris Johnson resisted lockdowns practiced in some parts of Europe, worrying that "coronavirus will trigger a panic and a desire for market segregation that go beyond what is rational to the point of doing real and unnecessary economic damage" (Johnson, 2020). Italian government's strategy of lockdown was also considered by the British government as "useless populism" and "unscientific" (Peston, 2020), and the media revealed that the British government did not even ask scientists to simulate whether lockdown could be a solution (Calvert et al., 2020).

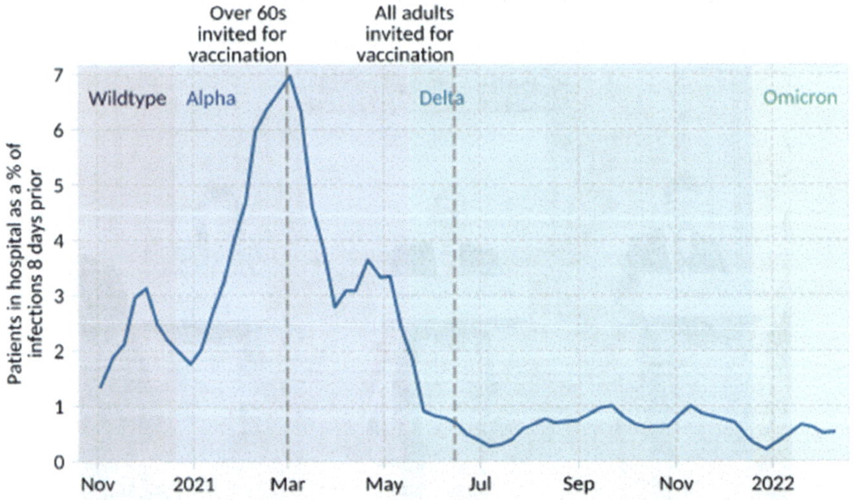

Different variants have brought new challenges to pandemic control, and the vaccine can act a critical role in controlling infection rate. (Source: UK Cabinet Office)

As the outbreak developed, the British government told the public to be prepared for pandemic expansion, while the government remained open to restricting citizens' freedom to gather in groups. The concept of herd immunity was first mentioned by the British government's chief scientist in the early days of the pandemic (Lintern, 2020). According to the number provided by the UK government, more than 720 thousand patients have been admitted to the hospital with more than 160 thousand COVID-19 death cases. As the number of confirmed cases skyrocketed and the NHS system became overwhelmed, the UK government eventually decided to impose three lockdowns or travel restrictions in February, April 2020, and January 2021. The Home Secretary even advocated closing Britain's borders (Murphy, 2021). At the stage, there was no vaccine or specific medicine to prevent and cure COVID-19, domestic travel restriction would be the best approach to cope with the widespread. UK citizens also show its agreement with the government's decision. Research published in Nature shows that the political trust of the UK government has risen at the onset of the UK lockdown imposed in March 2020 (Davies et al., 2021).

However, the mass and long-time lockdowns and travel restrictions triggered demonstrations and protests in the UK. The "Unite for Freedom" series of protests began in April 2020 against the government's nationwide lockdown and has continued across the country (Özdüzen et al., 2021). Mandatory travel restrictions have not been widely accepted by society in the interaction of strategic behaviours, especially long-term and multiple restriction orders. In addition, the trust of the government gradually declined in coming months due to political scandals and the effects on citizens' everyday life (Davies et al., 2021). Government's decision is facing the challenge of consent losing and the state capacity fading.

Policy-decisions of the government directly affect the national capacity and the consent of the government, the Liberal approach believes (Hau, 2016). Given the infeasibility of strong restrictions on civil liberties and the liberal philosophy of minimal government intervention, lockdowns and travel restrictions are not realistic. The British government has directed their pandemic control strategy more on vaccinations and medication. The UK became the world's first country to have invented and authorized the well-recognized COVID-19 vaccines. The UK government adjusted the policy into "live with COVID," removing domestic restrictions, vaccinating vulnerable people and deploying targeted testing, mass vaccination, and investing more in the life science sector (Cabinet Office, 2022). As of February 2022, 91% of the population over 12 years of age in the UK had received at least one dose of the vaccine and 66% had received three doses.

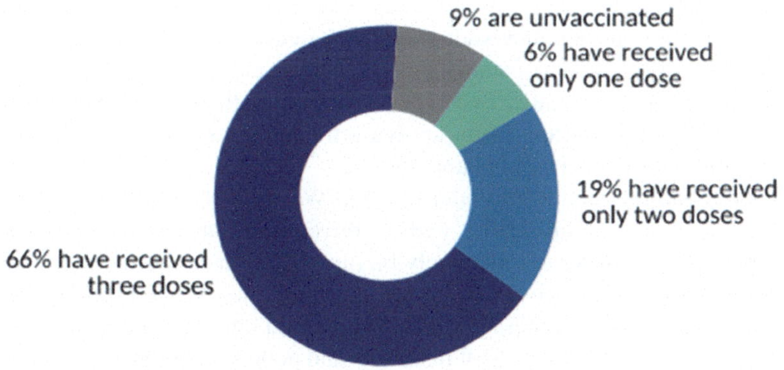

UK population COVID-19 vaccine coverage, by dose, of those aged 12 and over as of 16 February 2022 (Source: UKHSA, Vaccination in the UK, 18 February 2022.)

The change of pandemic control policy shows the state actor's attempt to ensure the authorisation from people and the consent to state power with the rise of the demonstrations and protests to the pandemic control policy and its effect on the economy (Özdüzen et al., 2021). To ensure the consent and legitimacy of the state, the Liberal approach would conceptualise it into the constituents' respect to the state policies. Hau (2016) considers that two conditions should be met, the people's recognition of the state-provided public goods and services, and the procedural fairness of decision making. "Live with COVID" is a case policy to demonstrate this thought.

Firstly, this policy pivots from the previous restriction on citizens' freedom to the provisions of more specific services on vaccination and medication accessibility, and precise pandemic surveillance (e.g., targeted testing and vaccination of vulnerable people). By using vaccine and medication, new policies remove restricted measures in response to people's appeal and the pandemic control demand. Boris Johnson considers this policy "will restore freedom" (O'Connor & Cursino, 2022). Secondly, the thought of procedural fairness of liberalism would require the government not to limit most people's freedom to prevent the potential infection of certain people. Law scholars published an article in "King's Law Journal", asking why the UK people should comply with the lockdown law and questioning the compliance of the lockdown. (Halliday et al., 2022) Therefore, the "live with COVID" provides the procedural fairness that the state consent and legitimacy require after the loss of public trust of UK government. The "live with COVID" type policy might be the only solution for the states like the UK to act under the Liberal approach of state theory.

5 Conclusion

To conclude, the state theory provides us with a new perspective to read the different pandemic control policies adopted by different countries. This chapter firstly reviews different approaches of state theory and the discussions in political sciences academia. Then, it investigates two types of the state theory, the Neo-Weberian approach, and the Liberal approach, applying them onto analysis of selections of pandemic control policies in China and the UK.

We have found that countries under the Neo-Weberian approach, like China, as a set of autonomous administrative and coercive organisations, the state has the right of monopoly in making collectively binding decisions in accordance with the organisational logic of the government. The state legitimacy is embedded into the state performance, collectiveness, nationalisation, and citizens' socialisation. State power is presented by the leverage of resources and popular mobilisation. In the case of the Chinese pandemic control policy, Neo-Weberian can explain its consideration and rationale.

Firstly, the Neo-Weberian thought provides the rationale for the Chinese government's decision-making of its zero COVID strategy. The state executes the monopoly of making collectively binding policy with its logic. Then, the state leverages its propaganda system to explain its policy logic to win support from society, and the power to mobilize national resources to cope with the pandemic. The Chinese government has organised a series of filmography, news series, and other measures to explain the necessity of zero COVID strategy and encourage collectiveness to support government measures that might limit personal freedom (lockdowns, quarantine, and mass testing).

Secondly, the Chinese government makes good use of its organisational uniqueness to strengthen the state capacity. State organisms, including the Communist Party of China and party members, are being mobilised to ensure the executive capability of zero COVID strategy in Chinese territory. Also, this organisational arrangement provides the Chinese government with strong resource scheduling. Numbers of inter-regional resource support and mobilisation prove its capacity. In addition, the special neighbourhood management in China equips the state with in-depth population mobilisation to cooperate with the government's pandemic control policy. Thirdly, the zero COVID strategy helps China keep a very low number of confirmed cases and death cases, giving the state and the public a great incentive in maintaining this strategy, especially in the competition on the state system with many Western countries under the Liberal approach. Conversely, China government's practices on its state monopoly autonomy (decision-making), holistic executive capability to employ zero COVID strategy within Chinese territory and resource scheduling (state capacity), and the capability to ensure adequate resource in maintaining the zero COVID strategy and popular mobilisation (consent to state) tender itself motivation, legitimacy, and support in making "zero COVID strategy and maintain it.

Conversely, the Liberal approach conveys a different idea on state. We detect that the Liberal approach does not consider the state own power naturally. Power is conferred by the people and society through a set of social contracts, such as division of power between the governed and the governor, and the arrangements like elections. The power's transfer from people to state is also limited. "Live with COVID" strategy is therefore implemented under such political considerations. The chapter illustrates this point by using the example of the United Kingdom.

The UK government's power to make pandemic control policy is granted by the UK Parliament, formed by members through popular elections. The UK government made the travel restriction decision based on its emergency power and the circumstances of the overwhelming state medical resources (NHS). People supported it in the preliminary state and provided trust to the government. However, the support and trust lost due to the government and the long-time restriction on citizens' freedom. Protests against lockdowns broke out nation-wide, and people started questioning the rationale of the restriction decisions. State capacity and legitimacy are being challenged. For the Liberal approach, the government soon gave up the lockdown and established the "live with COVID" strategy, developing and rolling out vaccination and medication treatment. The new policy removes most restrictions and deploys more specific services and measures in exchange of citizens' consent to policy (source of state legitimacy). Therefore, the indicators of mass recognition to state-provided public goods and services, and the procedural fairness link directly to the consent of people to the state, and the legitimacy of the state.

References

Anter, A. (2014). *Max Weber's theory of the modern state. Origins, structure and significance.* Palgrave Macmillan.

Aristotle. (1995). I.2. In T. J. Saunders (Ed.), *Politics: Books I and II* (1st ed.) (p. I252). Clarendon Press.

Cabinet Office, T. U. G. (2022). *Guidance: COVID-19 response: Living with COVID-19 live with COVID.* [Online] Available at: https://www.gov.uk/government/publications/covid-19-response-living-with-covid-19/covid-19-response-living-with-covid-19#covid-19-data-and-impacts. Accessed 6 Aug 2022.

Calvert, J., Arbuthnott, G., Leake, J. & Gadher, D. (2020). *22 days of dither and delay on coronavirus that cost thousands of British lives.* [Online] Available at: https://www.thetimes.co.uk/article/three-weeks-of-dither-and-delay-on-coronavirus-that-cost-thousands-of-british-lives-05sjvwv7g. Accessed 10 Aug 2022.

CCTV. (2021). *"Dynamic zero clearing" is not absolute "zero infection": Pandemic prevention and control measures will be continuously optimized and upgraded according to the situation* ("动态清 零"不是绝对"零感染" 将因时因势对防控措施不断优化升级). [Online] Available at: https://news.cctv.com/2022/04/29/ARTITleHrhLcJpbkuk0tSq5R220429.shtml

Davies, B., et al. (2021). Changes in political trust in Britain during the COVID-19 pandemic in 2020: Integrated public opinion evidence and implications. *Humanities & Social Sciences Communications, 166*(8), 1–9.

Department for Transport, U. G. & Public Health England, U. G. (2020). *COVID-19: Guidance for staff in the transport sector.* [Online] Available at: https://www.gov.uk/government/publications/covid-19-guidance-for-staff-in-the-transport-sector/covid-19-guidance-for-staff-in-the-transport-sector. Accessed 10 Aug 2022.

Economist Intelligence Unit, T. E. (2021). Asia's transition away from "zero COVID". *The Economist.*

Economist, T. (2021). *Asian countries are at last abandoning zero-covid strategies.* [Online] Available at: https://www.economist.com/asia/2021/10/09/asian-countries-are-at-last-abandoning-zero-covid-strategies. Accessed 6 Aug 2022.

Edelman Holdings, I. (2022). *Edelman Trust Barometer 2022,* s.l.: Edelman.

Evans, P. B. (1995). *Embedded autonomy: States and industrial transformation.* Princeton University Press.

Evans, P. B., Rueschemeyer, D., & Skocpol, T. (1985). *Bringing the state back in.* Cambridge University Press.

Gellner, E. (1983). *Nations and nationalism.* Cornell University Press.

Halliday, S. et al. (2022). Why the UK complied with COVID-19 lockdown law. *King's Law Journal,* Vol. 2022.

Hau, M. . V. . (2016). State theory: Four analytical traditions. In S. Leibfried (Ed.), *The Oxford handbook of transformations of the state* (pp. 131–151). Oxford University Press.

JHU, J. H. U. (2022). *Johns Hopkins university coronavirus resource center.* [Online] Available at: https://coronavirus.jhu.edu/map.html. Accessed 8 Aug 2022.

Johnson, B. (2020). *Boris Johnson: Britain must become the superman of global free trade.* [Online] Available at: https://www.spectator.co.uk/article/boris-johnson-britain-must-become-the-superman-of-global-free-trade. Accessed 10 Aug 2022.

Levi, M. (1988). *Of rule and revenue.* University of California Press.

Liang, W., et al. (2022). The dynamic COVID-zero strategy on prevention and control of COVID-19 in China. *National Medical Journal of China, 102*(4), 239–242.

Lintern, S. (2020). *Coronavirus: 60% of UK population need to become infected so country can build 'herd immunity', government's chief scientist says.* [Online] Available at: https://www.independent.co.uk/news/health/coronavirus-herd-immunity-uk-nhs-outbreak-pandemic-government-a9399101.html. Accessed 10 Aug 2022.

Meng, Y., 2020. *The number of nucleic acid testing institutions in Beijing reached 67 (北京核酸检测机构数量达67家)*. [Online] Available at: http://www.xinhuanet.com/politics/2020-05/12/c_1125971501.htm. Accessed 6 Aug 2022.
Murphy, S. (2021). *Priti Patel says she wanted UK borders shut last March.* [Online] Available at: https://www.theguardian.com/politics/2021/jan/20/priti-patel-says-she-wanted-uk-borders-shut-last-march-over-covid. Accessed 10 Aug 2022.
O'Connor, M. & Cursino, M. (2022). *Covid: Living with covid plan will restore freedom, says Boris Johnson.* [Online] Available at: https://www.bbc.com/news/uk-60455943. Accessed 12 Aug 2022.
Özdüzen, Ö., Ianosev, B. & Ozgul, B. A. (2021). *Freedom or self-interest?: Motivations, ideology and visual symbols uniting anti-lockdown protesters in the UK.* [Online] Available at: https://www.psa.ac.uk/psa/news/freedom-or-self-interest-motivations-ideology-and-visual-symbols-uniting-anti-lockdown. Accessed 10 Aug 2022.
People's Daily, P. (2021). *Win the war against the pandemic, The Communists led the charge (打赢疫情防控阻击战 共产党员冲锋在前)*. [Online] Available at: http://cpc.people.com.cn/GB/67481/431707/. Accessed 6 Aug 2022.
Peston, R. (2020). *Response from senior government source is "the Italians did several of the populist - non-science based - measures that aren't any use. They're who not to follow"*. [Online] Available at: https://twitter.com/Peston/status/1236739529303830530 Accessed 10 Aug 2022.
Silver, A. (2021). Covid-19: Why China is sticking to "zero tolerance" public health. *BMJ, 375*(n2756), 1–3.
Simmons, A. J. (2006). Theories of the state. In D. Rutherford (Ed.), *Early modern philosophy* (pp. 250–273). Cambridge University Press.
Sobhan, S. (2022). *What is China's zero-COVID policy?*. [Online] Available at: https://www.dw.com/en/what-is-chinas-zero-covid-policy/a-61736418. Accessed 8 Aug 2022.
UK Government, U. G. (2022). *How government works.* [Online] Available at: https://www.gov.uk/government/how-government-works. Accessed 6 Aug 2022.
UK Parliament, U. P. (2020). *Coronavirus act 2020.* [Online] Available at: https://www.legislation.gov.uk/ukpga/2020/7/section/100. Accessed 6 Aug 2022.
Weber, M. (1978). *Economy and society: An outline of interpretive sociology.* University of California Press.
Wimmer, A. (2002). *Nationalist exclusion and ethnic conflict: Shadows of modernity.* Cambridge University Press.

Chaofan Chen studies Globalisation and Development at Beijing Normal University-Hong Kong Baptist University United International College. Chen's main research interests include International Relations, Diplomacy, Chinese Politics, Chinese Foreign Policy and Comparative Politics. Chen has founded the Sustainable Development Research Group at UIC, an organisation that promotes student research on UN SDGs and global governance. Chen is also a member of World Ocean Day Youth Advisory Council since September 2022.

Jiawei Chen studies globalisation and development at BNU-HKBU UIC. Chen's main research interests focus on geopolitics and global governance issues. Chen's collective research project on the successful development experience of small countries was supported by the college faculty. Chen also actively participates in government research; In 2022, he partook in the investigation of public infrastructure resources in Foshan City.

Bo Yan received Ph.D. in History from Nanyang Technological University, Singapore. He is a Research Fellow in the Faculty of Humanities and Social Sciences. Yan's main research interests include modern Chinese history, postwar United States, and modern Southeast Asian history, and the history of the global 1960s. He has published in the *Journal of Asian Studies, Twentieth*

Century China, and others. Yan's Ph.D. dissertation was listed as the International Convention of Asia Scholars (ICAS) Book Prize 2021-Best Dissertation in Humanities. Bo Yan is currently conducting studies and research on global pandemic of COVID-19.

COVID-19 Pandemic as the Cosmopolitan Moment of World Risk Society

Siu Han Chan

1 Introduction: COVID-19 Pandemic and a New Beginning of the 21st Century

The new coronavirus COVID-19 and the public health crises associated with it have thrown the whole world into great disarray. The most important lesson the pandemic teaches us is normal is after all not that *normal*. Things we are doing every day, things we take for granted can be ruthlessly interrupted overnight. We may wake up with our "natural attitude," to borrow the concept from social phenomenology, towards the world shattered. Borders were closed and cities were locked down without forewarning. Transnational or even trans-city mobility was suddenly called to a halt. Production activities suspended and diesel oil prices even fell below zero. "Social distancing" was mandatory and face-to-face interaction was greatly limited or even prohibited. The online space slipped in to fill the lacuna and became the domain we conducted most of our daily routines, from school to work, from marriage to parenting, from information to grocery-shopping. Our world has been transformed by this very pandemic—most likely permanently. But to what end?

Is our world going to become utterly different from before as Arundhati Roy (2020) suggests that the pandemic "is a portal, a gateway between one world and the next?" Are we going to live in a "post-normal" world in which the "normal" state of being is being perpetually called into question? When will a "new normal" arrive, if it would arrive at all? (Matthewman & Huppatz, 2020; Sardar, 2019; Zinn, 2021) No one has a definite answer to these solemn questions. But sociologists usually believe our future is relatively open, depending upon both our imagination and collective will to it. Our world has been reset yet again by the pandemic, only two decades after

S. H. Chan (✉)
Beijing Normal University-Hong Kong Baptist University,
United International College, Zhuhai, China
e-mail: siuhanchan@uic.edu.cn

© The Author(s), under exclusive license to Springer Nature Singapore Pte Ltd. 2023

S. X. B. Zhao et al. (eds.), *Comparative Studies on Pandemic Control Policies and the Resilience of Society*, https://doi.org/10.1007/978-981-19-9993-2_8

we entered the new millennium and projected our millennial hopes. Not unlike the collapse of the World Trade Centre in New York in 2001, the first media stunt of the new era, COVID-19 pandemic is a catastrophe that obliges all of us to rethink the future of humanity. Just like any catastrophe, it is a nadir of a society and an abrupt end to a particular state of being. But it also spells the arrival of a new, better or worse, state of being. It is now the time we start to (re)imagine another new beginning of the twenty-first century, and to reflect on the emerging formation of world order after the pandemic shock.

One salient theme in this regard is the future of our globalised world as a whole: (Re-)rigidification of national borders, the disintegration of pre-pandemic global supply chain, the strengthening of the role of the state in pandemic control and thereafter, and so on all point to the reverse process of globalisation. Nonetheless, I argue that COVID-19 pandemic has not only not reversed globalisation, but brings forth an ever more globalised world. The pandemic is indeed the first palpable *global* phenomenon in (modern) human history (Chernilo, 2021; Hanieh, 2022). The *globality* of the pandemic describes not just its variegated impacts on a planetary scale. More importantly, the COVID-19 pandemic lays the foundation for a fundamental transformation of collective consciousness about the globe we are inhabiting. The pandemic functions as "global risk," other than climate change, financial risk and terrorism as Ulrich Beck predicts, which plagues and connects the whole world simultaneously. It portends the arrival of "world risk society" more than 20 years after Beck first introduced this idea into the English-speaking academic discourse (Beck, 1999). The pandemic demarcates an opportune moment that Beck calls "risk society's 'cosmopolitan moment'" that world society transfigures fully into "world risk society" (Beck, 2008, 2009). It is a belated moment Beck himself missed narrowly by a few years because of his precocious death in 2015.

This article will engage with Ulrich Beck's theory of world risk society and interpret COVID-19 pandemic as the episode that instigates an "anthropological shock" necessary for a wide recognition of world risk society, i.e., the world as composing of "cosmopolitan communities of global risk" (Beck, 1999, 2008, 2015, 2016). In the following, I will first discuss the deglobalisation thesis as both *descriptive* and *reflexive* response to the deepening of world integration, especially one that is the driven by neoliberalism. Next, I will explicate the relation between global risks and world risk society in Beck's sociology before going on to interpret COVID-19 pandemic as the "cosmopolitan moment" of world risk society. I will shed light on the *pessimistic optimism* of Beck's sociology and world risk society in the concluding part.

2 Global Reflexivity: Deglobalisation as Descriptive and Reflexive Response to Deep Globalisation

The idea of deglobalisation is gaining attention in both public and academic discourses over the last two decades. The outbreak of the COVID-19 pandemic further casts the idea into the limelight. The deglobalisation thesis argues that a different image of the global world is taking shape—one that is less interconnected, and is likely to be the new normal state of being even after the end of the pandemic (Haass, 2022; Kornprobst & Paul, 2021; Nigrinová, 2022). If deglobalisation is taken as a *descriptive* term referring to the decrease in international economic volume and connection, quite a lot of corroborating evidences suggest it is happening during the pandemic and will likely persist for some time. The Davos World Economic Forum, the staunch supporter of economic globalisation and neoliberal ideology, advocated that our world should be "preparing for deglobalization" after the pandemic (Baker et al., 2021). Other phenomena also appear to support the reverse of the globalisation process is taking place—national border and nationality become real and consequential again after we imagine our world as a "borderless world" of free-flowing goods, information, capital and people, and Europeans to be living in the "post-national constellation" and the whole population forms a global citizenry for long (Habermas, 2001; Nussbaum, 1996; Ohmae, 1992; Parekh, 2003).

National vulnerability exposes when the global production chain becomes fully efficient and cost-effective. The intense competition for strategic medical materials like ventilators, masks, and vaccines during the initial period of the pandemic makes governments more aware of the potential risks of economic and production interdependence to national security and autonomy. The global supply chain will be revamped with strategic considerations and a sizeable part may take the form of back-shoring or near-shoring after the pandemic (Alicke et al., 2021). Politically, different states deploy draconian measures and declare state of emergency one after the other to deal with the pandemic crisis. In the process, they claw back much of the power they lose to the "runaway world" of globalisation, to borrow Anthony Giddens (2000). Giorgio Agamben (2020a) even controversially argues that the overreaction of different states to a flu-like virus reflects their "tendency to use a state of exception as a normal paradigm for government." The concerted political effort in staging the "invention of an epidemic" around the world has driven us crazy, and believed in nothing but "bare life" at the very early stage of the coronavirus outbreak. (Agamben, 2020a, 2020b). Though Agamben is widely reprimanded for underestimating the nature and impacts of the coronavirus outbreak, the forewarning he issues about the (re)centralisation of powers, especially in democratic societies, is hardly just paranoid.

The attenuation of place, border, state are important markers of globalisation. It seems nonetheless placeness, "the return of borders to a borderless world," "the return of the state" are in place, especially after the pandemic (Allen et al., 2021;

Ferguson & Mansbach, 2012). Besides being a descriptive concept, deglobalisation is, more importantly, invoked as a critical and *reflexive* concept to globalisation, particularly the neoliberal strand that sweeps through the world in the last few decades and culminates in the twenty-first century. Madhi Elmandjra (2000) and Walden Bello (2004, 2009, 2013) for example have long been arguing for "the need for a 'deglobalization' of 'globalization.'" They actively promote the "virtues of deglobalization" in hope of counteracting the hazards of the neoliberal globalisation started in the 1980s, including prevalent economism, hierarchisation, unilateralism of the world, and empowering the locals, the communities, and societies.

For several decades, many people and governments regard globalisation as bringing net benefits, and are generally content to let it run its course. But the darker side of globalisation soon emerges as the repercussions of deepened global integration and interdependence between places and peoples. The idea of deglobalisation thus enters into our consciousness in the twenty-first century as a form of *global reflexivity*.[1] The medium of *reflexivity*, as Beck (1992b) underscores, is *unawareness* and *non-knowledge*, as in *reflexes* that do not require the mediation of consciousness. Followed from Beck's line of reasoning, global reflexivity describes the process of "self-confrontation" of global modernity with the unintended and suppressed consequences—the global risks, it produces in the course of its development. Global reflexivity is by nature an integral part of the actions of the global system taken as a whole. One way or another, sooner or later, reverberations of actions and decisions of the system will come back. They are just bracketed and deferred in the first modernity. The practical implication of the self-confrontation of global modernity is that things no longer develop in the linear manner as people expect in industrial modernity. *Reflection* on the other hand refers to the possible *awareness* and *knowledge* of the consequences of modernity as risks and the inherent dynamism of modern society, and the latter's transformation into (world) risk society (Beck et al., 1994; Beck, 1992a, 1992b, 1999). Beck's theoretical project thus endeavours both to capture the reflexive development of (global) modernity and bridge the gap between *reflexivity* and *reflection*.

As a form of global reflexivity, deglobalisation discourse is primarily about the self-confrontation of the triumph of neoliberal globalisation with its variegated intertwined economic, political and social outcomes. The pillar of neoliberal globalisation is first and foremost global trade and international economic division of labour. Transnational trade streamlines global supply chain. It also provides better-paying jobs in export-oriented factories and agricultural sectors in developing economies, and more affordable consumer goods with higher quality in developed societies. International capital increasingly dis-embeds economy from local society, but leaves the cost of improved production and distribution capabilities to be borne by the

[1] Global reflexivity is not a term Beck directly deploys, but is extrapolated from his writings. Towards the later stage of his career, Beck has shifted his theoretical focus and does not talk much about reflexivity.

majorities who are relatively powerless and stifled by it. The invisible hand of global market and the open policy of neoliberal states work together to make the lower class accept "desocialized wage labor" and the instability the latter brings in its wake (Wacquant, 2009a). Fragmentation of labour and flexibilisation of work are more prominent in the last few decades, turning the Fordist regime of permanent, full-time, socially secure employment to a risk regime of work (Beck, 2000a). Economic neo-liberalism's attack on big government also works to dismantle the social safety net, striking even the welfare states (Abrahamson, 2010). Widespread feeling of instability and insecurity is looming in many societies, especially in the central zone of globalisation—Europe and America, leading to a sea change in the political and social landscape (Wacquant, 2009a, 2009b).

Needless to say, prevalent feeling of economic insecurity in society is the hotbed for right-wing thought, and breeds political distrust. Rightist politicians, who advocate anti-immigrant agenda and instigate xenophobic sentiments, are elected in different parts of the world. Populist or even extreme right-wing parties are on the rise even in the more affluent part of Europe (BBC, 2019; Millman, 2021). The victory of Donald Trump in the 2016 U.S. presidential election is also symptomatic of widespread economic and social insecurity felt in the American society (Green, 2017). The rising right-wing nationalism and Eurosceptic sentiment challenge further integration and solidarity of Europe (de Wilde et al., 2019). The Brexit campaign exemplifies squarely the anxiety of the British people about the adverse consequence of neoliberal globalisation—the feeling that they are losing control of their own lives and society—unfortunately projected onto Europe (Elliott, 2016; Tsoukalis, 2022). Neoliberal globalisation is a powerful force reshaping the political terrain of Western democracies, and producing less tolerant, fraternal and inclusive societies (Beck, 2002a; Duggan, 2003; Giroux, 2008).

More detrimental still is, as Loic Wacquant astutely points out, the changing nature of Western society. America is quickly turning from a "social state" into a "penal state" (Wacquant, 2001, 2009a). In the last three decades, more and more Americans are losing their stable jobs and falling out of the social security net. Lack of means of life and social security lead to increase in petty crimes like robbery, thief, burglary and so on. Stricter punitive measures are installed as a response to the general rise in crime rate. American prisons have almost turned into the "surrogate ghetto, encaging the black subproletarians" and the lower stratum who need to be cared by the society (Wacquant, 2009b). "Criminalization of poverty," particularly the coloured people, contributes to major social cleavages, and underlies general hostility to the law enforcement. A penal society tends to produce more police brutality and higher level of distrust of public institutions among civilians. This may well explain why the pandemonium sparked by the wrongful death of the African American George Floyd, in the midst of being caught using an alleged counterfeit 20 dollar note in 2020, quickly escalates into the *Black Live Matters* movement. The Black Live Matters movement for certain discloses the inherent racism in the American society. But the entrenched racial contradiction is severely compounded by the growing economic and social frustration induced by the neoliberal mode of globalisation. Still more unfortunate is American "penal common sense" is also exported to other parts of the world, plaguing even the European states with strong leftist and republican traditions, like France and Norway (Glotova & Raoult, 2017; Shammas, 2016; Wacquant, 2009a, 2009b).

Entering into the twenty-first century when all the changes along economic, political and social planes come back as reflexive process to the modernisation of the global social system, people increasingly find themselves leading lives departing greatly from the classic image of industrial modernity. The global reflexivity of the (global) modern society does not automatically imply the recognition and reflection of the reality of the "self-dissolution" of modern society and the sense of certainty and predictability that comes with it (Beck, 1999; Beck et al., 1994). But the two are closely related and global reflexivity will eventually lead to higher awareness of the profound societal transformation in the longer run. Denial or apathy to the changing human condition is often the common reaction when sociological imagination is absent (Beck, 2012; Mills, 1959). Hence, sociologists are usually the people who herald a new existential condition. Sociologists of globalisation have variously captured the dynamics of the globalisation process (Ritzer, 2007). Ulrich Beck is but *the sociologist* who unambiguously *names* our increasingly globalised world "world risk society."

3 Global Risks and the Rise of "World Risk Society"

Sociologists began to talk about new existential condition of the late(r) stage of modernity in the 1960s. Various notions come into intellectual sights, including post-industrialisation, post-modernity, post-history, post-materialism. In Beck's somewhat exaggerated expression: "'post' is the expression of sociological helplessness," and its message is "as excited and exciting as it is empty" (Beck, 1992a, 1999). The exciting part about the various "post-isms" is they entail the dissociation from a more well-defined social formation and a relatively definite and restrictive state of being. Yet, on the other hand, "post-ism" also connotes the uncertain feeling about the transition and transformation of our futures. Interpellation of the emerging condition is important to give our society a new identity, a new beginning and an appropriate orientation.

Contrary to various ambivalent "post-isms," Beck ventures to name the late modern society straightforwardly as "risk society" (Beck, 1992a, 1992b). Living in a risk society does not necessarily mean that our lives have become more dangerous than before. We are just more aware of the potential threats. Danger, threats and other forms of "otherness" are the things the first modern society attempts to keep at bay and obscure to maintain the facade of personal safety, and to construct social order and the sense of orderliness and security in social lives. Towards the end of the twentieth century, the suppressed in the first modernity returns reflexively as risk, "a product of the successes of civilization" (Beck, 2009, p. 4). Risk infuses the "normal" state of being and is the repercussion of regular life practices, a constituent of normal life we have long (un)wittingly overlooked. Risk is now the central structuring parameter of the emerging social order of the "second age of modernity," as Beck (1992b, 2000a, 2000b) puts it. It is something we can no longer ignore without suffering potentially devastating outcomes. Our society is hence not experiencing a further (linear) extension of modernisation, guided by the central value of social "progress." Its development is but, as Beck emphasises, the "modernization of modernization," or "reflexive modernization," a process as involuntary as reflex action, triggered by the risks produced by the first modernity (Beck et al., 1994; Beck, 1992a, 1992b). Reflexive modernisation represents a breakthrough from the *first* to the *second* modernity. Our society no longer takes the *pre-modern* as the contrasting image, but is transcending the condition of the first modernity, and transforms its organisation form and principles. Beck later coins a new term "metamorphosis of the world" to gauge the nature of such transformation (Beck, 2015, 2016). As prefix, *meta* connotes both transformation and a higher stage of development. *Morph* means form and structure. The etymological implication of *metamorphosis* is appropriated by Beck to underscore the transformation of the form and structure of the second modern world towards a more advanced stage of development that follows non-linear principles.

Beck argues that the risk logic of the second modernity is going to override the wealth logic of the first modernity. Wealth and risk follow rather different distribution principles: the inequality of wealth versus the eventual equalisation of risk. In the first modern society, social polarisation and concentration of wealth is the norm. In a risk society, risk, like radiation and pollution, will in the long run even out its impacts to all (Beck, 1992a, 1992b). Even though wealth and power are still the basis for minimising risks in risk society, as is evident in acquisition of medical supplies and vaccines during the COVID-19 pandemic, risk is at least a compatible consideration to wealth—an important utility of wealth is risk avoidance. Wealth may not fully eliminate risk, but can defer risk. However, in the last instance, risk comes back to those who produce or profit from it in the first place, eck calls the "boomerang effect" (Beck, 1992b).

The vivid "boomerang" image and the "debounding" of uncontrollable risks along the spatial, temporal and social dimensions explain why the basic unit of risk society is not one society nor one nation (Beck, 1992b, 2002b). It is the "world risk society" with all societies added and bound together by risks. "World risk society" is a theoretical development of Beck's "risk society" to the evolving reality of greater interconnection and integration of the global system, following the conclusion of the Cold War. Multi-directional and multi-dimensional interpenetration of influence and risks among nations is the way Beck interprets globalisation. "Globalisation" appears to Beck as a misleadingly abstract and simplistically linear idea that explains changes as originating from the global, and sees the global and the local as "cultural polarities" (Beck, 2002a). He prefers seeing globalisation as a "non-linear and dialectic process" of "cosmopolitanization" with different nations and localities function as "combined and mutually implicating principles," which transform the fabrics of everyday life in a national society into cosmopolitan (Beck, 2002a).

Widespread recognition of global risks will be the "cosmopolitan moment" of world risk society. In world risk society, the cosmopolitan perspective, "an imagination of alternative ways of life and rationalities, which included otherness of the other," reigns over the national perspective of the first modernity (Beck, 1999, 2002a, p. 18; 2006, 2008).

> Risk society's cosmopolitan moment means the *conditio humana* of the irreversible non-excludability of the distant and alien other. Global risks tear down national boundaries and jumble together the native with the foreign. The distant other is becoming the inclusive other—not through mobility but through risk. Everyday life is becoming cosmopolitan: human beings must find the meaning of life in the exchange with others and no longer in the encounter with like. We are all trapped in a shared global space of threats without exit. (Beck, 2008)

Global risks force us to rediscover the repressed dimension and otherness of the first modernity, and realise the interdependence of our existence. They are the *mediation* of world risk society. The world of otherness and enemies, in which we attempt to keep the others away from violation and contamination, has given over to the world of sameness and risks—the collective vulnerability to risks of the second modernity (Beck, 1999). Human beings are no longer bound together by shared aspiration for future prosperity or the normative idea of *Humanity*, but by the common *challenges*, *threats* or even *fate* in world risk society. The emergence of the anthropological condition of global risk, and, more importantly, the collective recognition of such new condition connects us all.

The three major types of global risks Beck perceives to be precipitating the cosmopolitan moment of world risk society are climate change, crisis of financial market and international terrorism (Beck, 2002a, 2002b, 2015, 2016). They are all transnational in nature. Precaution or solution to them can by no means be found in the effort of one or few nations. Closing the national border for example does not stop pollution from spreading around the globe as an ecological system. No established financial market can be fully shielded from the domino effect of financial turmoil from the Asian Financial Crisis in 1997, Financial Tsunami in 2008 and the 2009 Euro Crisis. Terrorism and international terrorist networks cannot be fully kept out from homeland even with stringent border controls. Second generation immigrants or even nationals can be in touch with foreign terrorist groups and sympathetic with their cause (Beck, 1999, 2002a, 2002b). As seen from the European standpoint, the three types of global risks are cataclysmic enough to provide convergent focus for the rise of world risk society. They nonetheless have not been functioning adequately as *the* cosmopolitan moment as Beck predicts. These threats are hitherto either being deferred, averted or remain skewed in distribution that fail to incite the fateful sense of sameness and collective vulnerability, indispensable for invoking cosmopolitan consciousness at the planetary level.

Scientists, environmentalists and other relevant claim-makers may have successfully convinced us about the potential devastating effect of ecological risk. It is high on the agenda of the world, but the threat is not yet immediate enough and evenly felt by all. Many still believe climate change is going to take long time to materialise and irreversibly overturn our way of life (Unger, 1992, 2012). Besides perceived non-imminence, uneven distribution of contribution, responsibility, and adverse effect of climate change prevents it from serving as a unifying factor for the world population. On the one hand, major discourse about limiting emission of pollutants is driven by the affluent world, which is less directly affected by climate change thus far. On the other hand, countries in the global south are worst hit and their people concern more about the impacts of climate change (Stokes et al., 2015). They however find it hard to protect themselves because they have limited influence over climate policies (Khadka, 2021). In front of ecological risk, common ground and interest at the global level are still far from sight.

For the risk of global financial market, it appears to be a global phenomenon. But its globality is again not globalised enough. Despite financial meltdown sends ripple down to different corners of the world, globalisation of financial turmoil follows centre-periphery dynamics. The shockwave from the financial centres is usually much more powerful than that from the emerging markets, unless they affect asset markets of the centres. The spillover of financial crisis from the periphery is relatively regional in nature (Kaminsky & Reinhart, 2003). Moreover, financial turbulence affects economies better integrated to the global market far more than those with relatively low level of financialisation or excluded. Even though it is found that the different waves of financial crises have speeded up the financialisation of the periphery states and fostered greater financial globalisation (Barradas et al., 2018; Rato, 2020; Santos & Teles, 2021). A more integrated global financial market for certain implies greater financial risk and instability in the future. Further financialisation at the present stage indicates that financial risk can at least be partially averted.

For terrorist risk, international terrorism or "superterrorism" with global reach, organisational network, agenda and supranational ideologies is ascending. But terrorism remains relatively regional in perception and in scope at the moment. Terrorism is fundamentally asymmetric violent conflict and political in nature. It is bred by major power and status asymmetries and ideological and structural disparities in the encroachment of powers. From the perspective of the perpetrators of terrorist acts, who are always the disadvantaged party in a lopsided power relation, extreme violence is the repayment of violent extremism inflicted by the powerful that other forms of conflict resolution cannot do justice (Stepanova, 2008; Vertigans, 2011). Taking into consideration of the fundamentally political and asymmetrical nature, global terrorism affects more the (inter)national hegemons and power-wielders and much less those who are powerless to begin with. The developed world is nonetheless not the worst victim of terrorism in terms of casualty, perhaps because of better intelligence and counter-terrorist squads and networks. But they are definitely the most vocal group and prepared to use terrorism as pretext to widen political imbalance.

Beck is aware that global risk does not bring about homogenising effect nor are the undesirable consequences evenly distributed.

> But that doesn't change the principle that everyone is affected, and everyone can potentially be affected in a much worse manner. Thus…these problems endow each country with a common global interest…we can already talk about the basis of a global community of fate. (Beck, 2002b, p. 42)

It cannot be denied that climate change, economic crisis and international terrorism are all consequential as global risks, and will eventually threaten everyone on earth. But global risks along the axes of ecology, finance or terrorism are in the main bothering Europeans much more than non-Europeans at the moment. As global risks, they are hitherto *divisive*, manifesting and reproducing the hierarchy and uneven distribution of power of the world, instead of being *unifying* in causes among different parts of the world (Giddens, 2009; Harrison & Sundstrom, 2010; Stehr & von Storch, 2009). The global risks Beck discusses are hitherto all far from transgressing the social structure of the world, and serving as the baseline of social integration of the global social system. The "cosmopolitan moment" of world risk society thus has yet to transpire as Beck's theory predicts. It should be noted that Beck's cosmopolitan imagination is still blinkered by European, if not Eurocentric, inflection. In other words, Beck is still not cosmopolitan enough when he forecasts financial crisis, climate change, and terrorist threats, the engrossing concerns of Europe, to be the momentous turning point of world risk society (Robertson, 2012).

Nonetheless, Beck has succinctly formulated the general condition for the rise of world risk society, and the global risks Beck discusses at length all have hastened the cosmopolitan consciousness of the world subsequently (Beck, 2002b, 2008, 2015, 2016). What Beck does not correctly anticipate is the ultimately rallying other comes from the microscopic level. Our microbiological enemy—the coronavirus—does bring about a new form of cosmopolitan consciousness.

4 COVID-19 Pandemic as the "Cosmopolitan Moment" of World Risk Society

The potential globality of pandemic is inscribed in the etymology of the word *pandemic*. *Pandemic* comes from the Greek word *pandēmos* meaning literally all the people. To borrow a quote Giorgio Agamben takes from Thucydides' *History of the Peloponnesian War*, "the plague marked for the city the beginning of corruption... No one was any longer disposed to persevere in what he had previously judged to be the good, because he believed that perhaps he would die before achieving it" (Agamben, 2020b). In front of death, survival outweighs all other risk and moral considerations. The fact that the COVID-19 pandemic reaches deep down to the very baseline of collective survival of the world population as a whole, not just some pockets of them, nor merely their well-being, warrants it to be the "cosmopolitan moment" of world risk society.

The planetary implications of the coronavirus as a global risk should be discussed at greater details. First, it is the worldwide *reach* of the COVID-19 virus along the social, cultural and geographical lines. The coronavirus definitely respects no established division of our world. Socially, the coronavirus breaks down the barrier of class. It infects indiscriminately the top leaders and echelon of the world, from the

British and Danish Queens and Spanish King to the British Prime Minister and American President, and the bottom of the society, the outcast in slum areas. Culturally, the virus cuts through cultural lines and plagues both the most *civilised* groups living in metropolitans and the indigenous peoples down to the Brazilian Amazon basin, leaving no corners of the world immune (Kim, 2021; Nuorgam, 2022). Geographically, the coronavirus travels from the more liveable subtropical areas to the Arctic and even the Antarctica (Godbold, 2020). At the same time, global *inclusivity* is also the feature of the spread of the various variants of the coronavirus that sustains the momentum of the pandemic. The evolving variants of SARS-CoV-2 from Alpha, Beta to Omicron are first found in both the less and more developed parts of the world (Otto et al., 2021; WHO, 2020). This powerful zoonotic disease has revoked our inveterate image that deadly viruses like Ebola, Zika originate and widely circulate in the relatively underdeveloped, less hygienic areas. Different parts of the world are making *fair contribution* to the continuous spread of the COVID-19 pandemic and death toll. Some observers even argue that the coronavirus is an "epidemic of the affluence" that is why it catches the world's attention instantaneously and containment measures are immediately put in place (van de Pas, 2020). The disease of the affluence hits the poor no less heavy and drives them out of jobs and deepens social inequalities.

The COVID-19 pandemic also provides a telling example of the not-so-subtle difference between *interdependence* and *interconnection*, and the two may at times be driving divergent directions without compromising global integration. The interconnection between places and the flow of people and goods are largely suspended by the various pandemic-related immigration restriction and quarantine policies, especially during the first two years of the pandemic. Yet, the interdependence of the global system as a unit per se, which different constituent components are interacting with each other and mutually implicating, at least when the contagion of the virus and the fighting of the pandemic are concerned, becomes more apparent. COVID-19 pandemic as a global risk expresses a new form of global interdependence that we are not completely aware of before—our health is mutually dependent. The pandemic cannot be quelled at one country or one region, without spilling over to others. The end of the pandemic and the popular spread of virus can only be achieved at a global level with literally the inputs of all.

Most importantly, what certifies COVID-19 pandemic as a global risk is that it may well be the very first global phenomenon, a "global total social fact" that is impinging upon the world population, that all people are actually aware of collectively and experiencing simultaneously. All social institutions in the world are brought either into movement or a standstill unprecedentedly for the similar set of reasons (Vandenberghe & Véran, 2021). For more than two years, despatialised simultaneity, Benedict Anderson (2006) deems essential for the formation of national imagined community, has extended to the globe and produced a realistic sense of collective fate at the global level, turning the whole world into a global imagined community. The pandemic has affected everybody on this planet in some ways and synchronised collective experience, no matter one has caught the disease or not. The COVID-19 pandemic is thus the very first truly *collectively experienced global illness and risk*.

The pandemic "brings home the realisation that we have *reached the globalisation of the very globalisation processes that started in the 1990s*" (Chernilo, 2021, p. 164). While we have been talking about globalisation for decades, the COVID-19 pandemic is the epiphany of the profound *globality* of the global world.

More than other global risks, the COVID-19 pandemic, as Beck does not expect himself, presents as the other that recklessly "violates *sacred* (unwritten) norms of human existence and civilization" and causes "anthropological shock" (Beck, 2015, 2016). The shockwave transcends national boundary and bonds the world together into one plague-ridden planet. The pandemic exposes the vulnerability of human species as *one* species. For many centuries, nation carries the divine mission of (re)producing society, transforming lives with collective sense of national purpose by linking them to collective past and future. It is the furthest stretch of collective consciousness that the world population is used to. Going beyond nation requires level of abstraction and imagination that most would find unfathomable (Shils, 1997; Weil, 2002). It takes anthropological shock with the magnitude of the COVID-19 pandemic for the transgression of the threshold of national consciousness. The collective experience and the devastating impacts the pandemic brings catalyse the emotional impacts of this global epidemic risk in people's perception and experience, such that collective energy of the world society converges, leading to the palpable emergence of a new modality of supranational—cosmopolitan—consciousness.

For the very first time, we are living not only in a world risk society *in itself*, but a world risk society *for itself*. The pandemic subjects the world population to certain form of common experience, suffering, focal attention and orientation. The whole world is compressed into "imagined communities of global risk" by the coronavirus (Beck, 2011). It is a form of, to use Beck's notion, "enforced cosmopolitanization," driven not by choice or cosmopolitanism in a normative sense. For Beck, cosmopolitanism and cosmopolitanisation are practical concepts expounding the interpenetration and influence of different nations and localities. Enforced cosmopolitanisation by the pandemic means we are compelled to take into considerations of what is happening outside the boundary of our national society as "global risks activate and connect actors across borders who otherwise don't want to have anything to do with one another" (Beck, 2009, p. 61).

During the pandemic, we come to recognise that we must act together to save ourselves. Individual and national self-preservation is far from enough to protect individuals and nations from infection. The building of immunity barrier, the key to fighting the pandemic and turning it into an endemic, depends on "herd immunity," also called "population community" or "community immunity," of not just a national society, but the planetary society, either by high level infection and the resulted immunity or vaccination (D'Souza & Dowdy, 2021; WHO, 2020). Herd immunity is a shared mission no nation can be spared or achieve alone. No nation can fully shield itself from a pandemic with high virality and quickly evolving virus variants. Pockets of immune communities can only be momentarily maintained before new, potentially more fatal variants from beyond the communities emerge and trespass the communities' carefully guarded boundaries. From an epidemiological point of view, only when the pandemic becomes endemic at different parts of the world will no new viral or fatal variants be plaguing. The most important message of the COVID-19 pandemic is we are a *herd* and we are one *community*, despite the divide in colour, class, wealth, gender, nationality, culture, and etc., in front of our microbiological enemy, something we almost forget. Immunity barrier is the ultimate *common denominator* that determines the fate of the human population before the collapses of global financial market, the massacres caused by international terrorism and the environmental disasters. The COVID-19 pandemic is hence the "cosmopolitan moment" and it is riper to talk about world risk society of collective fate and destiny in the post-pandemic time.

5 Conclusion: The Pessimistic Optimism of World Risk Society

At an earlier stage when Beck theorises world risk society, he focuses more on conceptualising risk as the "bads" produced by the "common goods" of modernity. Towards the later stage, he introduces the idea of "emancipatory catastrophism" to supplement his early vision. The essence of this latter introduction is common "bads" can generate the normative horizon of the "common goods" (Beck, 2015, 2016). This concept "emancipatory catastrophism" epitomises the *pessimistic optimism* of Beck's sociology and worldview, as he once describes himself as a "pessimistic optimist" (Beck, 1999, p. 8). Beck vividly illustrates the integrative function of catastrophe that we may see from global risk like COVID-19 pandemic: "Global risks have the ability to press-gang…an unlimited number of actors who want nothing to do with one another, who pursue different political goals and who may even live in incommensurable worlds," and bundle them together into "imagined communities of global risks" (Beck, 2009, pp. 191–192; 2011). Beck's vision of collective fate and vulnerability bringing people together in world risk society appears to be gloomy. Risk is taken as the mediation and *negative* force of social integration—people come together only when their collective survivals are at stake. For long, sociologists, particularly those working in the Durkheimian tradition, believe the basis of social integration and social order is normatively *positive* with shared values, sense of cultural identities, and collective conscience. Risk integration is thus a form of *asocial* solidarity, and implies a pessimistic view of humanity.

Beck comes from the German sociological tradition, which is well-known for its "tragic consciousness" of modernity from Georg Simmel, Max Weber, Karl Mannheim to Walter Benjamin (Lenk, 1987). The tragic consciousness of German sociology manifests itself in the deep realisation of the intricate nature of modernity and in its incessant quest for *clarity*, as is evident in Weber's (1978, 2004) sociology. In Mannheim's words, "German sociology is the product of one of the greatest social dissolutions and reorganisation, accompanied by the highest form of self-consciousness and self-criticism" (quoted by Lenk, 1987, p. 58). Working within this tradition, Beck's sociology is similarly imbued with an intensified tragic sense of self-consciousness of the complexity of modernity when he writes about the new human condition at the turn of the twenty-first century, which is permeated by global risk consideration. Moreover, Beck's theorisation also reflects the "complexity turn" of sociological thinking, in which linearity and predictability give way to non-linearity, convolution and uncertainty (Urry, 2005). As a result, Beck alerts us to the sober reality of complexity, an inevitability of our new anthropological condition, following the further development and deepening of modernisation.

Despite the prophetic realistic image of the human future he divulges, Beck is hardly a prophet of wrath and his sociology is as *optimistic* as it is *pessimistic*. Beck does not interpret global risk as spelling the doom of human future. Catastrophe does not herald the apocalypse of humanity. Quite the contrary, catastrophe invokes "emancipatory catastrophism" (Beck, 2015). Beck takes catastrophe as a probable form of liberation from the naively simplistic and sanguine worldview characterising the first modernity that things and lives are thought to be calculable, predictable and controllable. For Beck and critical sociologist like Bruno Latour (1993), the (first) "modern constitution" serves to "obscure" no less than "clarify." The norms and "modern constitution" of the first modernity steer us to believe in the immense potential of social progress with the advance of science and technology. Human beings can master our planet and control our destiny. But the COVID-19 pandemic eventually functions as a rude-awakening, waking us abruptly from the carefully curated facade of the (first) modernity. The obscurantism of modernity can no longer sustain itself when the repressed, denied and excluded all come back to haunt humanity. The anthropological shock caused by catastrophe, no matter it is climate change as Beck theorises or the COVID-19 pandemic as argued in this paper, is hence emancipatory. Emancipatory by no means suggests that we can be set from risks and threats, nor perturbation on the ontological level—a risk-free world is a misguided premise to begin with. Despite the calamitous consequences the COVID-19 pandemic brings, global risk is the mediation for liberating us in terms of cognition, ideation and cultural horizons. It is about emancipating us from a set of anachronistic presumptions and concepts of the first modernity, and ushering into new form of cultural consciousness and recognition of our globe as composing of "cosmopolitan communities of risk" and fate (Beck, 1999, 2008, 2015, 2016). Human beings thrive and perish together in "world risk society." If we do something good (bad) for the world, we will benefit (harm) each other. Our collective future and destiny is up to all of us.

I shall end this paper with a note of Beck's pessimistic optimism when we rethink the new beginning of world risk society and the third millennium of the Anno Domini after the end of the pandemic.

> Global risks open up a moral and political space that can give rise to a civil culture of responsibility that transcends borders and conflicts. The traumatic experience that everyone is vulnerable and the resulting responsibility for others, also for the sake of one's own survival, are the two sides of the belief in world risk. (Beck, 2009, pp. 56–57)

Acknowledgments Some ideas of this paper were previously shared in my public lecture titled "Rethinking globalisation" on 15th July, 2020. The lecture is a part of the General Education Office's 2020 Summer Lecture Series at BNU-HKBU United International College. I would like to thank the audience for their stimulating questions, and as always I am grateful to my teacher Prof. CHAN Hoi Man at the Chinese University of Hong Kong for his teaching and introducing Ulrich Beck's theory to me.

References

Abrahamson, P. (2010). European welfare states beyond neoliberalism: Toward the social investment state. *Development and Society, 39*(1), 61–95.

Agamben, G. (2020a). Coronavirus and philosophers: The invention of an epidemic. *European Journal of Psychoanalysis.* https://www.journal-psychoanalysis.eu/coronavirus-and-philosophers/

Agamben, G. (2020b). Giorgio Agamben on health care and the religion of science. *Inscriptions, 3*, 2.

Alicke, K., Barriball, E., & Trautwein, V. (2021, November 23). *How COVID-19 Is Reshaping Supply Chains.* McKinsey. https://www.mckinsey.com/business-functions/operations/our-insights/how-covid-19-is-reshaping-supply-chains

Allen, P., Konzelmann, S. J., & Toporowski, J. (2021). *The return of the state: Restructuring Britain for the common good.* Agenda Publishing.

Anderson, B. (2006). *Imagined communities: Reflections on the origin and spread of nationalism.* Verso.

Baker, G., Trumka, R. L., Evenett, S., & Javorcik, B. (2021, January 26). *Preparing for deglobalization.* World Economic Forum. https://www.weforum.org/events/the-davos-agenda-2021/sessions/preparing-for-deglobalization

Barradas, R., Lagoa, S., Leão, E., & Mamede, R. (2018). Financialization in the European periphery and the sovereign debt crisis: The Portuguese case. *Journal of Economic Issues, 52*(4), 1056–1083.

BBC. (2019, November 13). Europe and nationalism: A country-by-country guide. *BBC News.* https://www.bbc.com/news/world-europe-36130006

Beck, U. (1992a). Modern society as a risk society. In R. V. Ericson & N. Stehr (Eds.), *The culture and power of knowledge: Inquiries into contemporary societies* (pp. 199–214). De Gruyter.

Beck, U. (1992b). *Risk society: Towards a new modernity.* SAGE Publications.

Beck, U. (1999). *World risk society.* Polity Press.

Beck, U. (2000a). *The brave new world of work.* Polity Press.

Beck, U. (2000b). The cosmopolitan perspective: Sociology of the second age of modernity. *British Journal of Sociology, 51*(1), 79–105.

Beck, U. (2002a). The cosmopolitan society and its enemies. *Theory, Culture & Society, 19*(1–2), 17–44.

Beck, U. (2002b). The terrorist threat: World risk society revisited. *Theory, Culture & Society, 19*(4), 39–55.

Beck, U. (2006). *The cosmopolitan vision.* Polity Press.

Beck, U. (2008). Risk Society's "Cosmopolitan Moment." *ComCiência*, 104. http://comciencia.scielo.br/pdf/cci/n104/en_a09n104.pdf

Beck, U. (2009). *World at risk.* Polity Press.

Beck, U. (2011). Cosmopolitanism as imagined communities of global risk. *American Behavioral Scientist, 55*(10), 1346–1361.

Beck, U. (2012). *Living in and coping with world risk society: The cosmopolitan turn* [speech transcript]. The Gorbachev Foundation. https://www.gorby.ru/userfiles/ulrich_beck_final_version_moscow.pdf

Beck, U. (2015). Emancipatory catastrophism: What does it mean to climate change and risk society? *Current Sociology, 63*(1), 75–88.

Beck, U. (2016). *The metamorphosis of the world.* Polity Press.

Beck, U., Giddens, A., & Lash, S. (1994). *Reflexive modernization: Politics, tradition and aesthetics in the modern social order*. Polity Press.

Bello, W. (2004). *Deglobalization: Ideas for a new world economy*. Zed Books.

Bello, W. (2009, September 3). *The virtues of deglobalization*. Foreign Policy in Focus. https://fpif.org/the_virtues_of_deglobalization/

Bello, W. (2013). *Capitalism's last stand?: Deglobalization in the age of austerity*. Zed Books.

Chernilo, D. (2021). Another globalisation: Covid-19 and the cosmopolitan imagination. In G. Delanty (Ed.), *Pandemics, politics, and society: Critical perspectives on the Covid-19 crisis* (pp. 157–170). De Grutyer.

Christaen, T. (2020, March 26). Must society be defended from Agamben? *Critical Legal Thinking*. https://criticallegalthinking.com/2020/03/26/must-society-be-defended-from-agamben/

De Wilde, P., Koopmans, R., Merkel, W., Strijbis, O., & Zürn, M. (2019). *The struggle over borders*. Cambridge University Press.

D'Souza, G., & Dowdy, D. (2021, April 6). *Rethinking herd immunity and the Covid-19 response end game*. John Hopkins Bloomberg School of Public Health. https://publichealth.jhu.edu/2021/what-is-herd-immunity-and-how-can-we-achieve-it-with-covid-19

Duggan, L. (2003). *The twilight of equality?: Neoliberalism, cultural politics, and the attack on democracy*. Beacon Press.

Elliott, L. (2016). Brexit is a rejection of globalisation. *The Guardian*. https://www.theguardian.com/business/2016/jun/26/brexit-is-the-rejection-of-globalisation

Elmandjra, M. (2000). The need for a 'deglobalization' of 'globalization.' In C. Pierson & S. Tormey (Eds.), *Politics at the edge: The PSA yearbook 1999* (pp. 29–39). Palgrave Macmillan.

Ferguson, Y., & Mansbach, R. (2012). *Globalization: The return of borders to a borderless world*. Routledge.

Giddens, A. (2000). *Runaway world: How globalization is reshaping our lives*. Routledge.

Giddens, A. (2009). *The politics of climate change*. Polity Press.

Giroux, H. (2008). *Against the terror of neoliberalism: Politics beyond the age of greed*. Paradigm Publishers.

Glotova, E., & Raoult, S. (2017). Penal legislation inflation and convergence in the West: A French example. *Droit Et Société, 97*(3), 571–594.

Godbold, J. (2020). Antarctica: Measures to address coronavirus pandemic. *The Polar Connection*. https://polarconnection.org/antarctica-coronavirus-measures/

Green, E. (2017, May 9). It was cultural anxiety that drove white, working-class voters to Trump. *The Atlantic*. https://www.theatlantic.com/politics/archive/2017/05/white-working-class-trump-cultural-anxiety/525771/

Haass, R. (2022). *Deglobalization and its discontents*. Council on Foreign Relations. https://www.cfr.org/article/deglobalization-and-its-discontents

Habermas, J. (2001). *The postnational constellation: Political essays*. MIT Press.

Hanieh, A. (2022). *This is a global pandemic—Let's treat it as such*. Verso Books. https://www.versobooks.com/blogs/4623-this-is-a-global-pandemic-let-s-treat-it-as-such

Harrison, K., & Sundstrom, L. (2010). *Global commons, domestic decisions: The comparative politics of climate change*. MIT Press.

Kaminsky, G. L., & Reinhart, C. (2003, February). The center and the periphery: The globalization of financial turmoil. *National Bureau of Economic Research*. https://www.nber.org/papers/w9479

Khadka, N. S. (2021, August 7). Climate change: Low-income countries 'can't keep up' with impacts. *BBC News*. https://www.bbc.com/news/world-58080083

Kim, C. (2021, February 9). *New COVID-19 variant threatens indigenous peoples in the Amazon—Rainforest foundation US*. Rainforest Foundation US. https://rainforestfoundation.org/new-covid-19-variant-threatens-indigenous-peoples-in-the-amazon/

Kornprobst, M., & Paul, T. (2021). Globalization, deglobalization and the liberal international order. *International Affairs, 97*(5), 1305–1316.

Latour, B. (1993). *We have never been modern*. Harvard University Press.

Lenk, K. (1987). The tragic consciousness of German sociology. In V. Meja, D. Misgeld, & N. Stehr (Eds.), *Modern German sociology* (pp. 57–75). Routledge.
Matthewman, S., & Huppatz, K. (2020). A sociology of Covid-19. *Journal of Sociology, 56*(4), 675–683.
Millman, N. (2021, July 8) Europe is becoming a right-wing continent. *The Week*. https://theweek.com/politics/1002381/europe-is-becoming-a-right-wing-continent
Mills, C. W. (1959). *The sociological imagination*. Oxford University Press.
Nigrinová, L. (2022). The impact of the spread of COVID-19 on globalization: The future of globalization. In G. Antošová (Ed.), *Innovative strategic planning and international collaboration for the mitigation of global crises* (pp. 181–199). IGI Global.
Nuorgam, A. (2022). *COVID-19 and indigenous peoples | United Nations for indigenous peoples*. United Nations https://www.un.org/development/desa/indigenouspeoples/covid-19.html
Nussbaum, M. (1996). *For love of country: Debating the limits of patriotism*. Beacon Press.
Ohmae, K. (1992). *The borderless world*. HarperCollins.
Otto, S., Day, T., Arino, J., Colijn, C., Dushoff, J., Li, M., Mechai, S., Van Domselaar, G., Wu, J., Earn, D., & Ogden, H. (2021). The origins and potential future of SARS-CoV-2 variants of concern in the evolving COVID-19 pandemic. *Current Biology, 31*(14), R918–R929.
Parekh, B. (2003). Cosmopolitanism and global citizenship. *Review of International Studies, 29*(1), 3–17.
Rato, J. (2020). *The European debt crisis: How Portugal navigated the post-2008 financial crisis*. Palgrave Macmilan.
Ritzer, G. (Ed.). (2007). *The Blackwell companion to globalization*. Blackwell Publishing.
Robertson, R. (2012). European cosmopolitanism and the global field. In R. Robertson & A. S. Krossa (Eds.), *European cosmopolitanism in question* (pp. 174–197). Palgrave Macmillan.
Roy, A. (2020, April 3). Arundhati Roy: "The pandemic is a portal." *Financial Times*. https://www.ft.com/content/10d8f5e8-74eb-11ea-95fe-fcd274e920ca
Santos, A. C., & Teles, N. (2021). *Financialisation in the European periphery work and social reproduction in Portugal*. Routledge.
Sardar, Z. (2019). *The postnormal times reader*. International Institute of Islamic Thought.
Shammas, V. (2016). The rise of a more punitive state: On the attenuation of Norwegian penal exceptionalism in an era of welfare state transformation. *Critical Criminology, 24*(1), 57–74.
Shils, E. (1997). *The virtue of civility: Selected essays on liberalism, tradition, and civil society*. Liberty Fund.
Stehr, N., & von Storch, H. (2009). *Climate and society: Climate as resource, climate as risk*. World Scientific.
Stepanova, E. (2008). *Terrorism in asymmetrical conflict: Ideological and structural aspects*. Oxford University Press.
Stokes, B., Wike, R. & Carle, J. (2015). *Global concern about climate change, broad support for limiting emissions*. Pew Research Center. Retrieved March 14, 2022, from https://www.pewresearch.org/wp-content/uploads/sites/2/2015/11/Pew-Research-Center-Climate-Change-Report-FINAL-November-5-2015.pdf
Tsoukalis, L. (2022). Brexit and globalization. *Policy Network*. https://policynetwork.org/opinions/essays/brexit-and-globalization/
Unger, S. (1992). The rise and (relative) decline of global warming as a social problem. *Sociological Quarterly, 33*(4), 483–501.
Unger, S. (2012). The artful creation of global moral panic: Climatic folk devils, environmental evangelicals, and the coming catastrophe. In S. Hier (Ed.), *Moral panic and the politics of anxiety* (pp. 190–207). Routledge.
Urry, J. (2005). The complexity turn. *Theory, Culture & Society, 22*(5), 1–14.
Vandenberghe, F., & Véran, J. (2021). The pandemic as a global social total fact. In G. Delanty (Ed.), *Pandemics, politics, and society: Critical perspectives on the Covid-19 crisis* (pp. 171–188). De Gruyter.

van de Pass, R. (2020). *Globalization paradox and the coronavirus pandemic: Hyper-globalization and an epidemic of affluence*. Clingendael Report. https://www.clingendael.org/sites/default/files/2020-05/Report_Globalization_Paradox_and_Coronavirus_Pandemic_May_2020.pdf
Vertigans, S. (2011). *The sociology of terrorism: Peoples, places and processes*. Routledge.
Wacquant, L. (2001). The advent of the penal state is not a destiny. *Social Justice, 28*(3 (85)), 81–87.
Wacquant, L. (2009a). *Prisons of poverty*. University of Minnesota Press.
Wacquant, L. (2009b). *Punishing the poor: The neoliberal government of social insecurity*. Duke University Press.
Weber, M. (1978). *Economy and society: An outline of interpretive sociology*. University of California Press.
Weber, M. (2004). *The vocation lectures*. Hackett Publishing.
Weil, S. (2002). *The need for roots: Prelude to a declaration of duties towards mankind*. Routledge.
World Health Organization. (WHO). (2020, December 31). *Coronavirus disease (COVID-19): Herd immunity, lockdowns and COVID-19*. World Health Organization. https://www.who.int/news-room/questions-and-answers/item/herd-immunity-lockdowns-and-covid-19
Zinn, J. (2021). Introduction: Towards a sociology of pandemics. *Current Sociology, 69*(4), 435–452.

Siu Han Chan is an Associate Professor in Sociology at BNU-HKBU United International College, China. She was trained in Hong Kong, UK, and Austria. She specialises in cultural sociology, historical sociology, sociology of knowledge, and social and political thought. Her research interest includes (post)colonial Hong Kong culture and society, Chinese tradition and modernity, and the indigenisation of Chinese sociology.

Politicization of Anti-Pandemic Measures in Europe: Cleavage Politics and Divided Publics

Michal Vavřík and Siyao Qin

1 Introduction

The coronavirus pandemic suggests that policy capacity is related to demonstrated the importance of policy capacity is related for social resilience, or, "the ability of social entities… to absorb the impacts of external and internal system shocks without losing the ability to function, and failing that, to cope, adapt, and recover from those shocks" (Tierney, 2014, p. 6). As one scholar put it, the "COVID-19 pandemic disclose[d] some of the central dimensions of state power in general, and the re-production of state power in the practices of health care and economic policy in particular" (Moisio, 2020). Even in European countries, the outset of the pandemic occasioned a previously unimaginable concentration of power in the hands of state administrators with the sole purpose of containing the spread of the virus, and seemingly little concern for legitimacy (Goetz & Martinsen, 2021).

By the end of 2020, however, public protests against the anti-epidemic emerged across Europe, with numbers in the following months reaching tens of thousands participants (Carnegie Endowment for International Peace, 2022). An originally marginal phenomenon of "COVID-skepticism" manifesting itself in statements by global populist icons and parliamentary eccentrics, took to the streets. According to some early observations, they were another manifestation of the "populist Zeitgeist": a syndrome of alienation targeting the "establishment," which can be traced back to the 1980s (see Mudde, 2004; Wondreys & Mudde, 2020). As the pandemic situation turned grave around the world, defying early COVID-skeptics' expectations, some commentators saw a silver lining: since the incompetence of populist leaders had thus been laid bare, with voters witnessing the sham, the populism itself might

M. Vavřík (✉) · S. Qin
Beijing Normal University-Hong Kong Baptist University United International College, Zhuhai, China
e-mail: michalvavrik@uic.edu.cn

© The Author(s), under exclusive license to Springer Nature Singapore Pte Ltd. 2023
S. X. B. Zhao et al. (eds.), *Comparative Studies on Pandemic Control Policies and the Resilience of Society*, https://doi.org/10.1007/978-981-19-9993-2_9

become the next victim of the pandemic. This prediction has not been fulfilled (see Katsambekis & Stavrakakis, 2020; Spilimbergo, 2021; Wondreys & Mudde, 2020).

As the pandemic mutated from a brief tsunami into a long-term inundation, and successive waves of the virus swamped societies around the world, the most spectacular occurrences continued to demonstrate that the governments especially in Europe and North America were far from enjoying social consensus over solutions. In many cases they encountered resistance when they tried to impose strict anti-pandemic measures on their societies. The examples featured careless spontaneous negligence of social distancing rules as well as open defiance of officially adopted measures (lockdowns, COVID passes and mandatory vaccinations). In not a few instances they involved mass protests and violent events across European, US, and Canadian cities (Carnegie Endowment for International Peace, 2022). Anti-pandemic measures became a matter of politics.

What interested us was the question whether politicization surrounding the public contestation about anti-pandemic measures in Europe reflected pre-existing divides in European polities. To address this problem, we used political party positioning and public opinion data to map the divides regarding anti-pandemic measures onto variables indicating existing political cleavages to assess to what extent they coincided.

2 Politicization and the Cleavage Theory

The concept of politicization has been central to multiple research programs interested in public contestation over complex issues, covering agendas as diverse as European integration, economic globalization, identity politics, and environment and climate change. At its core, the concept of politicization refers to "the demand for, or the act of, transporting an issue or an institution into the sphere of politics" (Zürn, 2019).

Public opinion has been regarded to be an important actor in politicization processes seen as "the result of the expansion of conflict from the political elites to the mass public" (Kriesi et al., 2012, p. 16). However, when approaching the concept analytically, we can distinguish three dimensions, namely, (1) an issue rises in political salience and receives public visibility; (2) the range of actors involved in public contestation increases as previously uninvolved actors join public debate; and (3) a polarization increases, as the actors take increasingly diverse positions with high intensity (Grande & Hutter, 2016; Hutter & Grande, 2014; Kriesi et al., 2012). Consequently, there are two aspects of politicization that accompany involvement of more actors, including the public: the issue's increased salience and differentiation of opinion.

In addition, we can ask whether the pattern of differentiation between positions in the political conflict displays path-dependency, due to which certain actors repeatedly find themselves on the same side of apparently little-related causes. Cleavage theory provides a tool to conceptualize and explain instances in which such path-dependent political conflicts occur, and to put them on a comparative basis.

Originated as an explanation of the development of national party systems in Europe, the cleavage theory related political parties to normative divides among citizens, and to social differences determined by position in the social structure. The result was an account based on the correspondence between the structure of social differences, the structure of normative beliefs, and the system of political parties representing them (see Lipset & Rokkan, 1990). In its classic form the cleavage theory identified four cleavages activated in the course of modernization of European societies in the nineteenth and twentieth centuries: center-periphery, state-church, urban–rural, workers-employers, which led to the emergence of major European party families, including Christian democratic parties and workers' parties etc. (Caramani, 2011, p. 239).

The class cleavage in Europe has been historically associated with the mobilization of the working class under the leadership of left-wing parties (see Bartolini, 2000). As one of the manifestations of a class conflict emerging from the Industrial revolution, it "structured politics around social-economic policy issues: the regulation of the market and the construction of social protection by the state" (Kriesi et al., 2006, p. 923). The result was a class-based politics with the dominant political conflict dividing left-wing and right-wing political parties, differentiated by opposing views of the role of government in economy. Even after an interaction with the cultural transformation discussed below, "on the social-economic dimension, the new conflict can be expected to reinforce the classic opposition between a pro-state and a pro-market position" (Kriesi et al., 2006, p. 924).

The image has become more complicated since the 1970s as a result of cultural change, globalization, and European integration that jointly undermined class-based politics. Some accounts emphasized the split within subjective classes along the different nature of work which involve radically different life experience, lifestyles and consequently also values (Kitschelt, 1994). Others expected emergence of "apartisan", sophisticated voters without interest in traditional party mobilization, and therefore disconnected from any fixed cleavage patterns (e.g., Dalton, 1984). Our own approach is informed by the so-called "neo-cleavage theory," which prefers "to conceive the weakening of traditional cleavages as a phase in the re-articulation of political conflict" (Marks et al., 2021, p. 175).

It accepts that politics has been transformed and has become more value-based and cultural. The materialism-postmaterialism divide has been identified as a result of generational change (Inglehart, 1971); a rise in numbers of specific professional groups, such as sociocultural professionals whose preferences could not be easily subsumed under the agenda of traditional parties, was observed (Kitschelt, 1994); and ultimately the globalization was seen as a factor that generated a split between "winners" and "losers" of globalization (Kriesi et al., 2012). The re-articulation of political conflict is not only manifested a transformation of the original class cleavage, but also in an emergence of new cleavage structuring political conflict: a integration-demarcation or, transnational cleavage: "the demarcation pole of the new cultural cleavage should be characterized by an opposition to the process of European integration and by restrictive positions with regard to immigration" (Kriesi et al., 2006, p. 924). Consequently, the neo-cleavage theory found a major field of application in the study of politicization of the European Union, and of the transformative effects of globalization (Hooghe & Marks, 2009, 2018; Kriesi, 2007; Kriesi et al., 2012).

Our conceptualization of party position on this new dimension included also another variable, which measures position of political parties on "social," non-economic issues, such as gender, environment, alternative lifestyles and the meaning of tradition and community. It is defined as a cleavage between two general orientations, labelled GAL ("green, alternative, liberal") and TAN ("traditionalist, authoritarian, nationalist"). Some authors consider GAL-TAN to be a third, or "social" dimension of party competition in Europe (e.g., Brigevich et al., 2017), while others see positions on this dimension closely associated with positions on European integration and globalization (e.g., Hooghe & Marks, 2018; Kriesi et al., 2012; Marks et al., 2021), whereby transnationalism and reforms it occasioned have been "at the core of the contemporary cultural divide" (Marks et al., 2021, p. 176).[1]

To summarize, in exploring data for this chapter, we bring together the two dimensions, and focus on two cleavages forming the two-dimensional political space in European societies today: class cleavage and transnational/cultural cleavage[2] (Kriesi et al., 2006, 2012). The first of these is conceptualized through subjective social class in case of public opinion and through positioning on government redistribution in case of political parties. Both these conceptualizations seek to be more sensitive to the way perspective on economic life is more reflective of choice than a mechanic reflection of one's relationship to the means of production. The transnational cleavage is conceptualized with reference to the attitudes to European integration and globalization for public opinion, and to European integration and GALTAN scale for party position.

3 Data and Methodology

Our research aimed at answering a descriptive research question: did the positions taken by political actors towards anti-pandemic measures coincide with existing cleavages? To answer this question, we conducted secondary analysis of existing datasets to investigate the positioning of two types of actors: political parties and the public. We wanted to know (1) whether there was a relationship between positions of political parties along the dimensions defined by the four theoretically relevant cleavages and their positions on coronavirus-related items; and (2) whether public

[1] Since GAL-TAN as a concept serves to classify political parties, we only explored it in relation to party positioning, but not when studying divisions in public opinion.

[2] The list could be expanded not only by inclusions of all original cleavages, but also some newer ones that we left unexplored, for example, the "cleavages of transformation," in the Central and Eastern Europe (see e.g., Hloušek and Kopeček, 2008; Sitter, 2002).

opinion in the EU in general, and in the member countries, was divided over the anti-pandemic measures in a way that paralleled divisions along the four selected cleavages.

Political party data came from the Chapel Hill Expert Survey (CHES). Specifically, our analysis used the *2019 Chapel Hill Expert Survey* (2019 CHES) dataset, which contained estimates of 277 European parties' positioning in relation to several policy issues and dimensions, such as government interventionism, immigration, European integration, or general left–right and GAL-TAN ideological orientation, collected before the coronavirus pandemic (Jolly et al., 2022). This dataset we merged with compatible data came from *Special Edition Chapel Hill Expert Survey* (SPEED CHES), which was conducted in June 2020 (Rovny et al., 2022).

The SPEED CHES contained party positions on four pandemic-related items, all measured on an 11-point scale (see Table 1). We then conducted regression analysis using ordinary least square (OLS) models to test the hypothesis that party positioning on pandemic response issues was related to predictor variables: redistribution *(REDISTRIBUTION)*, European integration *(EU_POSITION)*, and GAL-TAN *(GALTAN)*. For comparison, we reported also scores for two traditional-cleavage variables: state-church *(RELIGIOUS_PRINCIPLES)* and urban–rural *(URBAN_RURAL)* (see Table 2).

Unfortunately, the CHES dataset is only available for the estimates regarding the way European political parties reacted to the pandemic as of June 2020 when

Table 1 The pandemic-related variables in the 2020 *Special Edition Chapel Hill Expert Survey* (SPEED CHES)

Variable name	Meaning	Values (scale 0–10)
ECON_VS_CONTAIN	Position on responding to the COVID-19 since it emerged in 2020	0: Prioritizing keeping the economy open 10: Prioritizing containing the virus
FISCAL_UNION	Position on moving towards a European fiscal union	0: Strongly opposes moving towards a fiscal union 10: Strongly favors moving towards a fiscal union
GOV_VS_CITIZEN	Position on whether citizens cannot be trusted to follow public health advisory and thus governments should strictly enforce public health measures, or individual citizens should be trusted to self-enforce public health advisory on their own	0: Believes in strict government enforcement 10: Fully trusts citizen self-enforcement
SCIENCE_ESSENTIAL	The party's position on whether scientific expertise is essential for public policymaking	0: Scientific expertise is not essential for sound policymaking 10: Scientific expertise is essential for sound policymaking

Source *2020 Special Edition Chapel Hill Expert Survey* (Rovny et al., 2022)

Table 2 Predictor variables from the 2019 *Chapel Hill Expert Survey* (2019 CHES)

Variable name	Description	Values
REDISTRIBUTION	Position on redistribution of wealth from the rich to the poor	0: Strongly favors redistribution … 10: Strongly opposes redistribution
EU_POSITION	Overall orientation of the party leadership towards European integration in 2019	1 = Strongly opposed; 2 = Opposed; 3 = Somewhat opposed; 4 = Neutral; 5 = Somewhat in favor; 6 = In favor; 7 = Strongly in favor
GALTAN	Party position in terms of their views on social and cultural values	0: Libertarian/Postmaterialist … 5: Center … 10: Traditional/Authoritarian
RELIGIOUS_PRINCIPLES	Position on the role of religious principles in politics	0: Strongly opposes religious principles in politics … 10: Strongly supports religious principles in politics
URBAN_RURAL	Position on urban/rural interests	0: Strongly supports urban interests … 10: Strongly supports rural interests

Note The survey items refer to the party positioning as observed in 2019
Source *2019 Chapel Hill Expert Survey* (Jolly et al., 2022)

the SPEED CHES expert survey was conducted. As a result, we were not able to consider the developments that took place after this date, and neither can we make quantified argument about party positioning towards the restrictive measures as they were introduced and contested later in 2020 and in 2021.

In approaching the differentiation of public attitudes to the anti-pandemic measures, our main source was the data from Standard Eurobarometer surveys, namely, from *Eurobarometer 93.1* collected in July–August 2020 (European Commission, Brussels, 2022a) and *Eurobarometer 95.3* collected in June 2021 (European Commission, Brussels, 2022b). Our discussion relied mainly on the data from Eurobarometer 95.3. Eurobarometer 93.1 was explored for the sake of comparison, and as an indication of the state of public opinion at the time for which CHES party estimations were recorded.

Our analysis of public opinion focused on the association between items measuring attitude to coronavirus restrictions (see Table 4) and variables related to social differentiation and divides (see Table 3). The social differentiation variables

Table 3 Cleavage variables from the 93.1 and 95.3 Standard Eurobarometer surveys

Variable label	Description	Values
d63 (95.3) q132 (93.1)	Self-assessed social class	The working class of society; The lower middle class of society; The middle class of society; The upper middle class of society; The higher class of society; Other; None; Refusal; Don't know
d78 (95.3) q056 (93.1)	Perceived image of the EU	Very positive; Fairly positive; Neutral; Fairly negative; Very negative; Don't know
qb2_2 (95.3)	Perceived image of globalization[3]	Very positive; Fairly positive; Fairly negative; Very negative
d25 (95.3) q124 (93.1)	Type of community in which respondent lives[4]	Rural area or village; Small or middle-sized town; Large town; Don't know
qc7.10	Religion is an important EU value	

Source *Standard Eurobarometer 93.1 and 95.3* (European Commission, Brussels, 2022a, 2022b)

Table 4 Attitudes to coronavirus measures

Variable label	Description	Values
qa19 (95.3) q070 (93.1)	Respondent's opinion on whether coronavirus measures in the country were justified[5]	Absolutely justified; Somewhat justified; Not very justified; Not at all justified; Don't Know
qa21 (95.3)	Respondent's vaccination intent[6]	I have already been vaccinated; As soon as possible; Some time in 2021; Later; Never; Don't Know

Source *Standard Eurobarometer 93.1 and 95.3* (European Commission, Brussels, 2022a, 2022b)

were selected from the Eurobarometer data based on their theoretical relevance, i.e., to indicate socio-economic, transnational, state-church, rural–urban cleavages.

The items measuring respondents' attitudes to coronavirus measures were treated as dependent variables reflecting the normative divisions over anti-pandemic measures.

---

[3] *"Could you please tell for each of the following, whether the term brings to mind something very positive, fairly positive, fairly negative or very negative?"* [Globalisation].

[4] *"Would you say you live in a…?".*

[5] *"Thinking about the restriction measures taken by the public authorities in (OUR COUNTRY) to fight the coronavirus and its effects, would you say that they were…".*

[6] *"When would you like to get vaccinated against COVID-19 (coronavirus)?".*

4 Findings

4.1 Political Party Positioning

Country-specific accounts (see, e.g., Bobba & Hubé, 2021; Greer et al., 2021) suggest that early politicization of the COVID-19 pandemic had an idiosyncratic character. In some countries, path-dependent pattern of consensual governance played a role in keeping political conflict under control, such as in Austria or Switzerland, whereby especially Austria was capable of implementing a comparably harsh lockdown (Czypionka & Reiss, 2021; Mätzke, 2021). In Italy and Spain—two countries heavily affected in the early stages of pandemic—the opposition parties were apparently as disoriented as were the governments. Italian Far-Right, after initial skepticism about the threat, turned on the government for not being capable of a decisive enough response, and then shifted again to criticizing an undue length of the lockdown (Falkenbach & Caiani, 2021). In Spain, oppositional conservative Popular Party remained constructive throughout the spring, but moved in May 2020 to criticizing fifth extension of the state of alarm, and abstained in the parliamentary vote on it, denouncing government's "constitutional dictatorship." Spanish right-wing populist party VOX remained muted during the first weeks, but also gradually shifted to confrontational rhetoric (Dubin, 2021).

In general, early politicizations took place once opposition parties decided to criticize the government for ignoring factors that incidentally happened to be targeted by the opposition's long-term agenda, such as "open borders," "globalization," "immigrants," or "China" (see e.g., Barbieri & Bonini, 2020; Falkenbach & Caiani, 2021; Tyler, 2021).

In the Eastern Europe, originally a less-affected part of the continent, populist parties did not position themselves with respect to the pandemic in any systematic way (Katsambekis & Stavrakakis, 2020). For example, in contrast to Trump's or Bolsonaro's denialism, Czech, Hungarian and Bulgarian populists clearly prioritized public health over economic concerns, and in March 2020 implemented far-reaching anti-pandemic measures, including travel bans and lockdowns (Löblová et al., 2021). The politicization in Hungary, Poland and the Czech Republic, Central European countries with populist-led governments, took a peculiar form when opposition charged that anti-pandemic measures were being used as a political weapon to suppress opposition (Bobba & Hubé, 2021; Löblová et al., 2021). Centrist and liberal politicians in Central Europe were also wary of their governments accepting help from Russia or China, and voiced concerns that it would boost soft power of these countries in the region. In a most pronounced case, in Slovakia in March 2021, government's negotiation to receive Russian vaccines triggered coalition crisis, leading to a resignation of Prime Minister Igor Matovic.

The Chapel Hill Expert Survey 2020 data on party positioning are consistent with these anecdotic observations. They indicate that party positions on the cleavage variables (redistribution, EU position, GALTAN, urban–rural interests, and religious principles in politics) were not very strong predictors of their positions on pandemic items: prioritizing economy v. prioritizing containment, position to creation of the EU fiscal union, reliance on government enforcement or self-enforcement by citizens, and essential need for scientific expertise in policymaking (see Table 5).

Regression analysis of the CHES datasets (Jolly et al., 2022; Rovny et al., 2022) shows that among our predictors only redistribution supported a model in which party position on a predictor variable was to a meaningful degree associated with the party position on the most important pandemic-related issue: whether to prioritize keeping economy open or containing the virus (see Fig. 1).

On the face of it, it is an expected outcome. Faced with the spread of the virus, politicians encountered calls for containment measures that would protect citizens from what was understood to be a potentially deadly threat. Taking the class cleavage (operationalized through the "redistribution" variable) as a reference, these calls for containment were more likely to be resisted by "low-redistribution" political parties that represented views of the employers and self-employed.

The above findings do not suggest that there was no politicization in the early stage of the pandemic. They do, however, show that it could not be structured along the dimensions defined by the four national cleavages we investigated. Anecdotic evidence, as well as the regression model, suggests the important role of the idiosyncratic factors reflecting national circumstances.

4.2 Public Opinion

As the pandemic reached Europe, governments around the continent, led by prudential reasons, started introducing behavioral (non-medical) measures aimed at slowing down the spread of the virus. They included rules of social distancing, closures or bans on public events, mobility restrictions, testing, wearing protective equipment, or using mobile apps to track people's movement.

From the very beginning, public debates regarding appropriateness of the government policy occurred and kept their relevance. A 2020 study in seven European countries found that public approval of the measures widely varied both between and within countries, and the same can be said about health and economic anxieties (see Lazarus et al., 2020; Sabat et al., 2020). Higher endorsement of measures has been shown to be associated with better compliance behavior (Meier et al., 2020; Sabat et al., 2020). Another study focused on Italy linked prior political opinion to the attitude towards anti-pandemic measures, and to compliance behavior. It concluded that "political extremism and disillusion have the potential to undermine civic sense during the pandemic, making individuals less cooperative in decreasing mobility" and in contrast, "being more oriented towards the current legislation can enhance willingness to accept restrictions on one's freedom to contain the virus spread" (Barbieri & Bonini, 2020).

What does cleavage theory suggest about this potential for the development of politicization and divisive conflict? It suggests that there is a pre-existing, multilayered-sociological, normative, and institutional-structure behind political

Table 5 Priorities of coronavirus measures: linear regression models (R square, F, Beta coefficient, and p-values)

	Economy v. containment	Fiscal union	Government v. citizens	Science essential
Redistribution	0.226 59.627 (1,204)*** – 0.476***	0.011 2.329 (1,206) – 0.106	0.131 30.890 (1,205)*** 0.362***	0.008 1.583 (1,205) – 0.088
EU position	0.024 5.003(1,204)** 0.155**	0.626 344.841 (1,206)*** 0.791***	0.006 1.227 (1,205) 0.077	0.490 196.764 (1,205)*** 0.700***
GAL-TAN	0.050 10.671 (1,204)** – 0.223**	0.313 93.976 (1,206)*** – 0.560***	0.024 5.117 (1,205)* – 0.156*	0.341 105.887 (1,205)*** – 0.584***
Urban–rural	0.003 0.516 (1,204) – 0.050	0.199 51.185 (1,206)*** – 0.446***	0.046 9.954 (1,205)** – 0.215**	0.099 22.432 (1,205)*** – 0.314***
Religious principles	0.057 12.389 (1,204)** – 0.239**	0.111 25.773 (1,206)*** – 0.333***	0.005 0.976 (1,205) – 0.069	0.163 39.996 (1,205)*** – 0.404***

*p < 0.05; **p < 0.01; ***p < 0.001

Source: *Chapel Hill Expert Survey* (Jolly et al., 2022; Rovny et al., 2022)

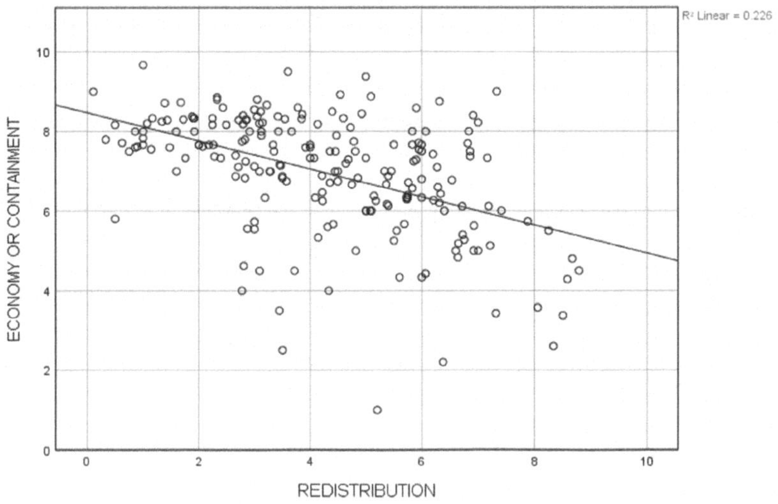

Fig. 1 Party positions on government redistribution and on priorities of coronavirus measures (Source *Chapel Hill Expert Survey* [Jolly et al., 2022; Rovny et al., 2022])

conflicts. Divisions in opinion are attached to values and through them to social structure. (Lipset & Rokkan, 1990; Marks et al., 2021) They become stable and resistant to change if they have merged with other existing divisions, with their argumentative ballast and with institutional support from political actors, and with underlying social differences providing the divisions with an air of naturalness. The following section explores to what extent the division of attitudes to coronavirus can be seen as an instance of this dynamic, that is, whether public attitudes to anti-pandemic measures overlayed differences along the four existing national cleavages.

Eurobarometer 95.3 in 2021 contained two items that were directly relevant for our research questions: justifiability of coronavirus restrictions (qa19) and vaccination intent (qa21). It was also these two issues that prominently resonated in public debates and became objects of political contestation. Justifiability of coronavirus restrictions was reported both in 2020 and 2021 surveys, while vaccination intent was only included in the 2021 survey.

Table 6 presents summary data for the EU countries. It shows that in 2021 the perceived justifiability of coronavirus restrictions was moderately associated with only two of our cleavage variables. They were the variables indicating transnational cleavage: positivity/negativity of the EU image ($V = 0.173^{***}$) and positivity/negativity of globalization image ($V = 0.151^{***}$). The association with other cleavage variables remained significant, but weak. It included association with subjective social class, which was moderate in 2020 (i.e., at the time we observed class cleavage partly reflected in differences among political parties in their preference for containment measures), but weakened in 2021.

It is notable that the moderate association produced a pronounced pattern showing that the view of coronavirus restrictions as either *not very* or *not at all justified* was

Table 6 Association (Cramér's V) between cleavage variables and attitude to coronavirus measures/vaccination intent: a whole EU perspective. Moderate associations highlighted.

	2020		2021	
	Coronavirus restrictions justified	Vaccination intent	Coronavirus restrictions justified	Vaccination intent
Subjective social class	**0.121*** **	N/A	0.062***	0.064
Type of community	0.038***	N/A	0.035***	0.034
Religion as EU value	N/A	N/A	0.035***	0.083***
EU image	**0.142*** **	N/A	**0.173*** **	**0.101*** **
Globalization image	N/A	N/A	**0.151*** **	**0.107*** **

*$p < 0.05$; **$p < 0.01$; ***$p < 0.001$
Source *Eurobarometer 93.1* and *95.3* (European Commission, Brussels, 2022a, 2022b)

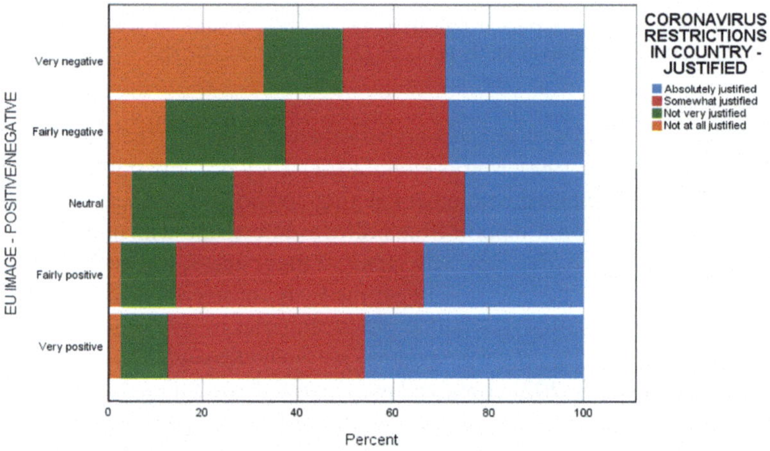

Fig. 2 Attitude to the European Union (EU image) and perceived justifiability of coronavirus restrictions (Source *Eurobarometer 95.3* [European Commission, Brussels, 2022b])

strongly concentrated among respondents holding very or fairly negative image of the EU. Among those with *fairly* or *very positive EU image*, this contrarian attitude was rare (see Fig. 2). Similar analysis will result if we substitute image of globalization for the EU image (see Fig. 3).

In line with quantitative observation, the association with subjective social class produced a more balanced pattern, with dissident voices more evenly distributed across (subjective) social classes (Fig. 4).

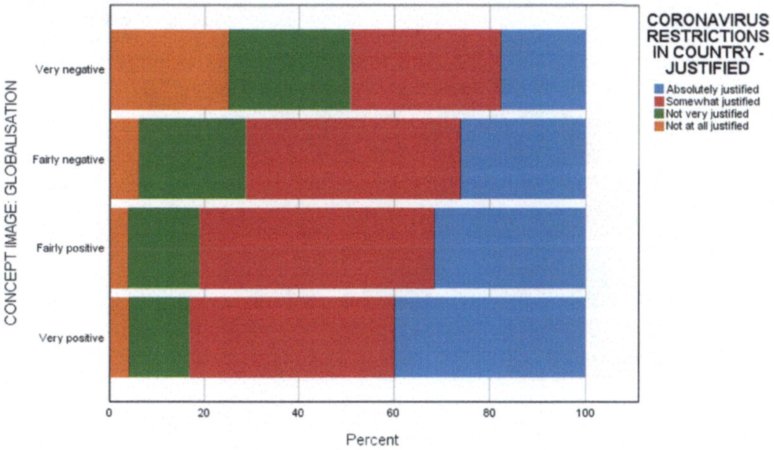

Fig. 3 Attitude to globalization (globalization image) and perceived justifiability of coronavirus restrictions (Source *Eurobarometer 95.3* [European Commission, Brussels, 2022b])

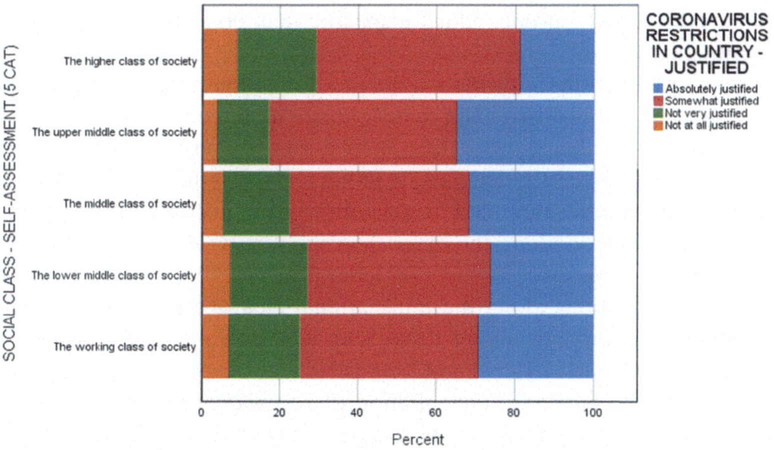

Fig. 4 Subjective social class and perception of justifiability of coronavirus restrictions (Source *Eurobarometer 95.3* [European Commission, Brussels, 2022b])

4.2.1 Vaccination Intent

By the mid-2021, vaccination had become the main source of hopes for extinguishing the pandemic, which otherwise threatened to persist in the form of repeated regional waves. Often intended to become a universal, and in some cases mandatory medical measure, it generated significant public debate and contestation. Eurobarometer 95.3 survey (European Commission, Brussels, 2022b), as well as numerous

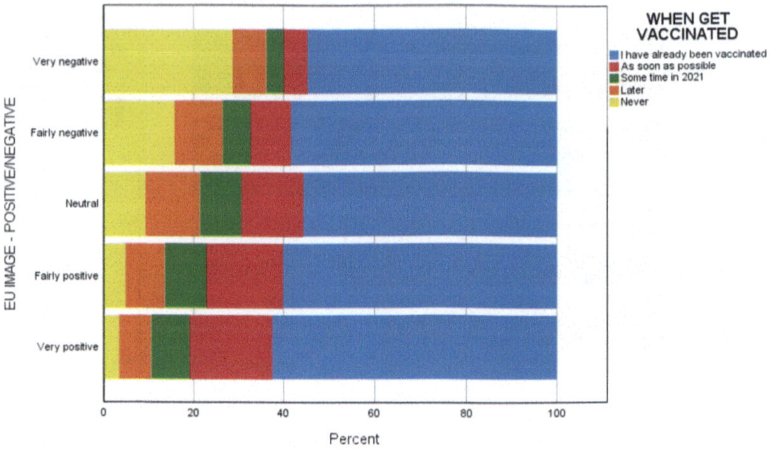

Fig. 5 Attitude to the European Union (EU image) and vaccination intent (Source *Eurobarometer 95.3* [European Commission, Brussels, 2022b])

national surveys across Europe, added a specific item concerning vaccination into their questionnaires.

Our analysis treated the information about vaccination intent as supplementary. Strictly speaking, the Eurobarometer questionnaire item reported actual or intended behavior of the respondents, rather than their normative position (even though the answer "never" may have included those wanting to express their categorical refusal). Nevertheless, the pattern resembled the one observed in justifiability of coronavirus restrictions. The division along the class cleavage was weak ($V = 0.064$***), just as urban–rural ($V = 0.034$***) and state-church (0.083***) cleavages. Moderate association with positions along the transnational cleavage generated a pattern of anti-vax attitudes dominant among those with a negative attitude toward the EU and globalization ($V = 0.101$*** and $V = 0.107$***, respectively). Exploring the relationship visually suggests concentration of respondents who "never" want to receive vaccination among people with a negative perception of the EU (see Fig. 5).

4.3 *National Perspectives*

Country-by-country analysis shows that justifiability of coronavirus restrictions was almost everywhere more strongly associated with the EU image than with subjective social class (see Appendix 1). In several countries, the strength of association was above 0.30, i.e., strong (see Healey, 2016, p. 303). These included: Latvia (0.383***), Bulgaria (0.343***), Slovakia (0.323***), Lithuania (0.311***), Estonia (0.309***), and Greece (0.304***). In contrast, it was weak in Portugal

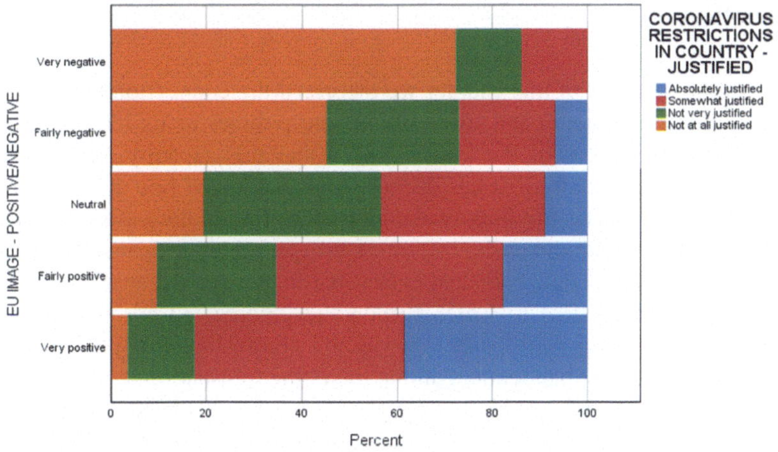

Fig. 6 Association between the EU image and justifiability of coronavirus restrictions in Latvia (Source *Eurobarometer 95.3* [European Commission, Brussels, 2022b])

(0.081***), Hungary (0.115***), Denmark (0.124***) and the Czech Republic (0.140***).

Figure 6 illustrates visually, using the case of Latvia, how a strong association may manifest itself. Some degree of refusal to see the restrictions as justified appears nearly emblematic for people with a negative view of the EU. However, an important caveat is at place. The interpretation needs to consider also absolute numbers. In the Latvian case, apparently massive resistance is mostly an illusion, because respondents declaring fairly and very negative image of the EU only accounted for 3.4% and 12.7% of the national sample, respectively. What the observation only shows is that the new coronavirus division overlays existing differences.

The association between perceived justifiability of restrictions and subjective social class was weak or moderate in all countries. Relatively strongest association was observed in the case of former East Germany (0.217***), Cyprus Republic (0.195***) and Romania (0.185***). The weakest was in Sweden (0.061***), Hungary (0.070***), Ireland (0.077***), Slovenia (0.084***) and the Czech Republic (0.090***). However, only in one country was the association with subjective social class stronger than the association with the EU image: in Portugal (0.165*** v. 0.081***).

It is notable that, as these and other observations recorded in the Appendix demonstrate, perceived justifiability was associated with class cleavage and transnational cleavage with varying strength, but the two tendencies were not communicating vessels. Several countries had low values of both, e.g., Belgium (0.059*** and 0.155***), Sweden (0.061*** and 0.159***), Hungary (0.070*** and 0.115***), or the Czech Republic (0.090*** and 0.140***).

5 Conclusion

In the opening sections of this chapter we theorized the connection between social resilience and government capacity to implement policies using its infrastructural power (Mann, 1984, 2008). We also assumed that this capacity can be affected by politicization, especially if it has deeper roots than an ad hoc differentiation of opinion. After reviewing the data, we suggest that this is the case—at least to a non-trivial degree—with the politicization of anti-pandemic measures in Europe.

Using public opinion data collected in Eurobarometer surveys, we found an association between the degree to which respondents' perceived coronavirus restrictions in their countries as justified and the variables indicating their placement along cleavages pre-dating the pandemic. Among them, the transnational or, transnational (integration-demarcation), cleavage (Hooghe & Marks, 2018; Kriesi et al., 2012; Marks et al., 2021) was the most important, with association moderate to strong. Skepticism about coronavirus restrictions, and refusals to accept COVID vaccination were concentrated among respondents on the "demarcation" side of the cleavage, who declared negative perception of European integration and globalization. These findings are consistent with the currently prevailing view (see e.g., Krastev & Leonard, 2020).

The division along class cleavage proved to have only a weak connection with coronavirus attitudes. This can be a sign that transnational cleavage is becoming increasingly important for political conflict in Europe, and policy activity of the governments should reflect on this.

Especially in countries in Central and Eastern Europe, however, we must be aware that our study does not explore cleavages that are typical for these post-communist countries (see e.g., Hloušek & Kopeček, 2008; Sitter, 2002; Whitefield, 2002). Moreover, Hungary is an extremely polarized country, with political conflict that is not easily subsumed under the four cleavages that we studied. Finally, the measure of association does not in itself testify to the strength or weakness of resistance to government policy, and this question should be answered with reference to actual numbers of its supporters or opponents.

With respect to political parties, it is notable that in the early stages of the pandemic, the polarization between keeping the economy open or containing the virus was relatively low. European parties' positions towards the issue of redistribution (chosen to represent the class cleavage) was better predictor of prioritizing keeping economy open or containing the virus than their positioning along the "cultural" GAL-TAN cleavage, or their position to European integration. However, country reports suggest that, in particular, populist parties gradually shifted their rhetoric towards libertarian values and anti-establishment sentiments (Bobba & Hubé, 2021; see Greer et al., 2021).

Our findings have bearing on current problems and discussions, confirming that critical views of coronavirus restrictions tended to be entrenched in certain social categories. Moreover, these categories were not mere collections of individuals, nor are they people connected by factors whose relevance is only temporary. Nor does this division appear to be based on pragmatic divisions between people differentially affected by the pandemic crisis. Rather, the divisions actualized by the coronavirus to a degree coincided with long-term social fractures. Intrinsically, there was no reason why opponents of the European integration should also oppose lockdowns or COVID-passes. This is where the task for future research remains.

Appendix 1 Association Between Cleavage Variables With the Attitude to Coronavirus Measures (qa19): A Country-by-Country Perspective

Country	Subjective social class	EU image	Community type	Religion as EU value	Globalization image
France	0.112**	0.229***	0.037	0.094	0.165***
Belgium	0.059	0.155***	0.058	0.114*	0.128***
The Netherlands	0.091	0.149***	0.062	0.076	0.134***
Germany (West)	0.110*	0.191***	0.050	0.075	0.088*
Italy	0.142***	0.291***	0.119***	0.057	0.220***
Luxembourg	0.123	0.178***	0.115	0.096	0.121*
Denmark	0.120***	0.124***	0.066	0.055	0.185***
Ireland	0.077	0.236***	0.064	0.073	0.176***
United Kingdom	0.120**	0.146***	0.037	N/A	0.145***
Greece	0.128***	0.304***	0.103**	0.058	0.173***
Spain	0.130***	0.151***	0.125***	0.051	0.145***
Portugal	0.165***	0.081*	0.150***	0.068	0.125***
Germany (East)	0.217***	0.253***	0.064	0.064	0.156***
Finland	0.104	0.176***	0.083*	0.104*	0.139***
Sweden	0.061	0.159***	0.053	0.056	0.134***
Austria	0.132***	0.288***	0.116**	0.072	0.136***
Cyprus	0.195***	0.266***	0.043	0.085	0.175***
The Czech Republic	0.090	0.140***	0.085*	0.107**	0.110***
Estonia	0.120***	0.309***	0.035	0.127**	0.211***
Hungary	0.070	0.115***	0.092*	0.056	0.110***
Latvia	0.090	0.383***	0.099*	0.053	0.185***
Lithuania	0.103**	0.311***	0.081*	0.155***	0.233***

(continued)

(continued)

Country	Subjective social class	EU image	Community type	Religion as EU value	Globalization image
Malta	0.103	0.132*	0.116*	0.044	0.113
Poland	0.118***	0.167***	0.147***	0.056	0.131***
Slovakia	0.146***	0.323***	0.062	0.149***	0.233***
Slovenia	0.084	0.199***	0.090*	0.082	0.145***
Bulgaria	0.139***	0.343***	0.120***	0.102*	0.285***
Romania	0.185***	0.283***	0.155***	0.076	0.218***
Croatia	0.074	0.181***	0.132***	0.081	0.131***

*$p < 0.05$; **$p < 0.01$; ***$p < 0.001$
Source *Eurobarometer 95.3* (European Commission, Brussels, 2022b)

References

Barbieri, P., & Bonini, B. (2020). Populism and political (mis-)belief effect on individual adherence to lockdown during the COVID-19 pandemic in Italy. *SSRN Journal*. https://doi.org/10.2139/ssrn.3640324

Bartolini, S. (2000). *The political mobilization of the European left, 1860–1980: The class cleavage, Cambridge studies in comparative politics*. Cambridge University Press.

Bobba, G., & Hubé, N. (Eds.). (2021). *Populism and the politicization of the COVID-19 crisis in Europe*. Springer International Publishing. https://doi.org/10.1007/978-3-030-66011-6

Brigevich, A., Smith, W. B., & Bakker, R. (2017). Unpacking the social (GAL/TAN) dimension of party politics: Euroscepticism and party positioning on Europe's "other." In *Presented at the 15th Biennial EUSA Conference*.

Caramani, D. (Ed.). (2011). *Comparative politics* (2nd ed.). Oxford University Press.

Carnegie Endowment for International Peace. (2022). *Global protest tracker* [WWW Document]. Carnegie Endowment for International Peace. https://carnegieendowment.org/publications/interactive/protest-tracker?gclid=EAIaIQobChMI-KXBk-bH9gIVZZhmAh3HwAOvEAAYASAAEgLbKvD_BwE

Czypionka, T., & Reiss, M. (2021). Three approaches to handling the COVID-19 crisis in federal countries: Germany, Austria, and Switzerland. In S. Greer, E. King, E. M. da Fonseca, & A. Peralta-Santos (Eds.), *Coronavirsu politics: Comparative politics and policy of COVID-19* (pp. 295–319). University of Michigan Press.

Dalton, R. J. (1984). Cognitive mobilization and partisan dealignment in advanced industrial democracies. *The Journal of Politics, 46*, 264–284. https://doi.org/10.2307/2130444

Dubin, K. A. (2021). Spain's response to COVID-19. In S. Greer, E. King, E. M. da Fonseca, & A. Peralta-Santos (Eds.), *Coronavirus politics: Comparative politics and policy of COVID-19* (pp. 339–360). University of Michigan Press.

European Commission, Brussels. (2022a). *Eurobarometer 93.1 (2020): Standard Eurobarometer 93 (COVID-19 Pandemic)*. https://doi.org/10.4232/1.13866

European Commission, Brussels. (2022b). *Eurobarometer 95.3 (2021): Standard Eurobarometer 95 (COVID-19 Pandemic)*. https://doi.org/10.4232/1.13826

Falkenbach, M., & Caiani, M. (2021). Italy's response to COVID-19. In S. Greer, E. King, E. M. da Fonseca, & A. Peralta-Santos (Eds.), *Coronavirus politics: Comparative politics and policy of COVID-19* (pp. 320–338). University of Michigan Press.

Goetz, K. H., & Martinsen, D. S. (2021). COVID-19: A dual challenge to European liberal democracy. *West European Politics, 44*, 1003–1024. https://doi.org/10.1080/01402382.2021.1930463

Grande, E., & Hutter, S. (2016). Beyond authority transfer: Explaining the politicisation of Europe. *West European Politics, 39*, 23–43.

Greer, S., King, E., Massard da Fonseca, E., & Peralta-Santos, A. (2021). *Coronavirus politics: The comparative politics and policy of COVID-19*. University of Michigan Press. https://doi.org/10.3998/mpub.11927713

Healey, J. F. (2016). *The essentials of statistics: A tool for social research* (4th ed.). Cengage Learning.

Hloušek, V., & Kopeček, L. (2008). Cleavages in the contemporary Czech and Slovak politics between persistence and change. *East European Politics and Societies, 22*, 518–552. https://doi.org/10.1177/0888325408315833

Hooghe, L., & Marks, G. (2009). A postfunctionalist theory of European integration: From permissive consensus to constraining. *British Journal of Political Science, 39*, 1–23.

Hooghe, L., & Marks, G. (2018). Cleavage theory meets Europe's crises: Lipset, Rokkan, and the transnational cleavage. *Journal of European Public Policy, 25*, 109–135. https://doi.org/10.1080/13501763.2017.1310279

Hutter, S., & Grande, E. (2014). Politicizing Europe in the National Electoral Arena: A comparative analysis of five west European countries, 1970–2010. *Journal of Common Market Studies, 52*, 1002–1018. https://doi.org/10.1111/jcms.12133

Inglehart, R. (1971). The silent revolution in Europe: Intergenerational change in post-industrial societies. *American Political Science Review, 65*, 991–1017. https://doi.org/10.2307/1953494

Jolly, S., Bakker, R., Hooghe, L., Marks, G., Polk, J., Rovny, J., Steenbergen, M., & Vachudova, M. A. (2022). Chapel Hill Expert Survey trend file, 1999–2019. *Electoral Studies, 75*, 102420. https://doi.org/10.1016/j.electstud.2021.102420

Katsambekis, G., & Stavrakakis, Y. (2020). *Populism and the pandemic: Introduction and preliminary findings* (POPULISMUS Interventions No. 7).

Kitschelt, H. (1994). *The transformation of European social democracy, Cambridge studies in comparative politics*. Cambridge University Press.

Krastev, I., & Leonard, M. (2020). *Europe's pandemic politics: How the virus has changed the public's worldview*. European Council on Foreign Relations, ECFR. https://ecfr.eu/publication/europes_pandemic_politics_how_the_virus_has_changed_the_publics_worldview/. Accessed 15 Mar 22.

Kriesi, H. (2007). The role of European integration in national election campaigns. *European Union Politics, 8*, 83–108.

Kriesi, H., Grande, E., Dolezal, M., Helbling, M., Höglinger, D., Hutter, S., & Wüest, B. (Eds.). (2012). *Political conflict in western Europe*. Cambridge University Press.

Kriesi, H., Grande, E., Lachat, R., Dolezal, M., Bornschier, S., & Frey, T. (2006). Globalization and the transformation of the national political space: Six European countries compared. *European Journal of Political Research, 45*, 921–956. https://doi.org/10.1111/j.1475-6765.2006.00644.x

Lazarus, J. V., Ratzan, S., Palayew, A., Billari, F. C., Binagwaho, A., Kimball, S., Larson, H. J., Melegaro, A., Rabin, K., White, T. M., & El-Mohandes, A. (2020). COVID-SCORE: A global survey to assess public perceptions of government responses to COVID-19 (COVID-SCORE-10). *PLoS One, 15*, e0240011. https://doi.org/10.1371/journal.pone.0240011

Lipset, S. M., & Rokkan, S. (1990). Cleavage structures, party systems, and voter alignments. In P. Mair (Ed.), *The West European party system* (pp. 91–111). Oxford University Press.

Löblová, O., Rone, J., & Borbáth, E. (2021). COVID-19 in central and eastern Europe: Focus on Czechia, Hungary, and Bulgaria. In S. Greer, E. King, E. M. da Fonseca, & A. Peralta-Santos (Eds.), *Coronavirus politics: Comparative politics and policy of COVID-19* (pp. 413–435). University of Michigan Press.

Mann, M. (1984). The autonomous power of the state: Its origins, mechanisms and results. *European Journal of Sociology/archives Européennes De Sociologie, 25*, 185–213.

Mann, M. (2008). Infrastructural power revisited. *Studies in Comparative International Development, 43*, 355. https://doi.org/10.1007/s12116-008-9027-7
Marks, G., Attewell, D., Rovny, J., & Hooghe, L. (2021). Cleavage Theory. In *The Palgrave handbook of EU crises* (pp. 173–193). https://doi.org/10.1007/978-3-030-51791-5_9
Mätzke, M. (2021). Political resonance in Austria's coronavirus crisis management. In S. Greer, E. King, E. M. da Fonseca, & A. Peralta-Santos (Eds.), *Coronavirsu politics: Comparative politics and policy of COVID-19* (pp. 280–294). University of Michigan Press.
Meier, K., Glatz, T., Guijt, M. C., Piccininni, M., van der Meulen, M., Atmar, K., Jolink, A.-T. C., Kurth, T., Rohmann, J. L., Zamanipoor Najafabadi, A. H., on behalf of the COVID-19 Survey Study Group. (2020). Public perspectives on protective measures during the COVID-19 pandemic in the Netherlands, Germany and Italy: A survey study. *PLoS One, 15*, e0236917. https://doi.org/10.1371/journal.pone.0236917
Moisio, S. (2020). State power and the COVID-19 pandemic: The case of Finland. *Eurasian Geography and Economics, 61*, 598–605. https://doi.org/10.1080/15387216.2020.1782241
Mudde, C. (2004). The populist zeitgeist. *Government and Opposition, 39*, 541–563. https://doi.org/10.1111/j.1477-7053.2004.00135.x
Rovny, J., Bakker, R., Hooghe, L., Jolly, S., Marks, G., Polk, J., Steenbergen, M., & Vachudova, M. A. (2022). Contesting COVID: The ideological bases of partisan responses to the COVID-19 pandemic. *European Journal of Political Research*. https://doi.org/10.1111/1475-6765.12510
Sabat, I., Neumann-Böhme, S., Varghese, N. E., Barros, P. P., Brouwer, W., van Exel, J., Schreyögg, J., & Stargardt, T. (2020). United but divided: Policy responses and people's perceptions in the EU during the COVID-19 outbreak. *Health Policy, 124*, 909–918. https://doi.org/10.1016/j.healthpol.2020.06.009
Sitter, N. (2002). Cleavages, party strategy and party system change in Europe, east and west. *Perspectives on European Politics and Society, 3*, 425–451. https://doi.org/10.1080/15705850208438844
Spilimbergo, A. (2021). Populism and COVID-19. https://voxeu.org/article/populism-and-COVID-19. Accessed 1 Nov 22.
Tierney, K. J. (2014). *The social roots of risk: Producing disasters, promoting resilience*. Stanford Business Books.
Tyler, C. (2021). The UK and COVID-19. In J. N. Pieterse, H. Lim, & H. Khondker (Eds.), *COVID-19 and governance: Crisis reveals* (pp. 125–138). Routledge.
Whitefield, S. (2002). Political cleavages and post-communist politics. *Annual Review of Political Science, 5*, 181–200. https://doi.org/10.1146/annurev.polisci.5.112601.144242
Wondreys, J., & Mudde, C. (2020). Victims of the pandemic? European far-right parties and COVID-19. *Nationalities Papers* 1–18. https://doi.org/10.1017/nps.2020.93
Zürn, M. (2019). Politicization compared: At national, European, and global levels. *Journal of European Public Policy, 26*, 977–995. https://doi.org/10.1080/13501763.2019.1619188

Michal Vavřík is Assistant Professor in the Globalisation and Development programme at the Beijing Normal University-Hong Kong Baptist University United International College. His current interests include research into social and political consequences of modernization, globalization and cultural change.

Siyao Qin is a postgraduate student at the University College London, majoring Global Governance and Ethics. Qin's academic interests are international organizations and global governance.

Understanding Sweden's COVID-19 Pandemic on Chinese Social Media: Public Perceptions Comparison Between Chinese Residing in Sweden and China

Liming Liu, Chally Chai Lee Lim, Chengxi Yao, and Zhiyi Liao

1 Introduction

Having to deal with the endless loop of COVID-19 outbreaks, the ongoing threat of new variants and the addition of vaccination booster shots make the world react to the coronavirus differently. Major policies such as eliminating any virus transmission without a single case report have been called the "zero COVID" policy (Aghion et al., 2021; Normile, 2021), as well as the one that treats the coronavirus as a flu-like illness and tolerates cases by claiming "to live with COVID" (Bennett, 2021). Sweden follows a non-lockdown approach from the start of the pandemic, which lies between two significant extremes. With largely open schools, industries, and communities, Sweden has followed a nudge strategy in the past two years (Pierre, 2020; Yan et al., 2020). Sweden's response differs from other European countries and the Scandinavian countries that share geographical, cultural, political, and economic similarities (Petridou, 2020). This unique response in the European and Scandinavian contexts has led to Sweden's liberal strategy being called the "Swedish experiment," an alias that means the non-lockdown approach represents a substantial social experiment (Esaiasson et al., 2021; Irwin, 2020). Sweden and China have adopted different measures based on their sociopolitical contexts. However, when a Chinese mainstream media outlet Global Times published a commentary regarding Sweden's strategy on March 13, 2020, it prompted an outcry among Chinese in China and Sweden. In calling on Sweden's abandonment of the pandemic that has hurt itself and the universe, millions of related discussions appeared on Chinese social media. Therefore, the polarized pandemic reflections in China and Sweden bring an essential undiscovered dynamic to understanding the pandemic through social media use

L. Liu (✉) · C. C. L. Lim · C. Yao · Z. Liao
Beijing Normal University-Hong Kong Baptist University, United International College, Zhuhai, China
e-mail: limingliu@uic.edu.cn

© The Author(s), under exclusive license to Springer Nature Singapore Pte Ltd. 2023
S. X. B. Zhao et al. (eds.), *Comparative Studies on Pandemic Control Policies and the Resilience of Society*, https://doi.org/10.1007/978-981-19-9993-2_10

in everyday life. This chapter aims to study how these related contents on Chinese social media impact Chinese people, especially in different resident contexts. This chapter explores the tension between Chinese social media and the understanding of Sweden's pandemic by answering the research question (RQ): How do Chinese residing in China and Sweden perceive Sweden's pandemic through Chinese social media?

Existing research about Sweden's pandemic has primarily centered on the government's strategies for curbing the pandemic with fewer strict regulations than in other countries (Hale et al., 2020) and the voluntary responses from citizens instead of mandatory regulations (Paterlini, 2020; Pierre, 2020; Strang, 2020). Meanwhile, different scholars are also researching the dynamics between Sweden's pandemic and the media domestically (Garz & Zhuang, 2021; Giritli Nygren & Olofsson, 2021; Giritli Nygren et al., 2021) and internationally (Irwin, 2020; Simons, 2020). Little research explores Swedish exceptionalism in multiple media representations to showcase the dynamics between Sweden's pandemic and social media. Therefore, this chapter is based on the Chinese social media context to explore and compare how two groups of Chinese perceive Sweden's pandemic. In this chapter, Chinese social media has been applied as an umbrella term for multiple social media platforms in mainland China, as well as one that takes account of all user experiences within the Chinese socio-political contexts. In this context, the term "Chinese social media" opens a much-needed space for complexities of practices.

This research adopts a critical discourse analysis approach (Fairclough & Wodak, 1997; Wodak & Meyer, 2009). We have conducted an entry-level or thematic analysis of key themes and topics in the interviews between two groups of Chinese. We also have given an in-depth analysis of arguments and discursive strategies (Krzyżanowski, 2010) deployed in the discursive construction of the perception of Sweden's pandemic by using Chinese social media. This chapter commences by reviewing the literature to offer context for positioning Sweden's pandemic in the global pandemic narrative and the dynamic between social media and the pandemic. Theoretically extending social media use, particularly im/migrant social media use in a pandemic context, this study also refers to empowering the im/migrants in Sweden. Next, we provide a brief overview of social media use in Chinese and Chinese immigrant contexts, allowing us to understand better the nature of the interviews we conducted.

2 Literature Review

2.1 Contextualizing Sweden's Strategy in the Global Pandemic Narrative

As one of the most significant global public health emergencies in human history, COVID-19 has tremendously impacted the world. Different countries have adopted a wide range of measures to curb the spread of COVID-19, while Sweden has taken a different approach compared to other countries. A combination of strict and lenient containment strategies can be found in most European countries to prevent and control infection spread (Yan et al., 2021). France and Germany implemented strict restrictions during the pandemic by having closed schools, industries, and society (Pierre, 2020). Compared to other countries, Sweden has established a largely non-lockdown pandemic strategy with limited gathering but inadequate supervision and a voluntary-based approach (Giritli Nygren & Olofsson, 2021; Liu, 2021), which focused on mitigation rather than directly containing the spread of the virus (Ludvigsson, 2020). In order to contextualize Sweden's strategy, it should bring Sweden's facts into this study. As Nilson (2021) discusses, Sweden's strategy to deal with COVID-19 is mainly based on interventions that adapt the capacity of health care to the degree of interference while protecting the elderly, vulnerable groups, and underserved populations. By maintaining the continuous opening of schools and industries and supplemented by crucial management of vulnerable groups, Sweden prevented the overload of the medical system by not completely blocking the society at the social policy level in response to the outbreak of the epidemic (Pierre, 2020). The Swedish strategy is said to be reproduced from "Swedish exceptionalism," where the Swedish government establishes its social-democratic welfare state on high levels of social trust, secularization, and individualism (Greve, 2007; West Pedersen & Kuhnle, 2017). As part of the society, Sweden's citizens have followed the strategy of exceptionalism in their daily lives. They perceived little or no change in their commutes and living arrangements. Their routine work schedule continued, and the government had little control over their lives (Blom et al., 2021).

As Sweden's exceptional strategy attracted global attention and criticism (Paterlini, 2020; Pierre, 2020), different media representations of Sweden during the pandemic turned Sweden into a newsworthy and media-interest story. The "Swedish model" is widely known through worldwide media coverage (Simons, 2020). Scrutinizing the official reports of the Swedish government on the recovery of economic activities, we have found that the recovery policies have focused on vulnerable groups and emphasized the importance of individual responsibility in the crisis as well as the dual concern of public health and economic health. However, Sweden has also been projected as a failing country in different media reports for employing a herd immunity strategy and not following expert advice and recommendations from World Health Organization (WHO) (Irwin, 2020). Meanwhile, the Swedish media has criticized the authorities' slow reaction to the pandemic, and the Swedes' high trust in the government, which can turn into blind faith and quickly leads to a palpable pandemic paralysis (Giritli Nygren & Olofsson, 2021).

2.2 *The Interaction Between Social Media and Pandemic*

In order to prevent the spread of coronavirus, mediated communication has primarily replaced face-to-face meetings in everyday life (Fuchs, 2020). Social media has become a pivotal place to create individual connections and develop social networks among individuals and organizations (Obar & Wildman, 2015), especially in the context of a pandemic. However, it is also becoming one of the most widely circulated sources of information and rumors about COVID-19 (Ali et al., 2020). The previous pandemic context of emerging infectious diseases involved social media users sharing information gathered through other platforms and sharing personal experiences, which led to an exchange of information and the dissemination of individuals' ideas (Tang et al., 2018). Using social media to read and browse relevant content, as well as writing posts and comments about the pandemic, people improved their awareness of the coronavirus and related restrictions in their countries (Eberl & Lebernegg, 2022).

As much as social media plays a crucial role in the epidemic, the multiplex of social media uses results in a dramatic differentiation. In addition to being an important tool for disseminating updates on the pandemic and relevant medical information to the public (Tang et al., 2018), social media is also a double-edged sword when it comes to effective social health communication during the pandemic as well as spreading fake news (van Dijck & Alinejad, 2020). Information overabundance during the pandemic has been called "infodemic," which refers to the difficulty of finding trustworthy sources and reliable information on physical and digital systems about the pandemic, including social media (Eysenbach, 2020; Tangcharoensathien et al., 2020). In the information age, social networking sites have become rife with misinformation due to information overload and underload among different social communities. For example, the inadequate information about COVID-19 in Asian languages has prompted Asian Americans to seek information through Asia-based social media platforms (Chong et al. 2022). Social media indeed disseminate information about the pandemic but can also influence people's perceptions of risk management (Tang et al., 2018). The commercial value of social media as a priority rather than public interest is to be considered since it can quickly drive information in an unpredictable direction (van Dijck & Alinejad, 2020).

2.3 Chinese Social Media Use in the Pandemic Context

With over 1.011 billion internet users in China (CNNIC, 2021), social media use is part of essential routines in Chinese everyday life. As a result of the pandemic context, more issues must be considered, such as coordinating information sources across social media, measuring the credibility of online information, and selecting suitable social media for users. To better understand Chinese social media use, we shall examine it from the general conditions of society to a pandemic context. This section can be categorized into four types based on the functions of the different Chinese social media: official social media, aggregated social media, public social media, and professional social media (Li & Liu, 2020). The official social media are state-run media platforms generally concerned with party ideology and national interests (Repnikova & Fang, 2019). For example, the different digital platforms of People's Daily and Xinhua Net are important sources of the latest policies and viewpoints from the official. The second type is aggregated social media, a novel news aggregator that integrates a range of media outlets to provide relatively high-quality news (Dellarocas et al., 2016). Tencent News, Sina News, and Toutiao are aggregated social media that provide higher readability and personalization than official social media (Li & Liu, 2020).

As public social media are decentralized, they facilitate the diffusion of news among users and facilitate users learning from each other. For example, WeChat, Weibo, and Tik Tok represent different user-generated applications. WeChat is a multipurpose platform that offers various services, such as messaging, socialization, and mobile payments (Montag et al., 2018). On the other hand, Weibo is a hybrid of Twitter and Facebook, where users can gain information, update statuses, share views, and interact with other users (Guan et al., 2014). Additionally, professional social media provides knowledge about specific areas. For instance, Ding Xiang Doctor provides health consultations online (Tennant et al., 2015). These types of Chinese social media serve different purposes in daily life. The pandemic affects public and professional social media usage to different extents.

Weibo is regarded as a public sphere (Zheng & Wu, 2005) and a breeding ground for mobilization (Yip & Jiang, 2011), which is also an effective channel for spreading rumors and anti-rumor during the pandemic. Research points out that information in public social media has better readability, familiarity, and humanity (Repnikova & Fang, 2019). While governments tend to use official social media to denounce rumors and handle gatekeeping, the effectiveness of which often depends on the cooperation of the public social media. The rumors and anti-rumor mechanisms also take place on WeChat. People's familiarity and acquaintanceship with WeChat made messages related to the pandemic more credible since they often accept and share messages from close friends, relatives, or other sources they have trusted before (Stawarz et al., 2019). In times of pandemic, professional social media like Ding Xiang Doctor provides a professional role in the public health and medical field that can reduce fears and seek specialists' advice on preventative measures they can take (Huang et al., 2020; Li & Liu, 2020). In parallel, DingTalk, an online business platform with messaging, files, video, and voice calls functions before the pandemic, became an online teaching platform based on its existing features to maintain education systems running during the pandemic (Chen et al., 2020).

2.4 *Social Media in the Context of Chinese Im/Migrants During the Pandemic*

Chinese im/migrants have received prejudices and discrimination due to the unproven news stating that the virus comes from China during the pandemic (Wen et al., 2020). For immigrants, social media are virtual channels for getting information about the pandemic in their motherlands (Croucher et al., 2020). Apart from that, social media is not only a tool to express people's reactions, thoughts, and opinions on current events but also a playground for racism, which means that people of different races have been subjected to discrimination online based on their race (Yang & Counts, 2018). Large-scale epidemics of fatal diseases present severe obstacles to social order (Strong, 1990) by instilling people's suspicion of others and the fear that others might transmit the disease, leading to prejudicial attitudes and behaviors (Adida et al., 2020). Also, pandemics inspire acts of unity, compassion, and solidarity (Aguirre et al., 2011; Drury et al., 2009), especially when there is a sense of shared fate (Drury et al., 2019). During COVID-19, "No bats," a euphemism for "No Chinese," appeared on social media, which reflected the prejudice about race and the origin of the virus (Li & Chen, 2021). Chinese immigrants and migrants always post anti-rumor comments to fight against prejudice (Macguire, 2020). For instance, researchers have demonstrated how the im/migrants showed their "authentic" personal COVID-19 experiences to the global Chinese-speaking audience based on the content produced by Chinese immigrants and migrants on YouTube (Zhang & Zhao, 2020). The contents of their videos always provide subtitles in Chinese, and most information in other languages was adequately translated into Chinese. Through storytelling, the YouTubers have tried to make sense of their experience during the pandemic in their host countries. However, their stories did not appear in their new country due to linguistic and cultural inaccessibility or home country due to platform inaccessibility (Zhang & Zhao, 2020). Instead, they seem to share stories with other transnational im/migrants who encounter the shared fate and can understand the two interwoven perspectives in the stories (Drury et al., 2019).

It is also distinctive that some catchphrases in Chinese mainstream social media (e.g., Weibo) are found in immigrants' or migrants' social media posts through their phrasing. They have responded to the mainstream discourses in their posts by saying that "it would be better if you follow the Chinese methods" (Zhang & Zhao, 2020). Others have posted that "China has an advantageous political system and outperforms other countries before and during the pandemic" (ibid). The posts indicate their appreciation of the Chinese pandemic management and their scorn of other countries' lack of preparation and ineffectual actions (Zhao, 2022). However, these im/migrants' voices can hardly be heard by the authorities in their destination countries (Zhang & Zhao, 2020). Social media are more than information sources for Chinese immigrants and migrants in the pandemic. They also reflect the features of racism or nationalism, presented explicitly as a confrontational and xenophobic posture (Zhao, 2021). As governments have already been using social media to promote better health practices to more effectively tackle the pandemic (Croucher et al., 2020), more studies should consider such views of immigrants and migrants.

Chinese social media usage is pervasive and vast; however, this strength should not blind us from its weaknesses. Issues such as censorship, rumourmongering, fake news, and online anonymity have marked a strong presence in China's online space. The unfolding of the COVID-19 pandemic has demonstrated a rapid spread of misleading and false information which can fundamentally distort people's perception of the virus and pandemic. Nonetheless, the Chinese social media platform remains the most accessible information source among the Chinese in China (Li &

Liu, 2020). While it also serves as a piece of essential information and aspirational source for Chinese im/migrants (Sa & Leung, 2016) owing to their language and cultural compatibility (Veenhof et al., 2008). Chinese social media, as an umbrella term covering a wide range of social media platforms in mainland China, offers a "social media sphere" to better understand the Chinese narrative of the coronavirus pandemic taking part in and through social media.

Therefore, to answer the RQ by positioning this chapter in the dynamic between (Chinese-) social media use and the pandemic in Sweden. The chapter proposes three sub-RQs to support the main research question:

Sub-RQ1: What are the impressions of Sweden's pandemic among two groups of Chinese based on their social media usage?
Sub-RQ2: How does Chinese social media help construct Chinese people's understanding of Sweden's pandemic situation?
Sub-RQ3: To what extent Sweden's pandemic impacts their impressions of Sweden's national image through their Chinese social media use?

3 Methodology

This study applies CDA as the methodological approach since CDA indicates that certain language users establish attitudes and practices by continually and selectively asserting specific attributes (i.e., social roles, behavioral characteristics, and physical appearance) of social groups (Chilton, 2005). Through "de-mystifying ideologies and power through the systematic and reproducible investigation of semiotic data" (Wodak & Meyer, 2009, p. 3), CDA provides a general framework for problem-oriented social research (Wodak, 2008). A key concept here is discourse (Wodak, 2001), understood as a social practice and implies a dialectic relationship between a particular discursive event and the situation, institution, and framed social structure (Fairclough & Wodak, 1997). In order to understand discourse, texts need to be captured because "texts comprise a presentation of discourse in a particular context and situation" (Krzyżanowski, 2010, p. 76). In this chapter, the analysis explores how two groups of Chinese produce discourse and how this discourse shapes their understanding of Sweden's pandemic.

The analysis can be described broadly as comprising two levels, which consist of entry-level and in-depth analysis (Krzyżanowski, 2018). First, the entry-level analysis, also known as thematic analysis (Krzyżanowski, 2010), of key themes and topics of the perceptions of Sweden's pandemic among two groups of Chinese maps the content of the analyzed texts. Discourse topics, the center of the primary analytical category, must be clarified and viewed as the most "important" or "summarizing" idea that underlies the meanings of a sequence of sentences (Van Dijk, 1984, p. 56). In this chapter, the entry-level analysis follows two basic analytical categories on primary and secondary discourse topics (Krzyzanowski, 2008). Primary topics are analyzed based on the available frames of interview questions posed to participants. Secondary topics are explored by developing a list of topics focused on matters framed by the primary topics and put forward by participants themselves (Krzyzanowski, 2008). Therefore, the following thematic analysis connects the thematic links between two said topics and "locate varying contexts and areas in which different issues become prominent and different experiences are gathered (at the 'real-life' level)" (Krzyzanowski, 2008, p. 175).

Second, we adopt the in-depth analysis strategy, which is argumentation-oriented and focuses on the strategies of ideas and discourses (Krzyżanowski, 2010). Discursive strategies constitute "systematic ways of using language" to explore self-versus-other representation (Reisigl & Wodak, 2001, p. 44). The in-depth analysis is used to analyze the discursive strategies including nomination strategies, predicational strategies, argumentation strategies, perspectivation, framing or discourse representation, intensifying strategies and mitigation strategies (Krzyżanowski, 2010; Reisigl, 2018; Reisigl & Wodak, 2001). Therefore, the discursive strategies assist a deeper exploration of discursive practices behind the texts.

The analyzed data derives from 20 semi-structured interviews (Appendix I), with ten interviewees living in China and the rest dwelling in Sweden. The interviews were conducted online under the circumstances of the ongoing COVID-19 pandemic. The interviewees signed consent forms electronically before the interviews, and all personal information was anonymous and untraceable. In the beginning, interviewees were recruited through personal networks, and then snowball samples were recommended by interviewees. During the recruitment process, we tried to ensure respondents had a variety of social settings and a range of professions to capture as many voices as possible. The interviews were conducted in Chinese and lasted 30 min on average. The interview questions were prepared in advance and individualized into two groups, but they were not strictly followed (see Appendices II and III). During the interview, respondents were encouraged to add additional details and clarifications about their understanding.

4 Analysis

The following analysis has two parts. This first section presents the results of thematic analysis (Krzyżanowski, 2010) that maps the content of the analyzed texts to give an overview of key thematic areas and topics related to the perceptions of Sweden's pandemic on Chinese social media by Chinese in Sweden and China. Based on the above thematic analysis, the second section presents the results of the in-depth analysis (Krzyżanowski, 2010) of arguments and discursive strategies deployed in discursive constructions of Sweden's pandemic by highlighting some key and general aspects of the discursive construction of Sweden's pandemic in the interviews.

4.1 Sweden's Pandemic: Thematic Analysis

The first COVID-19 case in Sweden occurred at the end of January 2020, when a woman in her 20s returned from Wuhan, China, and tested positive in Jönköping, a city in southern Sweden (The Local, 2020). Soon Sweden's pandemic started appearing on different media and social media platforms worldwide, providing an essential starting point to initiate the interview. Based on the perception-specific framing, the discourse on Sweden's pandemic is further classified into specific topics of discourse. Table 1 outlines the key thematic areas during the interviews with Chinese in China and the core information the respondents provided in relation to each theme.

As Table 1 shows, the key themes during the interviews with Chinese in Sweden are primarily on the discourse construct about: (1) the dynamic between Chinese social media and Sweden's pandemic, (2) Sweden's national image before and during the pandemic, (3) how to understand Sweden's pandemic, and (4) China is the centrality of their pandemic understandings. Each of these four themes contains sub-themes portraying how interviewees construct Sweden's pandemic through Chinese social media use. The following table summarizes the key themes and sub-themes discussed in the interviews with Chinese in Sweden, as well as the core information provided by the interviewees concerning each theme and topic.

As seen in Table 2, four themes construct and structure the discourse about Sweden's pandemic on Chinese social media among Chinese in Sweden: (1) Chinese social media in Sweden's pandemic context, (2) English and Swedish social media as an alternative information source, (3) how to understand Sweden's pandemic, and (4) Sweden's national image before and during the pandemic. Several sub-themes

Table 1 Thematic areas and sub-thematic topics of interviews with Chinese in China

Key themes	Sub-themes and topics
The dynamic between Chinese social media and Sweden's pandemic	• Daily Chinese social media use • Chinese social media as the source of Sweden's pandemic • Chinese social media construct perceptions of Sweden's pandemic
Sweden's national image before and during the pandemic	• Sweden as a non-mainstream country in pandemic • Sweden as a welfare state • Sweden as a western country
How to understand Sweden's pandemic	• Sweden has its uniqueness • Sweden has multiple negative non-lockdown images • Sweden's strategy has advantages
China is the centrality of their pandemic understanding	• China's pandemic performance as the benchmark • China as a good example during the pandemic

Table 2 Thematic areas and sub-thematic topics of interviews with Chinese in Sweden

Key themes	Sub-themes and topics
Chinese social media in Sweden's pandemic context	• Chinese social media as an emotional connection • Chinese social media as an information source • Chinese social media creates panic • Life (in Sweden) differs from what is stated on social media
English and Swedish social media as an alternative information source	• World is in chaos during the pandemic • English social media build their network in Sweden • Sweden is mainly free during the pandemic on Facebook and Instagram
How to understand Sweden's pandemic	• Sweden as a negative example on social media • The pandemic is not serious in Sweden
Sweden's national image before and during the pandemic	• Sweden is still the same as usual during the pandemic • Swedish values are reproducing during the pandemic

describe interviewees' perceptions of Sweden's pandemic. The above two thematic analyses assist the followed-up in-depth analysis in exploring discourse structures in more depth than the contents above (Krzyżanowski, 2010). Understanding that the thematic content oscillates between a discursive strategy of understanding Sweden's pandemic in Swedish and Chinese contexts is essential. Therefore, the following in-depth analysis applies a comparative perspective to further explore the discursive strategies on the perceptions of Sweden's pandemic in a dual context.

4.2 Positioning Social Media Use Within Sweden's Pandemic Context

There is a growing trend among Chinese people to use social media as a form of modern everyday life (McDonald, 2016). Chinese people in China and Sweden are regularly using Chinese social media, such as WeChat and Weibo. Chinese social media plays a crucial role in producing and spreading pandemic information (Li & Liu, 2020). As a result of providing informative content, Chinese social media became the dominant source for Chinese in China by sharing the individual stories of Sweden during the pandemic. However, Chinese social media have become an alternative source for those living in Sweden to learn about Sweden's pandemic. Chinese in Sweden use social media to explore Sweden's pandemic as part of their lives; they use these platforms as emotional connections to explore the real Sweden within their lives.

4.2.1 Dominant Source Versus Alternative Source

Sweden's response to the pandemic has obtained global attention because of its liberal approach to dealing with the COVID-19 pandemic. The extensive international media coverage has shaped Sweden's response as "the Swedish experiment" (Irwin, 2020). The pandemic in Sweden was not a mainstream topic in Chinese media outlets or social media, with little discussion or information offered on a limited number of pages. However, the commentary from Global Times on March 13, 2020, triggered millions of discussions on Weibo, which was the first time Sweden's pandemic attracted attention in the Chinese context. Afterward, Sweden's non-lockdown policies became widely popular on Chinese social media, as they differed from China's strict policies. Chinese people in China have claimed these Chinese social media as the dominant source for them to know and understand the pandemic in Sweden, especially for those who have never visited Sweden before.

Compared to those who live in China, Chinese people in Sweden have more diverse information sources to stay informed about the pandemic in their host country. Immigrants and migrants utilize different media forms to introduce themselves in a cross-cultural context, construct their identities in the host country, and establish a transnational connection to their homeland (Baffoe, 2012; Dalisay, 2012; Shumow, 2010). When they use mainstream media outlets in Sweden, they also adopt Chinese social media daily. Chinese social media provide content in their ethnic language and cultural background that may not be available in their host country (Veenhof et al., 2008). As a result of their usage and language preferences, they regard Chinese social media as an alternative source of information about the pandemic. Meanwhile, Swedish media outlets, English-language social media platforms, and Swedish-language-oriented social media platforms contribute to their information receiving. According to interviewee 16, Chinese social media and other Swedish media outlets are used in different functions to learn about the pandemic in Sweden.

Example 1: "I get my first-hand information about Sweden's pandemic from news apps like SVT,[1] also Dagens Nyheter,[2] or even SvD[3]This is because important notifications come via pushed notifications. I followed the pandemic by using WeChat as well. It is sometimes very late, but I can see its importance from a Chinese viewpoint." (SE-Interviewee 16)

[1] SVT is the shortened name for Sveriges Television AB, the Swedish national public television broadcaster.

[2] Dagens Nyheter is a Swedish daily newspaper. It is published in Stockholm and aims to cover national and international news.

[3] SvD, or Svenska Dagbladet, is a national magazine with publication in paper and digital channels.

4.2.2 Informative Content Versus Emotional Connection

Chinese social media, which has become the dominant source of information in China, has provided information about the pandemic in Sweden. According to the interviewees, millions of posts on social media platforms like Weibo and WeChat shared Sweden's pandemic information, and the interviewees regarded them as "informative." Interviewees trusted the posts as accurate and factual because the hashtag #Sweden's pandemic# was used along with the posts or posts with Sweden's location. The hashtagged information made the Chinese social media seem credible. In addition to transmitting information, Chinese social media posts helped shape the viewers' perception of Sweden's pandemic. Viewers read how the government responded to the pandemic through social media platforms and the solutions adopted. At the very least, viewers learned about the actual state of the pandemic in Sweden. At the same time, Sweden provided the Chinese people with a new perspective on dealing with the pandemic as a country on the periphery of the mainstream pandemic response. The interviewee highlighted the following:

Example 2: "Because of these WeChat accounts and Weibo posts, I can see something different. It lets me know that lockdown is not the only option in times of pandemic. I do not know what they are thinking, but it seems the pandemic is not a big deal in Sweden." (CN-Interviewee 5)

As an alternative source for Chinese living in Sweden, Chinese social media serves as a dual emotional connection for them. By sharing Chinese-written WeChat articles regarding the pandemic in Sweden with their families and friends, Chinese social media becomes an emotional connection for them to communicate with their homeland. At the same time, it also establishes an emotional connection to their Chinese cultural backgrounds in Sweden. Being a small ethnic group in Sweden,[4] the Chinese in Sweden as a minority is marginalized by white Swedish society. The contents on Chinese social media shared cultural perspectives with these immigrants and migrants, which were sometimes vastly different from the majority of voices in Swedish media outlets. For example, wearing a facial mask as part of self-protection during the pandemic is mandatory in China, but it was merely a recommendation in Sweden. In addition, this emotional connection to their native country, mother language, and racial/cultural backgrounds can further assist them in integrating into Swedish society. The interviewee elaborated:

Example 3: "It is not easy living in Sweden. That is why I feel more secure using these Chinese social media platforms. I can follow Sweden's pandemic news and contact my 'people' in Sweden and China on these platforms. It mitigates the loneliness and sense of exclusivity brought to me by cultural differences. More importantly, it can help me to adapt to this country in my way." (SE-Interviewee 12)

---

[4] Data from Statistic Sweden in December 2020. The Chinese immigrant community is one of the largest Asian communities in Sweden and occupies 0.3% of the Swedish population. See http://www.scb.se/en/finding-statistics/statistics-by-subject-area/population/population-composition/population-statistics/.

4.2.3 Mediatized Sweden Versus Real Sweden

What is the truth about the pandemic in Sweden? How accurate are the statements posted on Chinese social media? Interviewees indicated that they were aware of Sweden's pandemic through the informative content on these social media platforms. According to reports on Chinese social media, Sweden's death toll continued to rise during the pandemic outbreak in 2020. Medical conditions worsened, the medical staff incompetent, and the government irresponsible. As a result of these descriptions, the interviewees in China had no idea what reality was like in Sweden. Hence, one interviewee said that the social media representation of the pandemic in Sweden should be considered in light of commercial logic. Clickbait might have been the commercial entices behind Chinese social media platforms. She explained:

Example 4: "In my mind, Sweden has a very high degree of public health conditions, and it is not worth giving up. There are no neutral articles on WeChat because it needs a thriller title to draw users' attention. In order to achieve a high click rate, social media always creates a lot of shocking articles. However, the content is not as shocking as the title suggests, at least not in Sweden's pandemic case." (CN-Interviewee 4)

In addition, there is a lot of disinformation and fake news on social media platforms. The validity of the Swedish pandemic-related content has been doubted by interviewees, especially those who have never visited Sweden before. Thus, their attitude determines whether the images represent Sweden as it truly is. An interviewee said, "I keep a suspicious standpoint on these contents" (CN-Interviewee 5). In contrast, interviewees who live in Sweden said the posts are exaggerated, especially those that claim Sweden is the most dangerous country facing the pandemic, worse even than America and India. These exaggerated contents were widely shared on Weibo and the Chinese version of TikTok, which further intensified the feeling of anxiety among Chinese in Sweden. Several WeChat accounts had daily updates about confirmed cases and vaccination rates in Sweden, reminding viewers that the pandemic was not over yet. These daily reminders contrasted with the Swedish non-lockdown situations. Real life and life on social media showed significant disparities. As one interviewee remarked, the pandemic rarely impacted their daily life, and "it was not as serious as what the articles said on WeChat" (SE-Interviewee 11). Consequently, some interviewees started disconnecting from the pandemic information on Chinese social media after becoming tired of the oversensitive content. Although they continued to use Chinese social media as a signal light to determine how serious the situation was in Sweden, they began believing more in their own experience as the truthful source of information.

4.3 Constructing the Understanding of Sweden's Pandemic

Social media has helped construct the perception of Sweden's pandemic in the Chinese context. The discursive construction consists of two opposing images among Chinese in Sweden and China. On the one hand, some Chinese in China adopted a "China centrality" view by using Chinese medical and cultural values as the standards to evaluate Sweden's strategy. Following the social media posts and articles on Chinese social media, such viewers, who saw little scientificity in the herd immunity policy, have regarded Sweden as the most dangerous country. Conversely, the Chinese in Sweden adopted a "Sweden's centrality" standpoint to construct their understanding of Sweden's pandemic relying on Chinese social media accounts residing in Sweden. Furthermore, they proposed that Sweden's exceptionalism (Greve, 2007) was in fact, quite scientific and that Sweden's pandemic performance needed to be rewarded.

4.3.1 China's Centrality Versus Sweden's Priority

As Sweden's pandemic continued to change, its non-lockdown strategy became the main content on Chinese social media platforms. There was widespread criticism of Sweden on these platforms since Sweden was seen as a negative example on social media and in interviews. China's strict regulation was viewed as both a good example and a standard for evaluating other countries' anti-epidemic measures, particularly when the western world was struggling with a high mortality rate, a low cure rate, and an increasing number of new patients. Some social media users appreciated the Chinese government's measures to save people's lives by imposing a mandatory quarantine on international travelers who wished to enter the country and closing the borders to tourists. As a result, they upheld China as the benchmark for understanding the pandemic in Sweden. As the interviewee stated:

Example 5: "Sweden should consider learning from China's experience because our government has come up with effective solutions and positive attitudes, and these things are impacting mortality in China. To me, less death is desirable for common citizens." (CN-Interviewee 1)

Chinese in Sweden adopted the "Sweden's centrality" approach to criticize the Chinese social media's misrepresentation of the pandemic in Sweden. Specifically, they mainly relied on some Chinese social media accounts managed by immigrants and migrants in Sweden that translated Swedish news into Chinese. According to such posts, the pandemic was far less severe than what was presented in other Chinese posts on social media. To them, the pandemic had a limited impact on their ordinary life. At the same time, they admitted that Sweden was a bad example if evaluated by Chinese cultural and public health value standards. The interviewees, however, stressed that Sweden's centrality should be prioritized due to their residences and distance from their homeland. As such, they believed that maintaining a normal life in Sweden was something to be rewarded for. According to one interviewee:

Example 6: "Sweden's biggest and best achievement in this pandemic is that my way of life has not been impacted at all. It is impossible to accomplish this in China and other countries, but in Sweden, it is possible and attainable. It's true in Sweden and needs to be highlighted after two years of the pandemic." (SE-Interviewee 15)

4.3.2 Sweden's Little Scientificity Versus Sweden's Large Scientificity

Global society has referred to Sweden's non-lockdown strategy as "herd immunity," a medical term that describes how the majority of the population is immune to the disease and, therefore, can act as a buffer against further population infection surges if others are not (Orlowski & Goldsmith, 2020). Sweden's liberal policies are treated as ununderstandable with little scientificity among Chinese people in China. On the one hand, they constructed Sweden's herd immunity as a negative image during the pandemic because few countries followed this kind of herd immunity. The UK, for example, has already given up the practice (Pierre, 2020). When they read the news about other European countries, especially Scandinavian and Nordic countries, they blamed Sweden's irresponsibilities on ignoring the nearby neighbors due to the compact borders and the unification of the European Union and/or Nordic union. The herd immunity has sacrificed Sweden's citizens and economic expenses, which is an "unacceptable action" (CN-Interviewee 1).

Comparatively, the Chinese in Sweden regarded Sweden's pandemic control measures as scientific and positive because, as mentioned above, there is no serious impact on their daily life. The scientificity of Sweden's measures is based on its uniqueness. Sweden adopts the well-known exceptionalism to maintain a strong social-democratic welfare state, egalitarian values, secularism, individualism, and high social trust (Greve, 2007). Due to its welfare state system, the country has enough economic and medical capabilities to handle potential risks during the pandemic. Having a relatively small population, Sweden ensures a high social distance before and during the pandemic, further reducing the risks. Sweden's non-lockdown strategy follows its political, economic, and medical values through these Chinese in Sweden, as the following example demonstrates:

Example 7: "My understanding is that Sweden's (pandemic) strategy logic is relativistic. The government of Sweden is changing its strategies based on how many severe patients it has, especially considering the country's small population. This way, Sweden has enough medical capability to deal with the pandemic. Because of this, many people think that Sweden has given up. It has not." (SE-Interviewee 18).

4.4 Constructing Sweden's National Image

Our study on Chinese social media shows that Sweden is presented as a non-mainstream country before and after the COVID-19 pandemic among Chinese in China. Some interviewees in China may have developed a negative view of Sweden due to its marginality. For Chinese in Sweden, it is, however, their residing country. Many of the interviewees in Sweden appreciated their stable life there. Consequently, Sweden's national image is shaped by how the two groups of viewers evaluate the significance and values of the non-mainstream countries and the residing countries.

4.4.1 Non-Mainstream Country Versus Residing Country

Sweden, as a Nordic country, has long been understood as an outlier in Europe due to its exceptionalism (West Pedersen & Kuhnle, 2017). In our interviews, the Chinese in China nominated Sweden as a non-mainstream, European, western country. Compared with China, "Sweden has limited impact, so people are rarely concerned" (CN-Interviewee 1). Sweden's pandemic was the same as in other western and European countries in the minds of the Chinese in China. However, the Chinese in Sweden nominated Sweden as their residing country and chose to stay because they largely agreed with the Swedish system. Due to Sweden's liberalism, they had a relatively free life and extensive individualism during the severe pandemic in Sweden. As the example shown:

Example 8: "I really like Sweden; this is why I do not go back to China. I enjoy it here. Honestly, there is the pandemic, but the non-lockdown policy allows me to live freely. My life here is very comfortable. Hence, I am still here." (SE-Interviewee 14)

4.4.2 Negative Image Versus Stable Image

Based on its pandemic performance, Sweden's national image continued to change in the minds of the Chinese in China. Before the pandemic, the Chinese in China nominated Sweden as a small European country with beautiful sceneries, polar lights, and a high level of civilization and welfare. Its "high degree of developments in civic and nature are popular on Chinese social media" (CN-Interviewee 6), although Chinese people often confuse it with Switzerland. During the pandemic, Sweden's national image has gradually changed due to the little scientificity of its measures, dangerous conditions, and radical liberalism. Social media posts have helped construct such negative views. As the example shown:

Example 9: "I have changed my view of Sweden because of their lousy performance during the pandemic. These social media contents have increased my understanding of this country since I have more information to know about this country. This country does not seem like a good option for me to travel to in the future." (CN-Interview 3)

On the other hand, Chinese people in Sweden have experienced a slight change in Sweden's national image over the pandemic years. The national image of Sweden has not been divided into two parts as a result of the pandemic strategy of the country. In contrast, Sweden's national image has been nominated as a stable image because "Sweden is still the same Sweden" (SE-Interviewee 16). There is still growth in the country, the welfare state system is not bankrupt, and the civic and country lives are still stable and operating well.

5 Discussion and Conclusion

Through the two-level analysis of the two groups of Chinese people, we have answered the research questions by demonstrating their perceptions about Sweden's pandemic based on their Chinese social media usage and how their pandemic understandings impact Sweden's national image. Our analysis firstly answered the Sub-RQ1 by exploring Chinese social media usage among two groups of Chinese. Chinese in China utilize social media as their dominant source to gain informative content and further construct a mediatized Sweden in their pandemic understandings. In contrast, the Chinese in Sweden utilize Chinese social media as an alternative source due to their usage and language preferences. Chinese social media has become a dual emotional connection to their homeland and their minority group in Sweden. To answer Sub-RQ2, we have found that the Chinese in China have used Chinese policies as a benchmark to e valuate the scientificity and effectiveness of Sweden's pandemic measures. On the contrary, the Chinese in Sweden have prioritized Sweden's principles in interpreting Sweden's pandemic. They have tried to justify the scientificity of the Swedish strategy by using their real-life experiences to refute what is posted on Chinese social media. By exploring how Sweden's national image is constructed based on its social media usage during the pandemic, the Sub-RQ3 has been answered. Chinese in China construct Sweden's less favorable national image as a non-mainstream country in both Europe and the Western world. Conversely, the Chinese in Sweden construct Sweden's stable national image before and during the pandemic due to that being their residing country. We have therefore answered the main research question. Chinese residing in China and Sweden both have perceived Sweden's pandemic on Chinese social media to be mediatized, although in different ways by these two groups.

According to our analysis, the construction of Sweden's pandemic among the two groups lies in the different cultural and political values. The sense of self, of others, and interdependence is strikingly different between people living in different cultural contexts (Markus & Kitayama, 1991). Chinese in China and Sweden apply Chinese cultural beliefs on healthcare in understanding Sweden's pandemic, for instance, their perceptions on lockdown regulation and facial mask usage. However, the Chinese in Sweden employ Swedish cultural values when considering the pandemic as they claim Sweden's exceptionalism (Greve, 2007) on high levels of trust and individualism in their daily lives. At the same time, Swedish and Chinese political values have been applied separately in perceiving the pandemic. Volunteering social distance and self-restraint (Yan et al., 2020) may have laid the foundation for people to understand why Swedish non-lockdown policies can be acceptable. On the contrary, those who have benefited from the Chinese epidemic control policies under the centralizing regime (Wimmer, 2018) may find it hard to understand Sweden's policies to be scientific.

In light of the evolving perceptions of Sweden's pandemic, we suggest their understandings are shaped differently by two different processes of perceptions: a linear process in China and a floating process in Sweden. The linear process in China goes from having no knowledge of Sweden due to its non-mainstream position to perceiving it through the information provided in the Global Times and other negative discussions on social media. The floating process in Sweden refers to the fact that a proper understanding of the scientificity of Sweden's policy relies on a significant and positive presence of people's real-life experiences of residing in Sweden. Consequently, social media users have constructed two polarized images of the pandemic in Sweden on Chinese social media. Using social media as the singular information source and adopting their nationalistic standpoint, some viewers have criticized the "Swedish experiment" as being irresponsible and ignorant of people's lives. A failed Swedish image has been constructed (Irwin, 2020). Other users have constructed a rewarding Swedish image by justifying, explaining, and acknowledging Sweden's nudge strategy (Yan et al., 2020) based on multiple information sources, residence, real-life experiences and perceived Swedish cultural characteristics. A fluid dynamic exists between perceptions of the pandemic and the pandemic itself. Perceptions cannot be fixed as a unified image; they are affected by different aspects related to the standpoints on recognizing the pandemic in different contexts. Analyzing the two different processes of understanding Sweden's pandemic, we have showcased how Chinese im/migrants understand the same pandemic differently. The knowledge gained in the research argues that Sweden's officials, especially public health agencies, need to take the immigrants' viewpoints into their pandemic narratives and further include im/migrants as an integral part of Sweden's society in public and media agendas.

While this chapter is being written, the world is experiencing the Omicron variant of the COVID-19 crisis, which has attacked Sweden and China. In this case, this chapter helps add the Chinese and Swedish perspectives to the global pandemic narratives. Nonetheless, online interviews under the pandemic circumstance made it harder for ethnographic observation to be fruitful in this study. It explains the fluid dynamic between public perceptions, social media use, and the pandemic. As the COVID-19 crisis intensively impacts the world, future research can be conducted to explore further the dynamic between the linear process and the floating process. More work needs to be done to trace the changes in discourses to portray a comprehensive picture of how different pandemic situations and measures are understood and presented in Chinese social media.

Appendix I: Demographic Information About the Interviewees

There are twenty interviewees evenly distributed between China and Sweden.

Name	Age	Gender	Location	Profession
CN-Interviewee 1	29	Male	Beijing, China	Product Manager
CN-Interviewee 2	28	Female	Beijing, China	Journalist
CN-Interviewee 3	20	Male	Harbin, China	Undergraduate student
CN-Interviewee 4	26	Male	Beijing, China	Editor
CN-Interviewee 5	26	Female	Beijing, China	Internet practitioner
CN-Interviewee 6	23	Female	Chengdu, China	Marketing
CN-Interviewee 7	23	Male	Chongqing, China	Freelance
CN-Interviewee 8	27	Female	Beijing, China	Community operator
CN-Interviewee 9	37	Male	Guangzhou, China	Programmer
CN-Interviewee 10	24	Female	Jiaxing, China	Human resource intern
SE-Interviewee 11	23	Female	Uppsala, Sweden	Postgraduate student
SE-Interviewee 12	26	Female	Malmö, Sweden	Service
SE-Interviewee 13	25	Male	Lund, Sweden	Freelance
SE-Interviewee 14	24	Female	Uppsala, Sweden	Consumer marketing
SE-Interviewee 15	23	Male	Lund, Sweden	Postgraduate student
SE-Interviewee 16	26	Male	Stockholm, Sweden	Bank
SE-Interviewee 17	24	Female	Göteborg, Sweden	Research Assistant
SE-Interviewee 18	24	Female	Uppsala, Sweden	Postgraduate student
SE-Interviewee 19	31	Male	Vallentuna, Sweden	Data scientist
SE-Interviewee 20	22	Female	Uppsala, Sweden	Postgraduate student

Appendix II: Interview Guide for Chinese in China

Demographic information: age, gender identity, location (city in China), profession.

Social Media

1. What social media platforms do you usually use?
2. How often? How do you use these platforms?
3. Do you use these platforms for seeking information?
4. What kind of information do you seek? Please provide examples.

Sweden's Pandemic

1. Which social media platform did you use to know about Sweden's pandemic? How do you find these information regarding Sweden's pandemic?
2. Why do you know about Sweden's pandemic?
3. How would you describe Sweden's pandemic?
4. What kind of content did you read on these platforms? Please give one or two memorable examples.
5. What is your opinion about these contents? Why do you think so?
6. Have you ever posted related content? Why (or why not)?
7. During your understanding of Sweden's pandemic, what is your opinion about social media? In what ways does social media affect your understanding?

Sweden's National Image

1. After reading these contents, what is your impression of Sweden (as a country)?
2. What is your impression about Sweden before the pandemic?
3. Did your impressions of Sweden change before and after the pandemic? Why (or why not)?

Appendix III: Interview Guide for Chinese in Sweden

Demographic information: age, gender identity, location (city in Sweden), profession.

Social Media

1. What (Chinese, English, Swedish) social media platforms do you usually use?
2. How often? How do you use these platforms?
3. Do you use these platforms for seeking information?
4. What kind of information do you seek? Please provide examples.

Sweden's Pandemic

1. Which social media platform did you use to know about Sweden's pandemic? How do you find these information regarding Sweden's pandemic?

2. What kind of content did you read on these platforms? Please give one or two memorable examples.
3. What is your opinion about these contents? What is your attitude about these contents? Why do you think so?
4. Have you ever posted related content? Why (or why not)?
5. Are there any people around you who posted related content? If so, what was it like?
6. What other social media platforms did you use to learn about Sweden's pandemic besides Chinese social media platforms? Why (or why not)?
7. During your understanding of Sweden's pandemic, what is your opinion about social media? In what ways does social media affect your understanding?
8. What do you think of the pandemic in Sweden? How has it affected your life?

Sweden's National Image

1. After reading these contents, what is your impression of Sweden (as a country)?
2. What is your impression about Sweden before the pandemic?
3. Did your impressions of Sweden change before and after the pandemic? Why (or why not)?

References

Adida, C. L., Dionne, K. Y., & Platas, M. R. (2020). Ebola, elections, and immigration: How politicizing an epidemic can shape public attitudes. *Politics, Groups, and Identities, 8*(3), 488–514.

Aghion, P., Artus, P., Oliu-Barton, M., & Pradelski, B. (2021). Aiming for zero Covid-19 to ensure economic growth. *VoxEU. org, 31.*

Aguirre, B. E., Torres, M. R., Gill, K. B., & Lawrence Hotchkiss, H. (2011). Normative collective behavior in the station building fire. *Social Science Quarterly, 92*(1), 100–118.

Ali, S. H., Foreman, J., Tozan, Y., Capasso, A., Jones, A. M., & DiClemente, R. J. (2020). Trends and predictors of COVID-19 information sources and their relationship with knowledge and beliefs related to the pandemic: Nationwide cross-sectional study. *JMIR Public Health and Surveillance, 6*(4), e21071.

Baffoe, M. (2012). Projecting their own images: The role of the Black ethnic media in reconstructing the identities and images of ethnic minorities in Canadian society. *Journal of Alternative Perspectives in the Social Sciences, 5*(1), 28–52.

Bennett, C. M. (2021). Learning to live with COVID-19 in Australia: Time for a new approach. *Public Health Research & Practice, 31*(3). https://doi.org/10.17061/phrp3132110

Blom, V., Lönn, A., Ekblom, B., Kallings, L. V., Väisänen, D., Hemmingsson, E., et al. (2021). Lifestyle habits and mental health in light of the two COVID-19 pandemic waves in Sweden, 2020. *International Journal of Environmental Research and Public Health, 18*(6), 3313.

Chen, T., Peng, L., Jing, B., Wu, C., Yang, J., & Cong, G. (2020). The impact of the COVID-19 pandemic on user experience with online education platforms in China. *Sustainability, 12*(18), 7329.

Chilton, P. (2005). Missing links in mainstream CDA: Modules, blends and the critical instinct. In R. Wodak & P. Chilton (Eds.), *A new agenda in (critical) discourse analysis: Theory, methodology and interdisciplinarity* (pp. 19–51).

Chong, S. K., Ali, S. H., Đoàn, L. N., Yi, S. S., Trinh-Shevrin, C., & Kwon, S. C. (2022). Social media use and misinformation among Asian Americans during COVID-19. *Frontiers in Public Health, 9*. https://www.frontiersin.org/article/. https://doi.org/10.3389/fpubh.2021.764681

CNNIC. (2021, August). *The 48th statistical report on China's internet development*. China Internet Network Information Center. https://www.cnnic.com.cn/IDR/ReportDownloads/202111/P020211119394556095096.pdf

Croucher, S. M., Nguyen, T., & Rahmani, D. (2020). Prejudice toward Asian Americans in the COVID-19 pandemic: The effects of social media use in the United States. *Frontiers in Communication, 5*, 39.

Dalisay, F. (2012). Media use and acculturation of new immigrants in the United States. *Communication Research Reports, 29*(2), 148–160.

Dellarocas, C., Sutanto, J., Calin, M., & Palme, E. (2016). Attention allocation in information-rich environments: The case of news aggregators. *Management Science, 62*(9), 2543–2562.

Drury, J., Carter, H., Cocking, C., Ntontis, E., Tekin Guven, S., & Amlôt, R. (2019). Facilitating collective psychosocial resilience in the public in emergencies: Twelve recommendations based on the social identity approach. *Frontiers in Public Health, 7*, 141.

Drury, J., Cocking, C., & Reicher, S. (2009). Everyone for themselves? A comparative study of crowd solidarity among emergency survivors. *British Journal of Social Psychology, 48*(3), 487–506.

Eberl, J.-M., & Lebernegg, N. (2022). The pandemic through the social media lens: Correlates of COVID-19-related social media use in Austria. *MedienJournal, 45*, 5–15. https://doi.org/10.24989/medienjournal.v45i3.2037

Esaiasson, P., Sohlberg, J., Ghersetti, M., & Johansson, B. (2021). How the coronavirus crisis affects citizen trust in institutions and in unknown others: Evidence from 'the Swedish experiment.' *European Journal of Political Research, 60*(3), 748–760.

Eysenbach, G. (2020). How to fight an infodemic: The four pillars of infodemic management. *Journal of Medical Internet Research, 22*(6), e21820.

Fairclough, N., & Wodak, R. (1997). Critical discourse analysis. In T. A. van Dijk (Ed.), *Discourse as social interaction* (Vol. 2, pp. 258–284). Sage.

Fuchs, C. (2020). Everyday life and everyday communication in coronavirus capitalism. *tripleC: Communication, Capitalism & Critique, 18*(1), 375–399.

Garz, M., & Zhuang, M. (2021). Media coverage and pandemic behaviour: Evidence from Sweden.

Giritli Nygren, K., & Olofsson, A. (2021). Swedish exceptionalism, herd immunity and the welfare state: A media analysis of struggles over the nature and legitimacy of the COVID-19 pandemic strategy in Sweden. *Current Sociology*, 0011392121990025.

Giritli Nygren, K., Klinga, M., Olofsson, A., & Öhman, S. (2021). The language of risk and vulnerability in covering the COVID-19 pandemic in Swedish mass media in 2020: Implications for the sustainable management of elderly care. *Sustainability, 13*(19), 10533.

Greve, B. (2007). What characterise the Nordic welfare state model. *Journal of Social Sciences, 3*(2), 43–51.

Guan, W., Gao, H., Yang, M., Li, Y., Ma, H., Qian, W., et al. (2014). Analyzing user behavior of the micro-blogging website Sina Weibo during hot social events. *Physica a: Statistical Mechanics and Its Applications, 395*, 340–351.

Hale, T., Petherick, A., Phillips, T., & Webster, S. (2020, November). *Variation in government responses to COVID-19* (Blavatnik School of Government Working Paper, 31).

Huang, Y., Wu, Q., Wang, P., Xu, Y., Wang, L., Zhao, Y., et al. (2020). Measures undertaken in China to avoid COVID-19 infection: Internet-based, cross-sectional survey study. *Journal of Medical Internet Research, 22*(5), e18718.

Irwin, R. E. (2020). Misinformation and de-contextualization: International media reporting on Sweden and COVID-19. *Globalization and Health, 16*(1), 1–12.

Krzyzanowski, M. (2008). Analyzing focus group discussions. In R. Wodak & M. Krzyzanowski (Eds.), *Qualitative discourse analysis in the social sciences* (pp. 162–181). Palgrave Macmillan.

Krzyżanowski, M. (2010). *The discursive construction of European identities: A multi-level approach to discourse and identity in the transforming European Union*. Peter Lang.

Krzyżanowski, M. (2018). Discursive shifts in ethno-nationalist politics: On politicization and mediatization of the "refugee crisis" in Poland. *Journal of Immigrant & Refugee Studies, 16*(1–2), 76–96.

Li, H., & Chen, X. (2021). From "Oh, you're Chinese..." to "No bats, thx!": Racialized experiences of Australian-based Chinese queer women in the mobile dating context. *Social Media + Society, 7*(3), 20563051211035350.

Li, X., & Liu, Q. (2020). Social media use, eHealth literacy, disease knowledge, and preventive behaviors in the COVID-19 pandemic: Cross-sectional study on Chinese netizens. *Journal of Medical Internet Research, 22*(10), e19684.

Liu, L. (2021). Mediatization, digital space, online support: Exploring the transformational practices of the Swedish Church during the COVID-19 pandemic. In M. Dudžáková, M. Bílik, A. Telepčáková & D. Cséfalvay (Eds.), *The world after the pandemic* (pp. 13–23). Kritika & Kontext.

Ludvigsson, J. F. (2020). The first eight months of Sweden's COVID-19 strategy and the key actions and actors that were involved. *Acta Paediatrica, 109*(12), 2459–2471.

Macguire, E. (2020). Anti-Asian hate continues to spread online amid COVID-19 pandemic. *Aljazeera*. https://www.Aljazeera.com/news/2020/04/anti-Asian-hate-continues-spread-online-covid-19-pandemic-200405063015286.html. Accessed 6 May 2020.

Markus, H. R., & Kitayama, S. (1991). Culture and the self: Implications for cognition, emotion, and motivation. *Psychological Review, 98*(2), 224.

McDonald, T. (2016). *Social media in rural China*. UCL Press.

Montag, C., Becker, B., & Gan, C. (2018). The multipurpose application WeChat: A review on recent research. *Frontiers in Psychology, 9*, 2247.

Nilson, F. (2021). A biased analysis of the Swedish management of Covid-19. *Health Policy and Technology, 10*(3), 100542–100542. https://doi.org/10.1016/j.hlpt.2021.100542

Normile, D. (2021). 'Zero COVID' countries seek exit strategies. *Science, 373*, 1294–1295.

Obar, J. A., & Wildman, S. S. (2015). Social media definition and the governance challenge-an introduction to the special issue. *Telecommunications Policy, 39*(9), 745–750.

Orlowski, E. J., & Goldsmith, D. J. (2020). Four months into the COVID-19 pandemic, Sweden's prized herd immunity is nowhere in sight. *Journal of the Royal Society of Medicine, 113*(8), 292–298.

Paterlini, M. (2020). Closing borders is ridiculous': The epidemiologist behind Sweden's controversial coronavirus strategy. *Nature, 580*(7805), 574.

Petridou, E. (2020). Politics and administration in times of crisis: Explaining the Swedish response to the COVID-19 crisis. *European Policy Analysis, 6*(2), 147–158.

Pierre, J. (2020). Nudges against pandemics: Sweden's COVID-19 containment strategy in perspective. *Policy and Society, 39*(3), 478–493.

Reisigl, M. (2018). The discourse-historical approach. In J. Flowerdew & J. E. Richardson (Eds.), *The Routledge handbook of critical discourse studies* (pp. 44–59). Routledge Taylor & Francis Group.

Reisigl, M., & Wodak, R. (2001). *Discourse and discrimination: Rhetorics of racism and antisemitism*. Routledge.

Repnikova, M., & Fang, K. (2019). Digital media experiments in China: "Revolutionizing" persuasion under Xi Jinping. *The China Quarterly, 239*, 679–701.

Sa, E., & Leung, R. (2016). The enabling role of social media for Chinese immigrants' health: A pilot study. *The Journal of Nursing Care, 5*, 356.

Shumow, M. (2010). A foot in both worlds: Transnationalism and media use among Venezuelan immigrants in South Florida. *International Journal of Communication, 4*, 21.

Simons, G. (2020). Swedish government and country image during the international media coverage of the coronavirus pandemic strategy: From bold to pariah. *Journalism and Media, 1*(1), 41–58.

Stawarz, K., Preist, C., & Coyle, D. (2019). Use of smartphone apps, social media, and web-based resources to support mental health and well-being: Online survey. *JMIR Mental Health, 6*(7), e12546.

Strang, J. (2020, April 6). Why do the Nordic countries react differently to the Covid-19 crisis? *Nordics. info.* https://nordics.info/show/artikel/the-nordic-countries-react-differently-to-the-covid-19-crisis/

Strong, P. (1990). Epidemic psychology: A model. *Sociology of Health & Illness, 12*(3), 249–259.

Tang, L., Bie, B., Park, S.-E., & Zhi, D. (2018). Social media and outbreaks of emerging infectious diseases: A systematic review of literature. *American Journal of Infection Control, 46*(9), 962–972.

Tangcharoensathien, V., Calleja, N., Nguyen, T., Purnat, T., D'Agostino, M., Garcia-Saiso, S., et al. (2020). Framework for managing the COVID-19 infodemic: Methods and results of an online, crowdsourced WHO technical consultation. *Journal of Medical Internet Research, 22*(6), e19659.

Tennant, B., Stellefson, M., Dodd, V., Chaney, B., Chaney, D., Paige, S., & Alber, J. (2015). eHealth literacy and Web 2.0 health information seeking behaviors among baby boomers and older adults. *Journal of Medical Internet Research, 17*(3), e3992.

The Local. (2020, January 31). First case of coronavirus confirmed in Sweden. *The Local Sweden.* https://www.thelocal.se/20200131/first-case-of-coronavirus-confirmed-in-jonkoping-sweden/. Accessed 16 July 2021.

van Dijck, J., & Alinejad, D. (2020). Social media and trust in scientific expertise: Debating the Covid-19 pandemic in the Netherlands. *Social Media + Society, 6*(4), 2056305120981057.

Van Dijk, T. A. (1984). *Prejudice in discourse: An analysis of ethnic prejudice in cognition and conversation.* John Benjamins Publishing.

Veenhof, B., Wellman, B., Quell, C., & Hogan, B. (2008). *How Canadians' use of the Internet affects social life and civic participation.* Statistics Canada.

Wen, J., Aston, J., Liu, X., & Ying, T. (2020). Effects of misleading media coverage on public health crisis: A case of the 2019 novel coronavirus outbreak in China. *Anatolia, 31*(2), 331–336.

West Pedersen, A., & Kuhnle, S. (2017). The Nordic welfare state model: The concept of a 'Nordic model.' In O. Knutsen (Ed.), *The Nordic models in political science: Challenged, but still viable* (pp. 249–272).

Wimmer, A. (2018). *Nation building: Why some countries come together while others fall apart.* Princeton University Press.

Wodak, R. (2001). The discourse-historical approach. In R. Wodak & M. Meyer (Eds.), *Methods of critical discourse analysis* (Vol. 1, pp. 63–94). Sage.

Wodak, R. (2008). Introduction: Discourse studies-important concepts and terms. In R. Wodak & M. Krzyzanowski (Eds.), *Qualitative discourse analysis in the social sciences* (pp. 1–29). Palgrave Macmillan.

Wodak, R., & Meyer, M. (2009). Critical discourse analysis: History, agenda, theory, and methodology. In R. Wodak & M. Meyer (Eds.), *Methods of critical discourse analysis* (pp. 1–33). Sage.

Yan, B., Chen, B., Wu, L., Zhang, X., & Zhu, H. (2021). Culture, institution, and COVID-19 first-response policy: A qualitative comparative analysis of thirty-one countries. *Journal of Comparative Policy Analysis: Research and Practice, 23*(2), 219–233.

Yan, B., Zhang, X., Wu, L., Zhu, H., & Chen, B. (2020). Why do countries respond differently to COVID-19? A comparative study of Sweden, China, France, and Japan. *The American Review of Public Administration, 50*(6–7), 762–769.

Yang, D., & Counts, S. (2018). *Understanding self-narration of personally experienced racism on Reddit.* Presented at the twelfth international AAAI conference on web and social media.

Yip, N., & Jiang, Y. (2011). Homeowners united: The attempt to create lateral networks of homeowners' associations in urban China. *Journal of Contemporary China, 20*(72), 735–750.

Zhang, L.-T., & Zhao, S. (2020). Diaspora micro-influencers and COVID-19 communication on social media: The case of Chinese-speaking YouTube vloggers. *Multilingua, 39*(5), 553–563.

Zhao, X. (2021). A discourse analysis of quotidian expressions of nationalism during the COVID-19 pandemic in Chinese cyberspace. *Journal of Chinese Political Science, 26*(2), 277–293.
Zhao, X. (2022). Chinese nationalism during the COVID-19 pandemic: Conciliatory and confrontational discourses. *Nations and Nationalism*, 1–20. https://doi.org/10.1111/nana.12813
Zheng, Y., & Wu, G. (2005). Information technology, public space, and collective action in China. *Comparative Political Studies, 38*(5), 507–536.

Liming Liu with a master's degree earned from Uppsala University, is an Instructor II of Public Relations and Advertising at BNU-HKBU United International College (UIC), Guangdong, China. His research interests lie in social media use and discourse analysis, media cultural studies, gender in digital society. His work has been published in *New Media and Society*.

Chally Chai Lee Lim obtained both Bachelor of Communication (Hons) with First Class Honour and PhD in Communication from Northern University of Malaysia. She is currently an Assistant Professor at the Faculty of Humanities and Social Sciences, BNU-HKBU United International College (UIC), Guangdong, China.

Chengxi Yao is an undergraduate student in Public Relations and Advertising programme at BNU-HKBU United International College (UIC), Guangdong, China. Yao's research interests are the approaches to energize companies in cultural industries through effective marketing communication. Yao has published a book review in *Media International Australia*.

Zhiyi Liao is an undergraduate student in Public Relations and Advertising programme at BNU-HKBU United International College (UIC), Guangdong, China. She has interned at L'Oréal Group and Weber Shandwick. Her research interests are primarily in strategic communication. She has presented in the 2021 Asia-Pacific Communication Forum.

Sino–US Great Game Out of the Pandemic: A Conspiracy Theories Perspective

Bo Yan, Simon X. B. Zhao, Yizhou Dong, and Shihao Sun

1 Introduction

Conspiracy theories have been an integral part of almost every major crisis in world history. For instance, one conspiracy theory goes that the United States government knew in advance or even orchestrated the September 11th terrorist attacks (Sunstein & Vermeule, 2009). Since the outbreak of the AIDS epidemic, numerous conspiracy theories cropped up to speculate the origins of the diseases (Bennet, 2012) The ongoing pandemic of COVID-19 which has swept every corner of the world is not immune. According to one study cited by Rachael Rettner, "more than 2000 rumors, conspiracy theories and reports of disinformation" (Rettner, 2020) emerged following the outbreak of the coronavirus cases as of August 2020.

Amidst the plethora of COVID-related conspiracy theories emerge the following main subsets: origins of the coronavirus (Douglas, 2021) conspiracy theories on social media (Zeng & Schäfer, 2021), the effects of conspiracy theories on vaccination (McCarthy et al., 2022; Zelič et al., 2022), and how to combat the conspiracy theories (Nyika et al., 2021). This chapter examines COVID-19 related conspiracy theories in rarely observed aspects: international relations, democratic politics, and globalized market economy. It uses a data-featured approach to contemplate three understudied conspiracy theories during the pandemic. It must be clearly stated that not all unproven theses are conspiracy theories. This chapter debunks three conspiracy theories that are falsified with concrete and clear evidence. Particularly, this chapter focuses on the conspiracy theories in the context of the COVID-19 pandemic amidst the fluctuating great game between China and the United States. It exposes how conspiracy theories have been interwoven into the pandemic and used by various

---

B. Yan (✉) · S. X. B. Zhao · Y. Dong · S. Sun
Beijing Normal University-Hong Kong Baptist University,
United International College, Zhuhai, China
e-mail: BYAN002@e.ntu.edu.sg

© The Author(s), under exclusive license to Springer Nature Singapore Pte Ltd. 2023

S. X. B. Zhao et al. (eds.), *Comparative Studies on Pandemic Control Policies and the Resilience of Society*, https://doi.org/10.1007/978-981-19-9993-2_11

forces across the political spectrum in both the United States and China, and how the mobilizations of the conspiracy theories have only backfired. This chapter will debunk the three following conspiracy theories: one that wishes the pandemic to be "China's Chernobyl," one that worships the COVID as "God's gift to the left" by the United States leftists, and last one, popularly satirized as *DaQiLun* (playing a great chess), that sees the pandemic as China's strategic chance of "corner overtaking" the United States.

2 "China's Chernobyl?": An Old-fashioned Narrative against Communist Countries

A historical review of the narratives in the West will find that the metaphor of "Chernobyl" has been invoked to predict the downfall of communist regimes, particularly in China, after the dissolution of the Union of Soviet Socialist Republics (USSR), whenever a major social crisis was made public.

"Chernobyl" refers to the Chernobyl Nuclear Power Plant explosion in northern Ukraine (former USSR) on 26 April 1986. Its economic, social, and political fallout (for example, the Soviet's failed attempt to cover it from the public) is often analyzed by scholars, journalists, and historians as inextricably linked or directly attributable to the collapse of the Soviet Union.

Since then, the "Chernobyl" narrative has appeared in the West associating sudden events with possibilities of regime change or social transformation in still-standing communist states. These events are usually man-made disasters, characterized by governmental attempt to cover up and large masses' anger and frustration once exposed. Therefore, public health crises or scandals are often labelled by observers in the West as "Chernobyl" moments or predictions. For example, both the SARS in 2003 (Saich, 2003) and the Tianjin explosion in 2015 (OpenDemocracy, 2015) were labelled as China's "Chernobyl" moments.

The coronavirus outbreak at the turn of 2020 in China was the latest Chinese episode joining this Chernobyl-labelling list. The early stage of the pandemic during the beginning months of 2020 when China was main suffering nation of coronavirus witnessed various scholars, politicians, think tanks fellows consider the COVID-19 as China's "Chernobyl." Robert O'Brien, United States national security advisor, thought the communist party's cover-up (of the coronavirus outbreak in Wuhan) would go down in history along with Chernobyl (Reuters, 2020). O'Brien's statement represented a typical stream of thinking in the United States and the West in general. Arthur Herman, senior fellow at the conservative Hudson Institute, also used this Chernobyl narrative linking the coronavirus with China's global ambition. Herman claimed that it was "possible to speculate that Beijing saw the COVID crisis not as a tragedy but as an opportunity to disrupt economies and societies around the world, to China's advantage" (Herman, 2021). Marco Rubio, Senator from Florida, compared the coronavirus and China's early handling to Chernobyl in the Soviet Union after two years since the COVID first broke out in Wuhan (Gancarski, 2022)

The international atmosphere, China's own economic slowdown, and a suspected governmental cover-up of the coronavirus outbreak were factors leading the West to think that the coronavirus outbreak was China's Chernobyl. In the words of Liubomir K. Topaloff, "the situation in China resembles that of the mid-1980s Soviet Union" (Topaloff, 2020). The international atmosphere, particularly, the growing confrontations between Trump' s America and Xi's China, resembled the US-USSR rivalry when Ronald Reagan and Mikhail Gorbachev headed respective camps. Donald Trump was often compared or juxtaposed parallel to Ronald Reagan, particularly among conservative Americans (Adams, 2021; Blodgett, 2019; Frankovic, 2021; Smith & Reagan, 2021). Compared with US presidents in the intervening years between him and President Reagan, President Trump adopted a more US national interest-oriented approach, rather than a globalist one at the expense of US interest. His China policy was also more aggressive than his several predecessors.

Since the very first day that he announced his run for US presidency, Donald Trump unequivocally held high the flag of "America First," dealing a heavy blow to globalization. "We reject the ideology of globalism, and we embrace the doctrine of patriotism," (United Nations, 2018) said President Trump in his speech to the 2018 United Nations General Assembly. It is therefore quite easy to fathom President Trump's every move during his presidency. Diplomatically, Trump re-negotiated trade deals with traditional trade partners, forcing them to buy more American-manufactured goods. The North American Free Trade Agreement (NAFTA) was replaced by the United States-Mexico-Canada Agreement (USMCA). Tariffs were frequently used as leverage to reverse trade deficit of the United States. Trump asked the North Atlantic Treaty Organization allies to raise spending on their own military expenses to alleviate the US burden. He pulled the United States out of many international organizations to discontinue the financial burdens imposed on the United States, such as the Paris Climate Accord in the name of avoiding "nearly $3 trillion in reduced output, over 6 million industrial jobs, and over 3 million manufacturing jobs" (White House, 2017).

The United States under the administration of Donald Trump resembled Reagan's America. Before the US was hit by the pandemic in 2020, the US economy experienced more than 111 streak months of growth, unemployment continued to go down to a 50-year-old low, agreed by both political parties, at 3.5% (Beyer, n.d.), and real median household income grew to reach $63,179 by the end of 2018 (Beyer, n.d.). Unlike the Obama administration whose foreign policy led to social chaos via Arab Spring and achieved little in either Middle East peace or North Korean nuclear negotiation, Trump claimed many diplomatic victories: elimination of ISIS, signing of Abraham Accords normalizing relations between Israel and two Arab countries,[1] and direct talks between a sitting US president and North Korean leader significantly cooling off tensions on the peninsula. While starting no new wars, the Trump administration substantially strengthened the military capability of the United States.

[1] The United Arab Emirates and Bahrain.

Aerospace and defense scholar Loren Thompson stated that "Trump has done more in four years to shift the vector of U.S. military preparations than most presidents accomplish in eight" (Thompson, 2020).

Foreign and diplomatic policies adopted by the four years of Trump presidency posed significant strategic pressure to China (Zhao et al., 2021). Trade war with the United States detrimentally impacted China's export, technology transfer, and talent attraction. Before the Trump presidency, China enjoyed decades of continuous and widening surplus from trade with the United States. That trend, however, came to a halt when Trump became president. Donald Trump won the bid to presidency because his campaign focused on two central issues of the US society: immigration and trade (WBRC, 2016). China was Trump's main target when he set to regulate and reverse the trade relations.

From the data released by the Census Bureau of the United States, it was revealed that after a historic peak of trade surplus for China against the US at $418,232.9 Billion in 2018, trade surplus saw a big drop to $344,312.0 Billion in 2019 and a further reduction to $310, 263.5 billion (United States Census Bureau, n.d.). Multiple economic data emerging in early 2020 pointed to China's shrinking export vis-à-vis the US narrowing its trade deficit. For example, Trump's aggressive policy regulating trade with China had affected China's export, resulting in China's gross domestic product growth hitting a thirty-year low at 6.1% in 2019 (Xu, 2020). Thanks to the 15% decline of the US' annual trade deficit with China, the overall gap between the US imports and exports shrank to its lowest level in three decades in November 2019 (Inman, 2020).

China's economic slowdown was accompanied by its reluctant strategic retreat. Through the years, China's economic boom was boosted by China's generous and ambitious recruitment of leading international, overwhelmingly overseas Chinese, experts in science and technology. A variety of China's Communist Party-backed and government-sponsored talent recruitment schemes were instituted, the most well-known of which was the Thousand Talents Plan (Program, TTP) or Overseas High-Level Talent Recruitment Programs. Center for Security and Emerging Technology (CSET)at Georgetown University issued a report which found that from 2011 to 2018, the youth branch of the TTP alone has attracted nearly 3600 people into monetary awards and academic positions, two thirds of which (66%) were recruited were employed in the United States (Fedasiuk & Feldgoise, 2020). Although the same CSET report has found that just 8.3% of the recruited went to military or defense related institutions under the youth branch of the TTP, the recruitment scheme still attracted suspicion from the foreign authorities, mainly the United States. Trump Administration started high-profile crackdown and arrested some participating scientists, dealing a heavy blow to the Chinese efforts on high-tech transfer. As a result, China suddenly silenced promotion of the TTP, including hiding identities of top scientific recruits amidst growing US scrutiny (Nature Index, 2018), and censoring mentioning of the program online and in official documents and discussions (Keown, 2019).

In addition to this seemingly rival picture of the US going up and China stumbling down like the years when Ronald Reagan began to roll back the Soviet Union,

China's cover-up attempts had provoked observers connecting COVID to Soviet's Chernobyl. The coronavirus outbreak at the turn of 2020, and particularly the government cover-up and the eventual death of the silenced whistleblower Dr. Li Wenliang, ignited public anger and discontent over censorship of speech. A call for the freedom of speech broke out on China's cyberspace. All episode triggered commentators and political scientists' memories of the former Soviet Union's final years. Together with economic decline, Trump's hard policies, pressure from growing ideological confrontation, many people perceived the pandemic in China as the last straw that would break the camel's back. The label "China's Chernobyl" has thus entered scholarly discussions (Sautman, 2021).

However, the conspiracy seeing the coronavirus outbreak in Wuhan as China's Chernobyl was very soon debunked. China's unique success in containing positive cases of coronavirus and more importantly, China's rebound in its economic performance and the subsequent growing legitimacy of the Chinese Communist Party within ordinary Chinese people, crushed any speculation of a forthcoming regime change. The Western countries, in particular the United States, were severely hit by the rampant infections of the coronavirus. From 20 March 2020, the infected cases in the United States skyrocketed, in contrast to China's almost negligible infections. When it came to economic growth, after contraction in the early months of the pandemic, China soon stabilized the production. Standing out to the economy of the US (contracted by 3.6%) and the eurozone's (shrunk by 7.4%), China was the only major economy with net economic growth for 2020 (Cheng, 2020).

Meanwhile, China further debunked the Chernobyl narrative by maintaining social stability in stark contrast to waves and waves of social unrest in the West. The rampant looting, vandalism, and clashes with police in the #BlackLivesMatter protests and unrest in the United States following George Floyd's death were well-known. The chaos on the U.S. Capitol Hill in protest of suspected election fraud on 6 January 2021 was also made international headlines. The Global Peace Index (GPI) released by the think-tank the Institute for Economics & Peace has reported that North America had the largest deterioration of any region in the 2021 GPI, largely due to the unrest in the United States and rising hate crimes in both the US and Canada. The number of recorded hate crimes against Asians in Vancouver, Canada increased by 717 per cent (Global Peace Index, 2021).

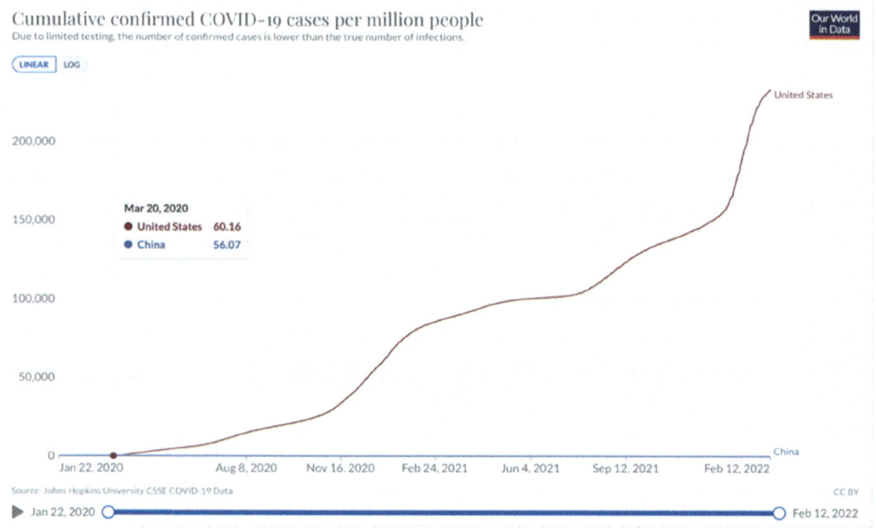

China's quick success in controlling the pandemic was marked by 20 March 2020, since which date the pandemic in the United States went rampant (Our World in Data, n.d.).

3 "God's Gift to the Left": How the United States Leftists Mobilized the Pandemic Narrative to Topple down the Trump Presidency and its Consequences

Narratives in the West portrayed the coronavirus outbreak in Wuhan as "China's Chernobyl." Radical leftist forces in the United States also mobilized the pandemic narrative to topple down the Trump presidency. Jane Fonda, actress and radical leftist activists, (in)famously known as "Hanoi Jane,"[2] openly called COVID-19 "God's gift to the left" in an online interview:

> We are people who can help determine which way humanity goes. What a great gift! What a tremendous opportunity, we're just so lucky…we just have to use it with every ounce of intelligence and courage and wherewithal we have… I just think COVID is God's gift to the left [giggles]…That's a terrible thing to say. I think it was a very difficult thing to send down to us, but it has ripped the Band-Aid off who [Trump] is and what he stands for and what is being done to average people and working people in this country… We can see it now, people who couldn't see it before, you know, they see it now and we have a chance to harness that anger. (Working Families Party, 2020)

---

[2] Jane Fonda was an anti-war activist during the Vietnam War. During her visit to North Vietnam, she was photographed with North Vietnamese troops on an antiaircraft gun that would have been used to shoot down American planes. This earned her the nickname "Hanoi Jane".

The radical leftism in the United States represented by Jane Fonda came under heavy pressures since Donald Trump was elected President of the United States. Echoing President Reagan, a staunch anti-communist, President Trump combated domestically the growing leftist forces in the United States. The growing inclination towards leftism in the United States led President Trump to categorically blazon his 2019 State of the Union address that America "was born free and will stay free...that America will never be a socialist country" (White House, 2019). The economic boost, around 7 million population freeing from food stamps,[3] jobs creation, and diplomatic achievements seemed to push Trump for a sound victory in his re-election. It was reasonably inferred that a second term of Trump presidency would have knocked the leftists, the globalization promoters, and globalization's beneficiaries (represented by China) even harder.

The coronavirus outbreak in the United States and its quick deterioration into a national pandemic, however, gave the leftists an edge to kick off a seemingly ensured Trump's victory. In addition to blaming Trump for the worsening pandemic in the United States, the left had maneuvered elections to their own benefit. The most prominent maneuvers included the rollout of mail-in ballots instead of insistence on in-person votes. In view of the deteriorating situation of the pandemic in the United States, the Democratic Party-states rolled out and enlarged large-scale mail-in ballots nation-wide. For example, the State of Massachusetts passed a law in July 2020, allowing all registered voters in Massachusetts to vote by mail in any 2020 election with no excuse needed (Galvin, 2020). President Donald Trump and the Republican Party, however, staunchly opposed the enlargement of mail-in votes (Burns & Kenneth, 2020). Their rationale was very simple: compared with in-person votes, mail-in ballots significantly increased possibility of voter fraud and damaged election integrity. A mail-in ballot in every state must arrive with a voter signature on the envelope to be counted. However, in real practice, verification of mail-in votes via matching signature went in a sloppy manner. States of North Carolina, Iowa, Wisconsin, Pennsylvania, and New Hampshire, all of which are battleground states in US presidential elections, have abandoned the requirement of signature-matching for mailed voting forms (Payne, 2020). This maneuver of mail-in ballots was called into further question when the Democrats' opposition to strict voter ID regulations. The Democratic Party has long been opposed to voter ID with the excuse that a strict voter ID would restrict people's rights to vote (Parks, 2021).

Because of the massive mail-in votes, votes continued to be counted after polls closed on the election night in the name of calculating the mail-in votes, adding suspicions that ballots could be fabricated in Biden's favor (Deliso, 2020). At the six battleground states whose results would have determined the election results, namely, Arizona, Georgia, Michigan, Nevada, Pennsylvania, and Wisconsin, calculations of votes were halted in the middle of the night. Shortly after the recount, mail-in votes were flocked in, and they were overwhelmingly for Biden. Thus, the #BidenCurve quickly climbed up the trending list on various social media since the election night.

[3] President Trump claimed so in his SOTU address 2020. Although political fact-checkers do not agree with the exact number, they agreed with the trend. See Samantha Putterman (2020).

The leftists insisted that mail-votes did not lead to voter fraud. However, their true attitude towards mail-in votes was revealed in their own elections. The National Labor Relations Board of Amazon.com Inc. asked its Alabama branch to consider having workers vote in person rather than by mail. The rationale was the raised risk of fraud (Day, 2021).

The political drama after the November 3 election night was well known. It is worth noting that the whole drama was well predicted and, with a hindsight, planned by prominent leftist activist, adding further speculations over a rigged election. Axios, a leftist-leaning news media that was established just after Trump started his presidency in 2017, published one article by its founder Jim VandeHei which precisely portrayed every dramatic scene following the election night all the way through the inauguration day on 20 January 2021. The author predicted that there would be a "Red Mirage" on the election night, namely, a landslide victory for Trump, but with waves and waves of mail-in votes coming in, Biden would gradually take the lead (VandeHei, 2020).

The controversies around the BidenCurve, mail-in votes, and election integrity triggered political repercussions from all ends of the political spectrum. The Democratic Party, without a doubt, denied any claims that elections were rigged or stolen, calling any such statements as right-wing conspiracy. Some Republican politicians joined the Democrats, calling for acceptance of election results. Most Republicans and President Trump lambasted the arrangement of mail-in votes and the continuing count of mail-in votes after polls were closed. Disputed claims that the Dominion Voting Systems, the voting machines used in 2020 US presidential elections switched votes from Trump to Biden, led Elections Canada to tweet and clarify that Canada had never used voting machines or electronic tabulators to count votes (Twitter, 2020). Demonstrated by a November 2020 Rasmussen poll shortly after the election, nearly half of America's likely voters, including 20–30% of Democrats, believed that it was likely that the Democrats stole votes or destroyed pro-Trump ballots in several states to ensure that Biden would win (Ramussen, 2020). An earlier Rasmussen poll also showed that just two weeks before the 2020 election day, 94% said their vote would be correctly recorded and counted, with 73% who said it was very likely; while following the 2020 election, those findings fell to 71% and 47% respectively (Rasmussen Reports, 2020).

As controversial and complex as it got, the election integrity was called into further question following a *Time* magazine article in February 2021. In *The Secret History of the Shadow Campaign that Saved the 2020 Election*, Molly Ball documented in great details of how a behind-the-scenes maneuver led by Mike Podhorzer, a Democrat insider, together with CEOs of big-tech companies, mobilized the coronavirus pandemic in arming election staff at local levels, in changing voting systems and laws, in instituting an enlarged and prolonged mail-in votes, in controlling social media to censor conservative voices in the name of fighting disinformation, and in pressing Republican officials in key battleground states to certify the controversial election results (Ball, 2020). Molly Ball might intend to write and publish the article in the leftist *Time* Magazine to cheer for these unsung heroes who "saved the 2020 election" from an alleged attack on democracy by Donald Trump. However, these sophisticated moves ironically validated all the criticism over election integrity in 2020.

The 2020 Presidential election of the United States came to an unpleasant climax on 6 January 2021. The chaos on the capitol hill of the United States, together with the belief that the election was compromised, damaged the soft power and national branding of the United States, when democracy, constitutionalism, and fair elections are its major charm offensives. The United States in 2020, shaken by coronavirus tragedy and haunted by election fraud accusations, surely witnessed its soft power eroded.

The United States' hard power was weakened afterwards. When Joe Biden became President of the United States, he immediately launched various de-Trump domestic and foreign policies. These policies not only worsened the coronavirus pandemic in the United States but plunged the once strong pre-pandemic US economy into what was satirically named "Bidenflation."

The coronavirus pandemic in the United States was worsened under Biden although Barrack Obama called Biden as having "the character and the experience to guide us through one of our darkest times (Jaffe et al., 2020)." Since the outbreak of coronavirus in the United States in January 2020, the number of the accumulated deaths under the Trump Administration was 412,893.[4] Under the Biden Administration, the number of deaths continued to skyrocket. By 28 December 2021, the deaths number of coronavirus under Biden's watch have surpassed those under Trump in just 11 months. Considering the Biden Administration rolled out large scale of vaccines that were invented during the Trump presidency, the high number of deaths during Biden's first year could only prove an even worse response to the coronavirus pandemic from Biden.

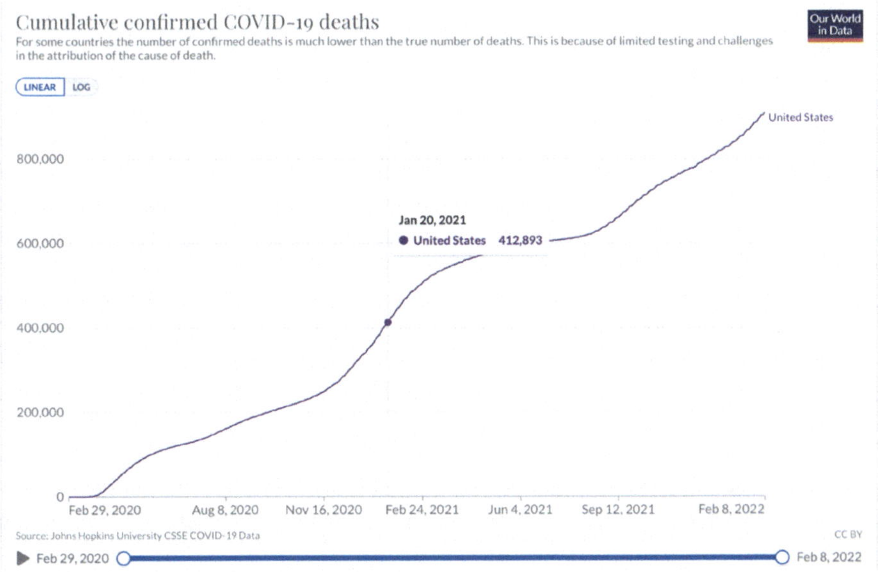

[4] "Cumulative Confirmed COVID-19 Deaths in the United States."

The number of the accumulated deaths under the Trump Administration was 412,893, according to Our World in Data.[5]

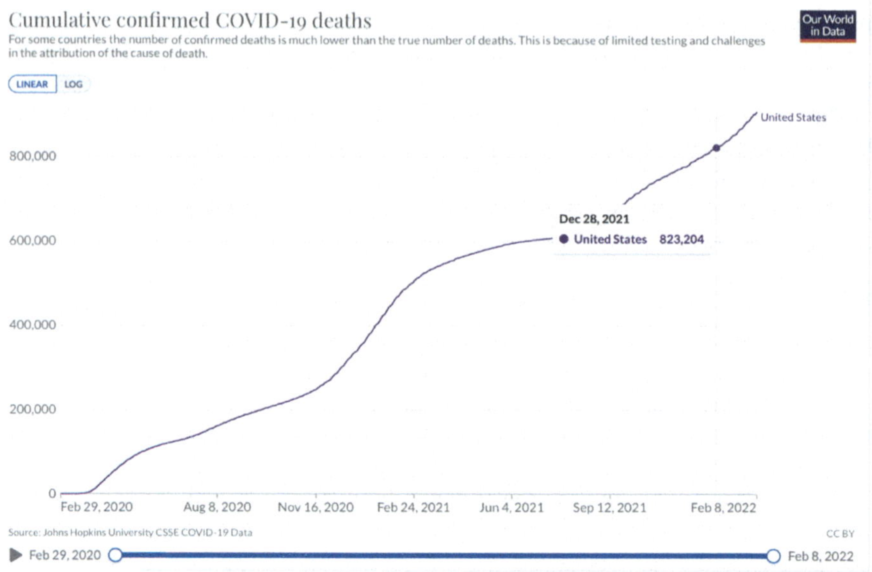

December 2021 marked that COVID deaths under Biden's watch started to surpass those under the Trump Administration.

On 10 January 2022, the United States reported 1.37 million coronavirus cases in one day, a highest daily total for any country since the pandemic.[6] The high number of infections was partly due to the high infection rates of the Omicron variant of the coronavirus. However, it was also attributable to Biden's perturbed approach to the pandemic situation. As early as December 2021, while still vowing to provide full federal support, mainly in the form of stimulus checks, Joe Biden ricocheted responsibility on combating coronavirus pandemic to the states. Attacking on the federal response to the pandemic by President Trump was a significant platform during Biden's presidential campaign. However, during a call with national governors on 26 December 2021, Biden said that federal response was not the key to combating the pandemic, he required resolution at the "state level" (Grossman, 2021).

In addition to pandemic-control fiasco, the de-Trump policies launched by Joe Biden resulted in economic inflation, crisis of shortages, and a worsening public security.

Under Joe Biden, the United States' inflation roared to the highest level in 40 years in December 2021. The consumer prices jumped by 7% for 2021. The surge was the highest annual increase since June 1982 for the Consumer Price Index (CPI), which

[5] Ibid.

[6] Ibid.

tracked the costs of goods and services including food, energy, apparel, gasoline, and vehicles. The same CPI also reported steep increases and record surges in prices of gasolines and vehicles (U.S. Bureau of Labor Statistics, n.d.).

The inflation was attributable to Biden's pandemic-control and other economic policies. As said, the main approach that Biden adopted in combating the coronavirus pandemic was to roll out USD $1.9 trillion "rescue package." However, the rescue package neither stimulated the economy nor focused on crushing the pandemic, but significantly boosted inflation (Smialek, 2021). The "Bidenflation" has caused the average US household to spend around USD $3500 more in 2021 to achieve the same level of consumption of goods and services as in recent previous years (2019 or 2020; Penn Wharton University of Pennsylvania Budget Model, 2021). A Gallup poll has shown that the surging prices and the inflation led 45% of Americans to report financial hardship, 10% to describe the hardship as threatening their current standard of living, and seven in ten lower-income Americans to experience hardship (Younis, 2021).

The Bidenflation was also the result of Biden's progressive and leftist social policies. Vowing to end fossil fuel (Steve Peoples, 2019), Biden has cut the production of coal, oil, and natural gas and driven up energy prices. During the Trump administration, the Justice Department was adamantly opposed to sanctioning consumption sites for narcotics. However, under the Biden administration, the Justice Department signals that it would allow opening of "safe injection sites," namely safe havens for people to use heroin and other narcotics with protections against fatal overdoses (Peltz & Balsamo, 2022). As this will require federal spending, the inflation will surely go high.

4 "Our State is Playing a Big Chess Game": How China's Nationalists Interpreted a Power Crunch When China Had the Upper Hand

The ouster of President Trump and the ascendancy of Joe Biden have relieved China's strategic pressure imposed by various Trump policies. Totally embracing the progressive and leftist agenda, Joe Biden is willing to push globalization and international cooperation. In this regard, the United States under Joe Biden found its common ground with China. The ascendancy of Joe Biden amidst the suspicion of election fraud in the pandemic mitigated the troubled Sino-US relationship. A DW News article analyzed that for Beijing the most difficult time had passed right after Biden's inauguration (Mu, 2021). Faced with a president who is willing collaborate with China for globalization and who is trapped in the suspicion of election fraud, China now has the capacity to play the upper hand in handling the Sino-US relationship.

China's having the upper hand was first evidenced by its different postures to Trump's and Biden's America. Faced with diplomatic, trade, and economic pressure brought by the Trump administration, China was occasionally fiery in diplomatic tone though, remained quite quiet and soft most of time. The tone of China media's reportage of Liu He's visit with President Trump in Washington and Yang Jiechi's visit with US Secretary of State Mike Pompeo was largely neutral and calm.

Since Joe Biden became the US President, China's diplomatic moves suddenly turned aggressive and forceful. Minutes after Joe Biden took office, China's Foreign Ministry announced decision to sanction 28 former Trump administration officials, including Mike Pompeo, Alex Azar, Peter Navarro, Robert O'Brien, Kelly Craft, Matthew Pottinger, John Bolton, and Steve Bannon.[7] The sanctions prohibited those individuals and their immediate family members from entering mainland China, Hong Kong, and Macao. They were also restricted from doing business with China, as were any companies or institutions associated with them (Treisman, 2021). Sanctioning these officials on the day when they left office and new officials were instituted was more symbolic rather than substantial. China's move was a head-on blow at the first encounter against Joe Biden. It also indicated that after Trump, China did not scruple to confront the United States.

Chinese officials and Chinese media reportage turned hawkish as well. The exemplary figure is Yang Jiechi, Director of the Central Foreign Affairs Commission General Office, China's highest diplomatic position. In his address to the National Committee on U.S.-China Relations (NCUSCR), Yang bombarded the Trump administration as it.

adopted misguided policies against China, plunging the relationship into its most difficult period since the establishment of diplomatic ties. Some in the United States, sticking to Cold War thinking, perceived China as a threat. Their rhetoric and actions have interfered in China's internal affairs, undermined China's interests, and disrupted exchanges and mutually beneficial cooperation between the two sides. There have also been attempts to seek "decoupling" and a so-called "new Cold War." Such moves, going against the trend of the times, have seriously damaged China-U.S. relations as well as the fundamental interests of the two peoples (National Committee on U.S.–China Relations, 2021).

Subsequently, in the speech given two weeks after Joe Biden started presidency, Yang demanded that:

1. China be seen as it is.
2. Normal interactions be restored, that the United States "remove the stumbling blocks to people-to-people exchanges, like harassing Chinese students, restricting Chinese media outlets, shutting down Confucius Institutes and suppressing Chinese companies."

[7] They were former Secretary of State, former Health and Human Services Secretary; former White House trade adviser; former national security adviser; former U.S. ambassador to the United Nations; former deputy national security adviser; former national security adviser, and former White House chief strategist.

3. Proper management of differences be called for, "the United States should stop interference in the affairs of Hong Kong, Tibet and Xinjiang, which all matter to China's sovereignty and territorial integrity, and stop attempts to hold back China's development by meddling in China's internal affairs." In Yang's words, this was the redline, and now it was China that defined the redline.
4. Mutually beneficial cooperation be broadened. The US should not abuse the concept of "national security." The two countries could cooperate on issues like climate change.[8]

Yang's forceful demands did not pause at the address to NCUSCR. Yang intensified bombardment at US officials in their face during Sino-US Alaska summit in March 2021 led by him and Wang Yi (Minister of Foreign Affairs), and Antony Blinken (Secretary of State) and Jake Sullivan (National Security Advisor). During the statements from the summit, Yang exuded a different tone when he met Secretary Pompeo. Instead, he angrily lectured and reprimanded the United States for pushing its democracy onto others and failing to deal with its own human rights:

> I think we thought too well of the United States. [我们把你们想的太好了!]We thought that the U.S. side will follow the necessary diplomatic protocols. So for China it was necessary that we made our position clear…So let me say here that, in front of the Chinese side, the United States does not have the qualification to say that it wants to speak to China from a position of strength.[我现在讲一句, 你们没有资格居高临下同中国说话] The U.S. side was not even qualified to say such things even 20 years or 30 years back, because this is not the way to deal with the Chinese people. [中国人不吃这一套!].[9]

As opposed to most in-door talks that were only downplayed by China's state media, many of Yang's straightforward and blunt expressions, whose emotions were particularly stronger and overbearing in Chinese, were broadly reported by Chinese media. The Chinese remarks became hotly searched trending words on Weibo. The *People's Daily* manufactured Weibo memes featuring these Chinese words which went viral. Online store soon began to sell merchandise featuring Yang's aggressive remarks.

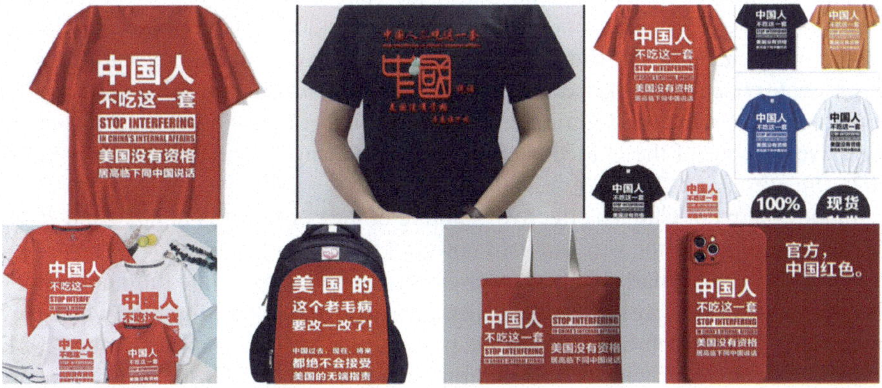

[8] National Committee on U.S.–China Relations (2021).

[9] The Chinese translations are invoked by the authors. For full transcript of the talk, see Nikkei Asia (2021).

Online store selling merchandizing t-shirts and mobile phone cases featuring Yang's words (Nan, 2021).

It may require another lengthy article to analyze why Yang could claim that "the United States does not have the qualification to say that it wants to speak to China from a position of strength" just two months after Biden took office. However, what followed in the Sino-US relations indeed reflected that the United States has been following and implementing ethos in Yang's demands.

In the first place, Meng Wanzhou, Chief Financial Officer of Huawei, the technology computer at the center of the China-US trade war and technology competition, was release in September 2021. Later in October 2021, the US Commerce Department issued more than USD $100 billion worth of export licenses for semiconductors and other productors to suppliers of Huawei (O'Keeffe, 2021). This was in line with Yang's request of the US to stop "suppressing Chinese companies." The once suppressed Thousand Talents Plan (Program) re-bloomed in China under Biden administration. China Initiative, a program launched by President Trump to counter China's absorption of the US intellectual property, was ended by Joe Biden's Justice Department (Lucas, 2022). Secondly, in November 2021, Jake Sullivan stated explicitly that the White House would not try to transform China as it would be the repetition of the errors of the past (the Trump presidency), nor to contain China in a new cold war; instead, the United States would be seeking coexistence with China (Wang, 2021). Even the habitually jingoistic and hawkish *Global Times* considered the remarks as a softer tone on China-US ties (Chen & Yan, 2021). Thirdly, when it came to trade, a highly contested issue in the recent Sino-US relationship, China scored well against the United States' trade maneuver. On 7 November 2021, General Administration of Customs of the People's Republic of China issued data which showed that trade between the China and the United States stood at CNY 3.95 trillion totally with an overall 23.4% increase; China's trade surplus to the US reported a 18.9% growth at 2.08 trillion (Welle, 2021). China's foreign exchange reserves saw a net increase for the first time since July 2021.[10] Considering that Biden's Commerce Secretary Gina Raimondo said in an interview that she would seek to improve US business ties with China (Davis, 2021), Trump's trade war was substantially ended by the Biden administration. Lastly, answering the call for tackling climate change, the United States joined China in announcing a joint declaration agreeing to cooperate on limiting carbon emissions and slowing climate change at the eleventh hour of the United Nations Climate Change Conference in November 2021 (Shapiro et al., 2021).

In the context of China's effective handling of the coronavirus in 2020 and 2021, the evident falling of the soft and hard power of the United States and other Western countries, and particularly, China's ever enlarged trade surplus, the third conspiracy theory came to the fore. The nationalists in China were the promoter when they interpreted reasons behind a nationwide power crunch in China in September 2021.

Since July 2021, provinces across China had been hit by power shortages. Yunnan, Zhejiang, Jiangsu, Guangdong, Liaoning, Chongqing, Inner Mongolia, Henan, and

[10] Welle (2021).

other provinces had to ration electricity. Related news mostly went unnoticed. However, a nationwide power crunch occurred abruptly in late September 2021 and instantly made China's 2021 power shortage an international news headline.

The reason for such a high visibility of China's power shortage was the September power crunch severely affected civilian lives in Northeast China. A power crunch in Liaoning province forced one local mining enterprise to a sudden halt. The resulting abrupt suspension of exhaust system in the mines caused 23 labors to get poisoned and lives threatened. Xinbei Water Supply Company in neighboring Jilin Province irritated netizens with their blunt announcement asking the public to get ready for electricity and water shortage on an irregular, uncertain, out-of-plan, and no-reminder basis (STCN, 2021).

Amidst the public outcry came an unusual online voice attributing the power crunch as a state-initiated maneuver to smartly shift inflation back to the United States. Creators of such narrative, mostly "We Media" bloggers on WeChat platform, used the metaphor of traditional Chinese chess game and interpreted this maneuver as "the state was playing a chess" (*guojia zaixia yipan daqi* 国家在下一盘大棋) to "corner overtake" (*wandao chaoche* 弯道超车) the United States. As in traditional Chinese cultural, playing a chess had the connotation of shrewd strategic maneuvering for eventual gains, often at the expense of the short-term interests.

The narrative contained the following logic flows. As China achieved unique effectiveness in controlling the pandemic, China became the first country to resume industrial production, especially in manufacturing section. The Western countries, such as the United States, according to these bloggers, set up a trap for China by pushing up prices for raw materials and commodities and shifting their inflation to the Chinese manufacturers as the West were in control of pricing right. The Chinese manufactures competed to woo overseas clients and orders to make up for their losses during the Sino-US trade war and the pandemic. Thus, an overstretched production resulted in lowered prices for and limited profits from manufactured Chinese goods bound for export overseas. These bloggers even worried that once the pandemic in the West neared its end and their manufacturing convalesced, manufacturing in China would be dealt an even heavier blow to. They interpreted this wave of power outage and rationing as Chinese state's response and solution to this dilemma: forcing enterprises and factories to shut down to reduce production output. The intended scenario, according to these bloggers, was illustrated below[11]:

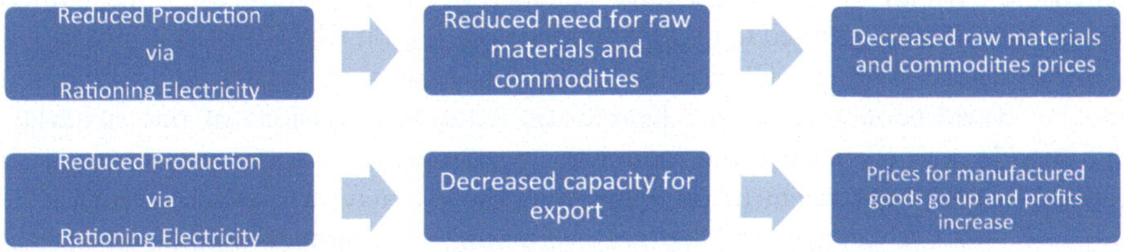

[11] In late September 2021, numerous articles published and circulated by WeChat platform emerged interpreting the power shortage and rationing as "playing a chess." After the repudiation by the state media, these bloggers deleted such articles. Therefore, only screenshots are available.

Such interpretations that the power crunch and rationing were an "international financial war;" and what was behind was a "battle for pricing rights on international commodities" were refuted by Chinese state and official media. Such narratives, as state media repudiated, indicated that the state had electricity but did not let industries and civilians use electricity. The true scenario, explained by the CCTV news, was that China lacked electricity. The power provision was faced with a severe deficit, and rationing was an act against the will (Yangshiwang et al., 2021).

Xia Ke Dao (侠客岛), the online sobriquet of the *People's Daily*, also debunked the conspiracy theory by explaining the reason for power cut at provincial level. China had initiated the policy of "dual control of energy consumption" for years. Since the pandemic and the rising costs of commodities, Chinese enterprises felt the urge to step up production. Meanwhile, the quick success of China's pandemic control and resumption of manufacturing paved the way for accepting orders from foreign markets. As China vowed to reach peak carbon dioxide emissions by 2030, these enterprises took the date as countdown for a "window period" and accelerated their production. Therefore, productions with high energy consumption and high pollution rushed on. All these resulted in the failure of many provinces to achieve their annual energy and pollution reduction goal. Consequently, the provinces hasted and imposed rigid power cut by the end of the year. That was the reason accounting for the nationwide power crunch.[12]

China's long-time practice of Dual-Track Electricity Price (DTEP) system was also behind this nationwide power crunch. The essence of the DTEP was that "the electricity price is determined by the market, while the transmission and distribution price is determined by the government with strict supervision" (Ding et al., 2020). Therefore, when China suffered from a coal shortage in 2021, the rising market-oriented electricity price could not be compensated by the government-controlled distribution price which remained low. Thus, electricity factories could not supply electricity at a widening deficit, leading to a shortage of electricity in late 2021.

It now came down to the question: why was there a coal shortage? The hyper-nationalistic bloggers were reluctant to admit that China's diplomatic feud with Australia, China's main source of coal importer, restricted China's access to high quality coal. In late 2020, China banned the import of Australian coal in retaliation for political tensions between the two states. The conspiracy theorists interpreted all the episodes: the power outage, China's victory in controlling the pandemic, resumption of economic production, and rising costs of materials in an imagined game in which the coronavirus boosted China's chance of "corner overtaking" the United States. However, they failed to touch upon the very reason why China was struggling with electricity shortage. China's economy has now been inextricably integrated into the world economy, relying heavily on external provisions of raw materials and outside markets. when other countries, particularly Western countries are still trapped in the predicament of the pandemic, the coronavirus pandemic cannot be seen as China's advantage. The "corner overtaking" was another thinking as wishful as the one seeing the pandemic as China's Chernobyl.

[12] Yangshiwang et al. (2021).

5 Conclusion

This chapter has analyzed three conspiracy theories during the pandemic concerning the Sino-US relations. When the United States was on the offensive by various Trump policies and forcing upon China strategic pressure, the coronavirus outbreak in Wuhan was interpreted as "China's Chernobyl." However, the radical left in the United States exploited this "God's gift to the left" to set up election agenda, ballot methods, and votes calculation, resulting in a presidency whose domestic and foreign policy have eased China's strategic pressure imposed by Trump. In a reversed power relationship, China seemed to play the upper hand against the United States evidenced by the United States' return to pre-Trump track to work with China on globalization, technology transfer, climate issues, and trade. China's unique success in pandemic control, production resumption, and China's strategic rebound after release from Trump pressure do not mean China could go alone. As China was haunted by global supply chain crisis closely associated with the pandemic and by the coal shortages, the third conspiracy theory, seeing China's nationwide power crunch imposed by electricity cut as the government's smart move to shift inflation back to the United States was as unreasonable as all the others with vicious wishful thinking. The pandemic is not any political force's opportunity, be they Western dragon slayers, US radical leftists, and Chinese hypernationalists. Mobilizing the pandemic narrative for a "I win, you lose" game will only be backfired.

References

Adams, N. (2021). *Trump and Reagan: Defenders of America.*

Bennet, N. (2012). The AIDS conspiracy: science fights back. *The Lancet Infectious Diseases 12*(8)596. https://doi.org/10.1016/S1473-3099(12)70195-0

Beyer, D. (n.d). Did Trump create or inherit the strong economy. Senate.gov. https://www.jec.senate.gov/public/_cache/files/2c298bda-8aee-4923-84a3-95a54f7f6e6f/did-trump-create-or-inherit-the-strong-economy.pdf?fbclid=IwAR1sxNZYkBAnuW3EmM3VA_Jw50pnNGCnSnBSktJh39RZd1O5PCo37Mk7xzI

Blodgett, T. (2019). Why 'the Donald' is the new 'Ronald': Parallels between Trump and Reagan. *Des Moines Register*, June 10, 2019. https://www.desmoinesregister.com/story/opinion/columnists/2019/06/10/parallels-between-trump-reagan-show-why-donald-new-ronald/1367579001/

Burns, P., & Kenneth, N. J. (2020). Coronavirus recovery: Trump campaign, GOP sues to block vote-by-mail; murphy says 'bring it on'. *PBS*, April 19, 2020. https://whyy.org/articles/n-j-coronavirus-recovery-gop-joins-trump-campaign-to-block-mail-in-election/

Chen, Q., & Yan, Y. (2021). Sullivan's remarks suggest softer tone on US-China ties, but 'words alone are not enough.' *Global Times*, November 8, 2021. https://www.globaltimes.cn/page/202111/1238448.shtml

Cheng, J. (2020). China is the only major economy to report economic growth for 2020. *Wall Street Journal*, January 18, 2021. https://www.wsj.com/articles/china-is-the-only-major-economy-to-report-economic-growth-for-2020-11610936187

Davis, B. (2021). Commerce secretary Gina Raimondo aims to strengthen business ties with China. *The Wall Street Journal*, September 24, 2021. https://www.wsj.com/articles/commerce-secretary-gina-raimondo-aims-to-strengthen-business-ties-with-china-11632475802

Deliso, M. (2020). Trump and Republicans have asked states to stop counting votes before. *ABC News*, November 7, 2020. https://abcnews.go.com/Politics/trump-republicans-asked-states-stop-counting-votes/story?id=74043767

Ding, W., Gao, F., Xu, Y., Zhu, J., & Hu, J. (2020). Economic impact assessment of the dual-track electricity price system in China. *IOP Conference Series: Earth and Environmental Science, 615*(1), 012053. https://doi.org/10.1088/1755-1315/615/1/012053

Douglas, K. M. (2021). COVID-19 conspiracy theories. *Group Processes and Intergroup Relations, 24*(2), 270–275. https://doi.org/10.1177/1368430220982068

Fedasiuk, R., & Feldgoise, J. (2020). *The youth thousand talents plan and China's Military.* Center for Security and Emerging Technology. https://doi.org/10.51593/20200041

Frankovic, K. (2021). Republicans used to call Ronald Reagan the best president in US history: Now it's Donald Trump. *YouGovAmerica*, February 15, 2021. https://today.yougov.com/topics/politics/articles-reports/2021/02/15/republicans-best-president-reagan-trump-

Galvin, W. F. (2020). Important elections updates. Secretary of the Commonwealth of Massachusetts. https://www.sec.state.ma.us/ele/covid-19/covid-19.htm

Gancarski, A. G. (2022). Marco Rubio says COVID-19 is China's Chernobyl. *Florida Politics*, January 13, 2022. https://floridapolitics.com/archives/486466-marco-rubio-says-covid-19-is-chinas-chernobyl/

Global Peace Index. (2021). *Measuring peace in a complex world.* Institute for Economic and Peace. https://www.economicsandpeace.org/wp-content/uploads/2021/06/GPI-2021-web.pdf

Grossman, H. (2021). Biden says states bear responsibility for COVID resolution, 'The five' reacts. *Foxnews*, December 27, 2021. https://www.foxnews.com/media/biden-states-responsibility-covid-resolution-the-five

Herman, A. (2021). Is COVID China's chernobyl? Hudson Institute. https://www.hudson.org/research/17109-is-covid-china-s-chernobyl

Inman, P. (2020). US trade deficit falls to three-year low in wake of China Standoff. *The Guardian*, January 7, 2020. https://www.theguardian.com/business/2020/jan/07/us-trade-deficit-falls-to-three-year-low-in-wake-of-china-standoff

Jaffe, A, Pace, J., & Barrow, B. (2020). Obama Endorses Biden as the best leader for 'darkest times.' *Associated Press*, April 15, 2020. https://apnews.com/article/donald-trump-ap-top-news-elections-barack-obama-joe-biden-16690d8c2cb5fc8503b33a32dde28359

Mu, Y. (2021). Beijing Zan Wu Yang Jiechi Fangmei Jihua de Sanda Liyou 北京暂无杨洁篪访美计划的三大理由 [Three reasons Beijing temporarily won't send Yang Jiechi to Visit America]. *DW News*, January 25, 2021.

Nan, X. (2021). Zhongmei Gaoceng Tanpan Yang Jiechi 'jinju' Baohoong Shangjia Yinzhi Duozhong Shangpin Zao Qianggou 中美高层谈判杨洁篪'金句'爆红 商家印制多种商品遭抢购[图] [Yang Jiechi's 'fiery Remarks' at US-China Summitt Went Viral, Merchandise Hotly Sought For]. *DW News*, March 22, 2021.

National Committee on U.S.–China Relations. (2021). A conversation with politburo member Yang Jiechi. *YouTube*, February 4, 2021. https://www.youtube.com/watch?v=k949eWl9a7M

Nature Index. (2018). China hides identities of top scientific recruits amidst growing US scrutiny. https://www.natureindex.com/news-blog/china-hides-identities-of-top-scientific-recruits-amidst-growing-us-scrutiny

Nikkei Asia. (2021). How it happened: Transcript of the US-China opening remarks in Alaska. https://asia.nikkei.com/Politics/International-relations/US-China-tensions/How-it-happened-Transcript-of-the-US-China-opening-remarks-in-Alaska

Nyika, A., Taponeswa Nyika, G., Tonderai Nyika, J., Tashinga Nyika, J., & Nyika, T. (2021). COVID-19 pandemic: Questioning conspiracy theories, beliefs or claims that have potential negative impact on public health interventions and proposal for integrated communication and information dissemination strategies (ICIDS). *Journal of Development and Communication Studies, 8*(1), 1–21. https://doi.org/10.4314/jdcs.v8i1.1

O'Keeffe, K. (2021). U.S. issued $100 billion in export licenses to suppliers of Huawei, SMIC. *The Wall Street Journal*, October 21, 2021. https://www.wsj.com/articles/u-s-issues-100-billion-in-export-licenses-to-suppliers-of-huawei-smic-11634869839

Day, M. (2021). Amazon requests in-person union vote in COVID-hit Alabama. *Bloomberg*, January 22, 2021. https://www.bloomberg.com/news/articles/2021-01-22/amazon-requests-in-person-union-vote-in-covid-plagued-alabama?sref=ZMFHsM5Z

McCarthy, M., Murphy, K., Sargeant, E., & Williamson, H. (2022). Examining the relationship between conspiracy theories and COVID-19 vaccine hesitancy: A mediating role for perceived health threats, trust, and anomie? *Analyses of Social Issues and Public Policy, 22*(1), 106–129. https://doi.org/10.1111/asap.12291

Ball, M. (2020). The secret history of the shadow campaign that saved the 2020 election. *Time*, February 4, 2021. https://time.com/5936036/SECRET-2020-ELECTION-CAMPAIGN/

OpenDemocracy. (2015). *Is this China's Chernobyl moment? OpenDemocracy*, August 26, 2015. 1707100409. ProQuest One Academic Research Library. https://www.proquest.com/magazines/is-this-chinas-chernobyl-moment/docview/1707100409/se-2?accountid=38789

Our World in Data. (n.d). Cumulative confirmed COVID-19 deaths in the United States. https://ourworldindata.org/explorers/coronavirus-data-explorer?facet=none&Metric=Confirmed+deaths&Interval=Cumulative&Relative+to+Population=false&Color+by+test+positivity=false&country=~USA

Parks, M. (2021). Democrats are now open to new voter ID rules. It probably won't win over the GOP. *NPR*, August 28, 2021. https://www.npr.org/2021/08/28/1031164994/democrats-are-now-open-to-new-voter-id-rules-it-probably-wont-win-over-the-gop

Payne, D. (2020). Key battleground states don't require signature-matching on mail-in ballots. *The Michigan Star*, October 16, 2020. https://themichiganstar.com/2020/10/16/key-battleground-states-dont-require-signature-matching-on-mail-in-ballots/

Peltz, J., & Balsamo, M. (2022). Justice dept. signals it may allow safe injection sites. *Associated Press*, February 8, 2022. https://apnews.com/article/business-health-new-york-c4e6d999583d7b7abce2189fba095011

Penn Wharton University of Pennsylvania Budget Model. (2021). Impact of inflation by household income. https://static1.squarespace.com/static/55693d60e4b06d83cf793431/t/61ba00dab7490056df65758d/1639579866872/Impact+of+inflation+by+Household+Income.pdf

Peoples, S. (2019). In intimate moment, Biden vows to 'end fossil fuel.' *Associated Press*, September 7, 2019. https://apnews.com/article/9dfb1e4c381043bab6fd0fa6dece3974

Putterman, S. (2020). Trump's state of the union claim about welfare, food stamp declines is off. *Politifact*, February 6, 2020. https://www.politifact.com/factchecks/2020/feb/06/donald-trump/trumps-state-union-claim-welfare-food-stamp-declin/

Ramussen. (2020). 61% think Trump should concede to Biden. *Rasmussen Reports*, November 19, 2020. https://www.rasmussenreports.com/public_content/politics/elections/election_2020/61_think_trump_should_concede_to_biden

Rasmussen Reports. (2020). Most decided their vote over a month ago, worry about voter fraud. https://www.rasmussenreports.com/public_content/politics/elections/election_2020/most_decided_their_vote_over_a_month_ago_worry_about_voter_fraud

Rettner, R. (2020). COVID-19 has fueled more than 2000 rumors and conspiracy theories. *Live Science*, August 12, 2020. https://www.livescience.com/covid-19-rumors-conspiracy-theories-infodemic.html

Reuters, S. (2020). Coronavirus 'cover-up' is China's chernobyl—White house adviser. *Reuters*, May 25, 2020. https://www.reuters.com/article/healt-coronavirus-usa-china-idUSKBN23106X

Saich, A. (2003). The real fallout from China's chernobyl. *FT.Com*, May 27, 2003, 1.

Sautman, B. (2021). China's Chernobyl: COVID-19 narratives of collapse and global domination. *Pacific Affairs, 94*(4), 683–706. https://doi.org/10.5509/2021944683

Shapiro, A., Brown, A., Caldwell, N., Venkat, M., & Jarenwattananon, P. (2021). U.S. and China announce surprise climate agreement at cop26 summit. *NPR*, November 11, 2021. https://www.npr.org/2021/11/11/1054648598/u-s-and-china-announce-surprise-climate-agreement-at-cop26-summit

Smialek, J. (2021). A regional fed analysis suggests biden's stimulus is temporarily stoking inflation. *The New York Times*, October 19, 2021. https://www.nytimes.com/2021/10/18/business/economy/fed-inflation-stimulus-biden.html

Smith, R. C., & Reagan, R. (2021). Donald Trump, and the Future of the Republican party and conservatism in America. *American Political Thought, 10*(2), 283–289. https://doi.org/10.1086/713662

STCN. (2021). Xiandian Yinfa! Liaoning Yi Qiye Meiqi Xielou, 23 Ren Zhongdu Songyi! Jilin Yi Shuiwu Gongsi: She Tingdian Tingshui Tongzhi Cuoci Budang 限电引发!辽宁一企业煤气泄漏,23人中毒送医!吉林一水务公司:涉停电停水通知措辞不当[Caused by Power Rattion! Enterprise in Liaoning had gas leaked, 23 poinsoned and hospitalised! Water enterprise in Jilin: announcement on power and water shortaged criticized as inappropriate]. https://news.stcn.com/sd/202109/t20210927_3716936.html

Sunstein, C. R., & Vermeule, A. (2009). Conspiracy theories: Causes and cures. *Journal of Political Philosophy, 17*(2), 202–227. https://doi.org/10.1111/j.1467-9760.2008.00325.x

Thompson, L. (2020). Love him or hate him, President Trump's defense legacy is profound. *Forbes*, December 15, 2020. https://www.forbes.com/sites/lorenthompson/2020/12/15/love-him-or-hate-him-president-trumps-defense-legacy-is-profound/?sh=77c5bec3795a

Topaloff, L. K. (2020). Is COVID-19 China's 'chernobyl moment'? *The Diplomat*, March 4, 2020. https://thediplomat.com/2020/03/is-covid-19-chinas-chernobyl-moment/

Treisman, R. (2021). China slaps sanctions on 28 Trump administration officials, including Mike Pompeo. *NPR*, January 20, 2021. https://www.npr.org/2021/01/20/958996415/china-slaps-sanctions-on-28-trump-administration-officials-including-mike-pompeo

Twitter. (2020). Elections Canada. https://twitter.com/electionscan_e/status/1328449565096366083

U.S. Bureau of Labor Statistics. (n.d.). 12-month percentage change, consumer price index, selected categories. https://www.bls.gov/charts/consumer-price-index/consumer-price-index-by-category-line-chart.htm

United Nations. (2018). US President Trump rejects globalism in speech to UN general assembly's annual debate. https://news.un.org/en/story/2018/09/1020472

United States Census Bureau. (n.d). Foreign trade: Trade in goods with China. https://www.census.gov/foreign-trade/balance/c5700.html#2018

VandeHei, J. (2020). A safe, sane way to navigate election night—and beyond. *Axios*, November 1, 2020. https://www.axios.com/election-night-how-to-watch-4731bd76-e0b8-4b4a-ba6f-324062e44de5.html

Wang, A. (2021). US wants coexistence not cold war with China, Jake Sullivan says. *South China Morning Post*, November 8, 2021. https://www.scmp.com/news/china/diplomacy/article/3155320/us-wants-coexistence-not-cold-war-china-jake-sullivan-says

WBRC. (2016). Transcript: Sen. Jeff Sessions speaks at the 2016 republican national convention. https://www.wbrc.com/story/32473757/transcript-sen-jeff-sessions-speaks-at-the-2016-republican-national-convention/

Welle, D. (2021). Zhongguo Waimao Baochi Qiangjing Shiyue Shuncha Chuangjilu 中国外贸保持强劲 10月顺差创纪录[Strong performance by China's trade export surplus for October set new record]. https://p.dw.com/p/42iof

White House. (2017). President Trump announces u.s. withdrawal from the Paris climate accord. *Trump White House Archives*, June 1, 2017. https://trumpwhitehouse.archives.gov/articles/president-trump-announces-u-s-withdrawal-paris-climate-accord/

White House. (2019). Remarks by President trump in state of the union address. *Trump White House Archives*, February 6, 2019. https://trumpwhitehouse.archives.gov/briefings-statements/remarks-president-trump-state-union-address-2/

Keown, A. (2019). As western fears mount, Chinese government silences promotion of thousand talents plan. *Biospace*, January 10, 2019. https://www.biospace.com/article/as-western-fears-mount-chinese-government-silences-promotion-of-thousand-talents-plan/

Lucas, R. (2022). The justice department is ending its controversial China initiative. *NPR*, February 23, 2022. https://www.npr.org/2022/02/23/1082593735/justice-department-china-initiative

Working Families Party. (2020). A conversation between activist & leader Jane Fonda and WFP National Director Maurice Mitchell. *YouTube*, October 3, 2020. https://www.youtube.com/watch?v=WuCvO7gNzfY

Xu, E. N. (2020). The U.S. trade war slowed China's 2019 economic growth to its weakest pace in nearly 30 years. *Fortune*, January 17, 2020. https://fortune.com/2020/01/17/china-gdp-growth-2019-weakest-30-years-trade-war/

Yangshi, S., Xiandian, L., & Nameduo, L. M. (2021). 'Daqi' 央视网:拉闸限电里没那么多大棋 [*CCTV News*: There is no 'chess game' in power rationing]. https://news.sina.com.cn/c/2021-09-29/doc-iktzscyx6952210.shtml

Younis, M. (2021). Inflation causing hardship for 45% of U.S. households. *Gallup*, December 2, 2021. https://news.gallup.com/poll/357731/inflation-causing-hardship-households.aspx

Zelič, Ž., Berič, M., & Kobal Grum, D. (2022). Examining the role of COVID-19 conspiracy beliefs in predicting vaccination intentions, preventive behavior and willingness to share opinions about the coronavirus. *Studia Psychologica, 64*(1), 136–153. https://doi.org/10.31577/sp.2022.01.844

Zeng, J., & Schäfer, M. S. (2021). Conceptualizing 'dark platforms'. Covid-19-related conspiracy theories on 8kun and Gab. *Digital Journalism* 9(9), 1321–1343. https://doi.org/10.1080/21670811.2021.1938165

Zhao, S, Simon, X. B., & Zhang, Z. (2021). US-China rivalry and its implications for the post-pandemic world. In S. X. B. Zhao, J. H. C. Wong, C. Lowe, E. Monaco, & J. Corbett (Eds.), *COVID-19 pandemic, crisis responses and the changing world: Perspectives in humanities and social sciences* (pp. 9–25). Springer. https://doi.org/10.1007/978-981-16-2430-8_2

Bo Yan received PhD in History from Nanyang Technological University, Singapore. He is a Research Fellow in the Faculty of Humanities and Social Sciences. Yan's main research interests include modern Chinese history, postwar United States, and modern Southeast Asian history, and the history of the global 1960s. He has published in the *Journal of Asian Studies*, *Twentieth Century China*, and others. Yan's PhD dissertation was a finalist in the International Convention of Asia Scholars (ICAS) Book Prize 2021-Best Dissertation in Humanities. Bo Yan is currently conducting studies and research on global pandemic of COVID-19.

Simon X. B. Zhao is Professor and Associate Dean at the Faculty of Humanities and Social Sciences at BNU-HKBU United International College (UIC). Professor Zhao used to teach in the University of Hong Kong as a founding Director of International Center for China Development Studies and he specializes at urban, regional and global studies and publishing in the broad areas of urban regional studies, geopolitical and economic development, spatial and global transformation, geography of international trade and finance, and development of international financial centers (IFCs). He is an editorial board member for four international journals and published more than 150 papers in international refereed journals, mainly SSCI/SCI/CSSCI Citation Journals, book chapters and consultative reports etc. Simon Zhao was a Member of the Central Policy Unit, HKSAR Government (Pan-PRD Panel) and chaired and co-led many substantial academic research projects, providing widely consultancy services to China national, provincial, and local governments as well as private sector. His latest edited volume *COVID-19 Pandemic, Crises Response and the Changing World: Perspectives in Humanities and Social Sciences* (by Springer) is well received globally.

Yizhou Dong is currently reading food science and technology at the Beijing Normal University-Hong Kong Baptist University United International College. Dong's main research interest falls within nutrients studies.

Shihao Sun earned a master's degree in Development Studies from the City University of Hong Kong before a bachelor's degree in finance from the University of International Business and Economics. Sun's main research interests include public policy and regional economy.

Cultural and Social Media Responses: Perceptions, Emotions, and Narratives

Body Politics and the Collective Well-Being: A Comparative Study of the Cultural Motives of Mask Wearing During COVID-19

Qiaoyun Zhang and Yushan Wang

1 Introduction

Taking mask wearing as an example, this chapter examines how culture influences people's risk perception and response. Mask wearing has become one of the most striking representations of the Covid-19 pandemic. Although masks are widely advocated by medical specialists for its effectiveness in virus transmission prevention, along with the rising demand for facemasks comes the even more heated debate about the use and meaning of them. In policy making and everyday practice, governments and groups across the world have adopted very different approaches to mask wearing as well as many other disease prevention and control measures. Clearly, risk perception and response is no longer only a matter of technology or knowledge.

From political leaders to the general public, people tend to account for the difference in pandemic control measures in terms of culture. But what does culture actually do to shape people's risk response logic and models? Such a question is not only at the center of the risk culture theory proposed by anthropologist Mary Douglas and her collaborators, but also pivotal for sociological research on risk perception and management. Crook (1999) summarizes in a probably oversimplified manner that the cultural theory of risk offers a "sociocultural account of risk processes." For Douglas and Wildavsky (1982), risk perception is largely shaped by specific groups' interest demands, cultural types and value orientation. The different ideas towards individual and group identities can lead to people's varied and sometimes contradictory understanding of what to be accepted as risk, its degree of harm, and how to deal with it.

Q. Zhang (✉) · Y. Wang
Beijing Normal University-Hong Kong Baptist University,
United International College, Zhuhai, China
e-mail: qiaoyunzhang@uic.edu.cn

© The Author(s), under exclusive license to Springer Nature Singapore Pte Ltd. 2023

S. X. B. Zhao et al. (eds.), *Comparative Studies on Pandemic Control Policies and the Resilience of Society*, https://doi.org/10.1007/978-981-19-9993-2_12

Analyzing discourse and practice of mask wearing among Chinese and Euro-American citizens since 2020 through interviews and case studies, the chapter proposes that cultural perceptions of risk can be respectively yet relationally analyzed in the individual and collective perspectives. The view of individual including ideas of death, reincarnation, body boundary and pollution reveals the effects of identity politics on risk response. The collective view pertaining to classification of social groups and the blame of "the other" exposes how the construction of groupness and order influences the evaluation of risk. People's different decisions of wearing a mask or not practically and symbolically are extensively influenced by their perceptions of the risk of Covid-19, which in turn, are shaped by the particular socio-political structure and tradition with which they identify themselves.

The chapter starts with introducing risk as collective construct known and managed through the principles of social classification so as to achieve and maintain social order. It follows a discussion of the history of mask use in the medical field to show that scientific and even "civilizing" measures such as mask wearing may not be cohesively accepted in social policy. The chapter then analyzes how the ideas of individual identity and collective well-being influence different groups' mask use during the Covid-19. Finally we provide advice on how public health policy-making can attend to people's particular sociocultural mapping of the risk of Covid-19 so as to be culturally sensitive and socially effective.

2 Risk as Collective Construct

As early as in the 1980s, Douglas and colleagues precisely pointed out the decisive role social structure and cultural tradition play in risk perception. In the seminal book *Risk and Culture*, Douglas and Wildavsky (1982) argue, "Risk should be seen as a joint product of *knowledge* about the future and *consent* about the most desired prospects." Risk is collectively constructed based on what kind of people we are and what we think we should do as a group. The fear of risk is largely due to uncertainties towards knowing. At risk refers to a situation where there is no social consensus of how to deal with it. According to the following chart (Fig. 1), it is easy to discern the real risk of the Covid-19. On the one hand, we hardly have obtained certain knowledge of the virus' origin or developmental paths scientifically. On the other hand, there lacks complete consent of the virus' impact on social life within and across societies.

The lack of knowledge and consent leads to danger, which, according to Douglas' notable argument (1966), is fundamentally caused by things "out of place" in the classified schema. Douglas (1966) regards classification as the foundation of human experience and relations despite that categorization is not always precise and inclusive of all factors. Risk remains dangerous when people fail to identify such an alien threat in the existing classification system which provides intellectual and moral guidance for proper reaction. Like a pair of dirty shoes put on the dinner table, pollution, as

Four Problems of Risk

Knowledge

Consent	Certain	Uncertain
Complete	Problem: *Technical* Solution: *Calculation*	Problem: *Information* Solution: *Research*
Contested	Problem: *(dis)Agreement* Solution: *Coercion* or *Discussion*	Problem: *Knowledge* and *Consent* Solution: ?

Fig. 1 Adapted from Douglas and Wildavsky (1982: 5)

things misplaced, becomes risk in society as it threatens social order. The failure of classification and ordering the world gives individuals a sense of insecurity.

Risk, danger and impurity are thus by-products of society's systematic ordering and classification (Rippl, 2002). Douglas (1986) interprets risk perception process as "coding," where people choose to recognize and respond to risk using familiar socio-cultural references such as political system, economic structure and moral standards. They classify and rank danger according to a desired value system and its corresponding institutional form. The standard for the social "coding" can be inferred from people's perceptions of the body as a symbol of their socio-political identity and of the collective as a solidary whole of social relations. Douglas and Wildavsky (1982) assert that the human body is in continuity with the social organism and that social structure and cosmology are revealed through the human body. The danger to the human body results from anxiety about the margins, a consciousness that is deftly projected into culture. There is more impure presence on a symbolic level where structures are blurred. The clearer the structure, the clearer the classification, the more obvious are the anomalies of disorder and transgression. Impurity is both a boundary required to maintain the stability of the social system and an expression of the idea of order. When boundaries are crossed, and order broken, the taboos or rituals associated with them will provide a way to re-establish order and maintain boundaries. From this point on, risk is no longer an objective, explicit consequence of danger. Together with impurity (danger), it is rooted in the conceptual world constructed by the categorization system, expressing the collective beliefs and values of a group about danger. Any attempt to provide a simple "functional explanation" of risk would be futile.

Our argument is inspired by the "grid/group analysis" proposed by Douglas and Wildavsky (1982). Group refers to the "outside boundary that people have erected between themselves and the outside world," while grid means "all the other social distinctions and delegations of authority that they use to limit how people behave to one another" (ibid.: 138). Two sets of idea types of "high group" and "low group" as well as "high grid" and "low grid" are proposed by the researchers. According to Lupton (1999: 69), "the high group ethos emphasizes cohesion among group members and makes strong distinctions between 'us' (members of the group) and 'them' (the world outside the group), while the low group ethos emphasizes individuality and weaker ties with others." Meanwhile, "Individuals in groups classified as ideal type 'high grid' are subject to a large number of cultural constraints, while those who are 'low grid' have few constraints shaping their actions" (ibid.).

Accordingly, four idea types and four approaches to risk can be identified. Following Douglas and Wildavsky (1982), Lupton (ibid.: 69–70) classified the four grid-group types as *hierarchists* (high group and high grid), *egalitarians* (high group and low grid), *individualists* (low group and low grid), and *fatalists* (low group and high grid). The following chart (Fig. 2) summarizes the four approaches to risk in the four idea types respectively. The chart is indeed too rigid to inclusively describe various forms of social grouping or personal preferences. Notwithstanding, the point is to "treat the concept of 'risk' as a given in suggesting that it is the worldview rather than the nature of the risk itself that is the source of differential responses to risk" (Lupton, 1999).

If risk perception and response are understood to be shaped by people's specific worldview and understanding of being, the sociocultural incentives promoting people's response to risk prevention and control measures such as mask wearing need to be closely analyzed. Particularly, the chapter discusses how the perspectives of individuality and collectivism across societies oblige people to respond to mask wearing in different ways. Admittedly, it is hardly possible to simply fit any existing society into the current "grid/group" schema due to the society's heterogeneric nature. The "grid/group analysis" is profoundly persuasive in revealing the

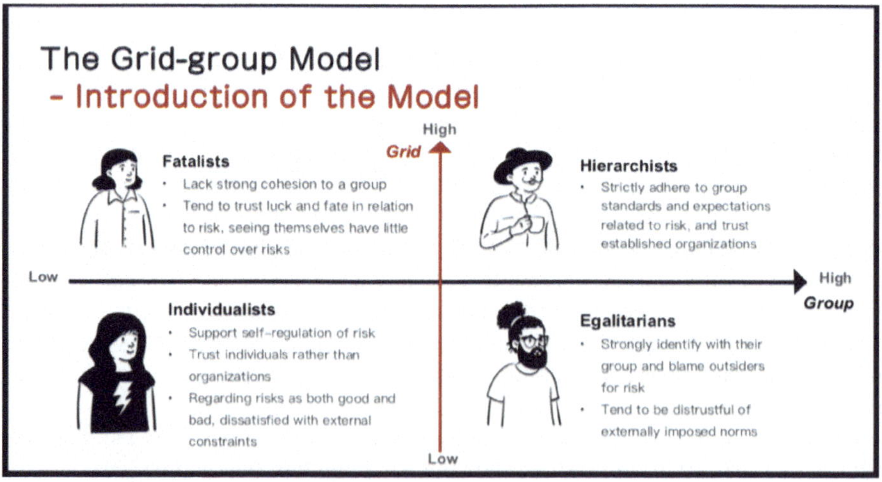

Fig. 2 Copyright of UIC students Sijia Yan, Zirou Zhang, Yifan Zhang and Zihao Zeng. Used with the authors' permission

effects of social norms, restriction, institutional power, and perceptions of selfhood on people's risk response. We thus propose to analyze how the Chinese and Euro-American people's different approaches to mask wearing are determined by varied conceptualization of individuality and self-regulation as well as group standards and expectations. Specifically, we argue that the debates of the mask wearing reveal the contestation of body politics where mask is differently regarded as concealment or containment. Mask wearing are also used as a measure to identify and distance the ill from the healthy so that the health of the community can be protected.

3 Research Methodology

The research is based on structured interviews with 10 Chinese citizens in their early 30s who cohabited with their parents and children in March–May 2020. As is discussed in latter sections, the co-habiting pattern of the parent–child–grandchild generations contributes a great deal to the mask wearing debates within Chinese families. We also conducted open-ended interviews with 20 Chinese and foreign residents about their opinions on mask use throughout the year of 2021. We used in-depth interview, focus group interview, and participant observation methods to complete the data collection. Due to travel restrictions, we collected data from primary studies on the European and American people's attitudes towards mask wearing during the Covid-19 outbreak for the comparative study. We purposely selected primary studies in Western European countries including the United Kingdom, France, Italy and Germany and in North American counties like the United States and Canada. In latter discussions, the countries surveyed are collectively called "Western countries" for they are historically considered as modern liberal states. While we are well aware of the risk of overgeneralization, we carefully sampled such countries as they fit the ordinary Chinese perceptions of "Western countries" which makes a comparative study between them and China meaningful. Comparing the interviewees' viewpoints of mask wearing with those of the Chinese, we examine how different perceptions of the individual and the collective impact people's risk response. Meanwhile, we collected official discourse, official policies, and social media coverage in China, the United States and the European countries since early 2020. Comments of the social media users are also an important source of references for us.

We identified four different scenarios according to the different developmental phases of the pandemic (see Table 1).

The same questions of mask-wearing behaviors were asked in the four scenarios so that we could gather information of people's changing ideas towards mask wearing along with the development of the pandemic situation. The following chart shows the key themes of the interviews (see Table 2).

Table 3 shows the basic demographics of our key informants. The first 10 Chinese interviewees were the ones co-inhabiting with their parents and children during the

Table 1 Covid-19 development phase

Timeline	Covid-19 development phase
2019.12.8 ~ 2020.1.19	The first case was confirmed by Hubei government, China. Hubei provincial health commission alerted WHO of several cases of pneumonia of unknown etiology
2020.1.20 ~ 2020.2.27	The Covid-19 outbreak was gradually brought under control in China after peaking on February 12, with outbreaks in other countries emerging
2020.2.28 ~ 2020.3.11	Covid-19 spread outside China. On March 11, the WHO declared Covid-19 a pandemic and called for global collaboration of the Covid-19 prevention and control
2020.3.11 ~	The pandemic continues to influence both the Western countries and China

Table 2 The key themes of the interviews

Key themes	Sub-themes and topics
The informants' mask-wearing frequency and scenarios	• The place and time when masks were worn • The relationship between the frequency of mask-wearing and the change of the pandemic situation • The relationship between mask-wearing behaviours and the actions of the surrounding masses
The influential factors of mask-wearing behaviours	• Political and economic aspects (Examples asked) • The influential degree of others' action (Examples asked) • The influential degree of the news or other social media (Examples asked) • The influential degree of the mask-wearing mandates (Examples asked)
Level of acceptance to wear a mask	• The processes in which people formed their approval or disapproval of the public mask wearing initiative
The relationship between people's mask-wearing behaviour and their management of kinship and friendship	• The relationship between the safety of family members and one's mask-wearing decisions • One's motivation to persuade family members to wear masks (if any) • The extent to which the decision to wear a mask was influenced by the one's family members and/or friends

pandemic outbreak. The rest 20 interviewees were key informants who shared with us their deep thinking on mask wearing.

Table 3 The basic demographics of key informants

Name	Age	Gender	Location(s)	Profession(s)
Interviewee 1 (China)	36	Female	Zhuhai, China	College professor
Interviewee 2 (China)	37	Female	Chengdu, China	Architect
Interviewee 3 (China)	36	Female	Chengdu, China	Civil servant
Interviewee 4 (China)	37	Female	Xiamen, China	Financial Adviser
Interviewee 5 (China)	35	Female	Beijing, China	Freelancer
Interviewee 6 (China)	34	Male	Hong Kong, China	Fresh PhD
Interviewee 7 (China)	33	Female	Beijing, China	NGO officer
Interviewee 8 (China)	31	Female	Beijing, China	Fresh PhD
Interviewee 9 (China)	36	Female	Changzhou, China	Nurse
Interviewee 10 (China)	36	Male	Zhuhai, China	Civil servant
Interviewee 11 (China)	34	Female	Zhuhai, China	Social worker
Interviewee 12 (China)	22	Female	Zhuhai, China	Undergraduate student
Interviewee 13 (China)	31	Male	Guangzhou, China	Manager (Accounting)
Interviewee 14 (China)	21	Female	Xi'an, China	Freelancer
Interviewee 15 (China)	17	Female	Xi'an, China	High school student
Interviewee 16 (China)	18	Male	Xi'an, China	Graduate student
Interviewee 17 (China)	23	Female	Freiburg, Germany	Graduate student
Interviewee 18 (China)	21	Female	Dongguan, China	Undergraduate student
Interviewee 19 (China)	75	Male	Weihai, China	Retirement (Teacher, Chinese)
Interviewee 20 (China)	82	Male	Heilongjiang, China	Retirement (Manager, food)
Interviewee 11 (UK)	21	Male	Edinburg, UK	Student
Interviewee 12 (US)	26	Female	Boston, US	Taxi driver
Interviewee 13 (Spain)	42	Male	Travelling alone in the Iberian Peninsula	Freelancer
Interviewee 14 (UK)	22	Female	Manchester, UK	Undergraduate student
Interviewee 15 (France)	26	Male	Paris, France	Painter
Interviewee 16 (UK)	28	Male	Glasgow, UK	Real estate agent
Interviewee 17 (UK)	30	Male	Birmingham, UK	Travel agent
Interviewee 18 (US)	17	Female	New York, US	High school student
Interviewee 19 (Canada)	56	Male	Quebec, Canada	Hotel manager
Interviewee 20 (US)	30	Female	Boston, US	PhD student

4 Medical Masks as Scientific Knowledge and Social Policy

The invention of medical masks is long regarded as a symbol of "science" and "civilization." In 1861, French microbiologist Louis Pasteur demonstrated the existence of pathogenic bacterium with the Gooseneck flask experiment. In 1895, Mikulicz Radecki, a German pathologist, recommended wearing a hood over the nose and mouth during surgery to prevent infection of the patient's wound from splashing saliva when the health care provider spoke. In 1899, French doctor Paul Berger invented a 6-layer gauze mask with a looped strap that could be hung on the ears, which became the prototype of the modern medical mask in Europe. Medical masks were created as a hygienic product with an isolation function in the field of health care. The primary function of the medical masks was to prevent the transmission of bacteria or viruses. In the medical field, mask wearing symbolizes "isolation," "shielding" and "protection" of the people from the infectious diseases (Elliott & Conneller, 2020).

In China, the masks' essential function has been reinforced in various protective scenarios, thus contributing to its recognition. For the Chinese, wearing masks reflects the health and hygiene habits that have gradually developed over the past century or so (Lynteris, 2018). In 1910, Chinese physician Dr. Lien-Te Wu modified the surgical mask into a more convenient form for the Chinese doctors and ordinary people during the plague in northeast China. It was the first time that the Chinese used masks to protect themselves in an epidemic. During the Republican era, when China was hit by outbreaks such as cholera, smallpox, typhoid and malaria, the simplest and least expensive measure to prevent from the infection was to wear a mask. In recent years, Chinese people have gradually developed the habit of mask wearing over the course of past disasters including the outbreak of the SARS in 2002, the global outbreak of bird flu in 2009, the major PM2.5 smog pollution in 2013 and the present Covid-19 pandemic (Tsang & Prost, 2021). The large outbreaks of contagious influenza have reinforced the perceived function of masks to prevent disease transmission. Masks have gained positive meanings of protection and safety.

In contrast, Western countries have had limited experience with infectious epidemics since the twentieth century. Masks played an important role in the 1918 pandemic influenza outbreak. During the epidemic, some American public hygienists recommended that everyone wear masks. Cities such as San Francisco and San Diego enacted the "Mask Order." Passengers who did not wear a mask would be refused boarding by trolley conductors. After 1920, masks were widely used worldwide as a preventive measure against respiratory infections. Advances in modern medical technology and historical reasons in Western countries have led to a restricted use of masks into special occasions such as medical treatment and to a gradual weakening of the public's perception of the protective function of masks. In addition, due to differences in the labor protection system and the concept of life, people in Europe and the United States are more likely to choose to take time off from work when they feel unwell. For instance, the current covid-19 prevention policies in European countries covid-19 suggest people "rest at home if you are unwell" (World Health

Organization Regional Office for Europe, 2020) rather than "wear a mask outside" (The UK Government, 2020).

Despite medical masks are largely accepted as a scientific method to prevent infectious diseases, there exist heated debates in public policy regarding the enforcement of mask wearing during the pandemic outbreak (Dai et al., 2021). On January 7, 2020, China confirmed that the pneumonia virus found in Wuhan was a new type coronavirus. On January 22, masks became a hot topic among Chinese residents after Chinese academician Zhong Nanshan, leading expert of National Health Commission (NHC) of China, seriously recommended on the national television program that everyone out in public needed to "wear a mask all the time, but not necessarily an N95 mask." On January 30, NHC of China issued mask wearing guidelines for the general public. In early February, 2020, every Chinese provincial Covid-19 prevention and control guidance office published a similar announcement requiring people to wear a mask in public places. In one of the official reports, mask wearing was deemed as "statutory obligations." Those who violated the public mask wearing mandate might receive corresponding administration penalty. On February 5th, Reuters used the imagery of "crazy masks" to describe China under the shadow of the Covid-19 when people of all walks of life rushed to purchase masks. Since then, mask wearing has become mandatory for people to take public transportation, to enter closed public space such as the hospitals, supermarkets and cinemas, and to participate in some large-scale group activities in the majority of Chinese cities. Current Chinese pandemic prevention and control policy insists on "closed-loop management" and "prevention of the contamination of people, materials and environment" which put a heavy emphasis on vaccination, health monitoring, nucleic acid testing, and cleaning and disinfection of exposed environments. These measures are still in effect at present.

Mask-wearing policy-making in the Western societies surveyed has witnessed a very different scenario. Take the United States for example. The Center for Disease Control and Prevention (CDC) followed WHO's suit which declared covid-19 was a "Public Health Emergency of International Concern" in February 2020. However until March 2020, infectious disease expert Dr. Anthony Fauci and other public officials did not encourage Americans to wear masks due to the deficiency of healthcare supplies during the early pandemic outbreak (Kemmelmeier & Jami, 2021). In America, a national mandate of mask wearing was not possible as public health powers belong to the states. Although a total of thirty-nine states issued statewide mask requirements at different stages of the pandemic, some have continued to challenge such mandates as violating the First Amendment for the fact that face covering have long been associated with political, religious and other symbolic meaning (Gostin et al., 2020). The mandatory policy encountered tremendous resistance. Some residents thought mask-wearing was a stigma that intruded their freedom. Others considered it as a certain political conspiracy (Finkelstein et al., 2020; Shepherd, 2020; Siegler, 2020; Thomson & Ip, 2020). Even when the CDC eventually advised Americans to wear masks, the White House reacted to the advice in different ways. In April 2020, then American President Donald Trump did recommend the public to voluntarily wear masks (Wu & Jackson, 2020). On May 21, Trump responded the media's question of why he refused to wear a mask by saying that he "did not want to give the press the pleasure of seeing it" (Carlisle, 2020). The myth of mask wearing was also increasingly politicized with 2020 being an election year (Ingraham, 2020; O'Connell, 2020). Until present, most of the Western countries surveyed have already lifted the public mask-wearing order. Similar public resistance of mask wearing for political, religious and other historical reasons has been found in countries including France, Great Britain, Italy and others.

5 The Healthy Body Vs. The Sick One: Mask Wearing as Protection or Concealment

The brief historical account of the use of masks in previous and current pandemics reveals that scientific knowledge and measures deemed "civilizing" or "advanced" may not be easily integrated into public policy to promote collective action (Dynel, 2021). Astonishingly, people in China and the Western countries surveyed have responded to the mask wearing mandates in almost contradictory manners. All Chinese people interviewed in this research admitted that they followed the mask wearing mandate and believed it to be critically helpful. Foreign interviewees however had mixed response to the mandates. The conflicting ideas prompts us to look into deep sociocultural reasons which influence people's attitudes towards mask wearing. We have noticed that the idea of whether mask wearing as protection or concealment became a prominent and recurring theme in Chinese and American interviewees' discussions of the masks.

One representative attitude of the Chinese interviewees is that masks can serve as an effective means of protection:

> When you asked me, I thought about it. It seems that I have never used the effectiveness as a standard of wearing a mask to decide whether to wear it or not. On the contrary, I often think about the troubles caused by not wearing a mask. I also imagined what my life plan would do if I was infected with covid-19.
>
> (Phone interview with Interviewee 13 on February 21, 2020)

> You can't go out without wearing a mask. It's government regulation. To enter the mall, you have to wear a mask. And we're all afraid of infection. The virus is so deadly and dangerous. Wearing a mask is important to save our life.
>
> (Phone interview with Interviewee 12 on February 21, 2020)

Interviewee 12's response resonates with that of many Chinese both home and abroad (Gerstein, 2021). Masks help prevent the infusion of the virus from entering into one's body. For the Chinese interviewed, the mouth and nose completely covered by a mask showed that the body's weakest porous areas were protected from the infection. The health of the individual body could be maintained. The completely covered and thus closed body with a mask on could no longer be infected by the interaction with others in public space. Also, when the mask was removed and the mouth and nose kept open, it brought damage and pollution to the public space and into other bodies through the saliva and droplets discharged from one's mouth and nose. In public and in private, the mask-wearing body symbolizes health, protection and precaution.

In contract, wearing mask might be regarded by the Western non-mask wearers as a way of concealing sickness or weakness. As one of the interviewees said:

> Why do people need to wear masks when they are not sick? You see, the infected people are the ones inside the hospital. It's the medical staff who wear masks to work ah. And wearing a mask is necessarily useful? The mask is just a thin layer. I do not understand why you must wear a mask. Don't you feel like you can't breathe?
>
> (Interviewee 11 on Instagram, March 23, 2020)

In most Western countries, mask wearing is regarded mainly as a means to restrict and limit the risk of disease transmission from the patients (Ali, 2021; Elliott & Conneller, 2020). The majority of the people who wear masks are patients and health care workers who come into contact with the patients. With the development of medical and health care and the improvement of the welfare system, Westerners generally accept the concept of resting at home when sick, and it is shameful to go out with sickness (Gerstein, 2021). Therefore, the act of mask wearing in public connotes two layers of meanings. On the one hand, the person wearing a mask is physically ill. On the other hand, the patient is going out without regard for the safety of others.

On a larger scale, while the preventive function of the masks is weakened in the minds of the Westerners, masks' concealing function is stressed in the socio-political environment. In the Western political context, mask wearing associates more with identity concealment particularly of illegal rallies, protests or terrorist incidents (Eli, 2020). Several Western countries including the US have enacted regulations to prohibit mask wearing in public except for special festivals or business occasions. Some countries have enacted ordinances prohibiting mask wearing or covering the face when participating in social events, and violators will be punished by fines or imprisonment. As Interviewee 17 said to us:

> I'm so ugly with the mask on. Don't you feel like a prisoner? I only wear a mask when I have a cold.
>
> (Interviewee 17 on February 17, 2021)

Embedded in the debate of mask wearing as protection or concealment is the different perceptions of the masked body as a healthy or sick one, which lead to positive or negative interpretations of the mask wearing in different cultural contexts. The body exposed to the social environment is a physical "fortress" of the individual. The body's surface is the frontier of "internal–external" and "personal-social" interactions. The holes and cavities of the body embody the pathways of cultural internalization and social integration (Kemmelmeier & Jami, 2021). In this regard, mask wearing sends out drastically different messages to the public in China and the Western societies. The Chinese masked body can be read as healthy, precautious, civilized, and responsible. For example, the younger generation Chinese interviewed (age 18–45) regarded mask wearing as a learned way to fight against virus. They have been educated to understand that mask wearing is much more scientific than many "local" or "traditional" way of infection prevention such as drinking vinegar or saline water.

Lynteris (2018) gives a detailed analysis of plague masks during the great Manchurian plague epidemic in 1910 in the Chinese-Russian border town of Manzhouli when "the Penang-born, ethnically Chinese, and Cambridge-educated" Wu Liande invented the anti-plague masks for the medical staff to wear. Masked medical staff were soon regarded as a "legion of 'plague fighters'" by the local residents as "a force of purification" (ibid.: 446–447). They formed a "mask unity" that distinguished the mask-wearing uniformed medical staff from the "supposed backwardness of victims" (ibid.: 446). More importantly, mask wearing constituted "a national body politic" (ibid.: 447). As Lynteris (ibid.: 449) argues, masks have become "an icon of agonistic medical rationalism" which also transform the social environment into "a performance of medical reason and hygienic modernity." Consequently, a mask-wearing body in public demonstrates a high level of hygiene and civilization, as it is a symbol of a scientifically well-protected healthy body (see also Zhang, 2021).

On the contrary, a masked body in the Western countries surveyed may be associated more with negative emotions that cause panic. In the same article, Lynteris (2018) shows that a masked doctor in late 18th and early 19th Century France was portrayed as a witch-hunter and a "cause of plague." Hence, the rationality of the devices or gestures of protection such as mask wearing depends upon "the redefinition of what pertains to reason and what not" (ibid.: 451). In Europe, the use of masks can be traced back to Zoroastrianism in the sixth century BC in Persia, when Zoroastrians considered the breath of the ordinary people unclean and therefore required believers to cover their faces with cloths during religious ceremonies. As for the oriental masks, it was first recorded in 1275 by the famous Italian traveler Marco Polo in his book "The Travels of Marco Polo" about the dining etiquette of the Yuan Dynasty palace. It wrote, "The numerous persons who attend at the sideboard of his Majesty, and who serve him with victuals and drink, are all obliged to cover their noses and mouths with beautiful veils or cloths of worked silk, so that his victuals or his wine may not be affected by their breath (Polo & Komroff, 2002: 87)."

In the primary studies we collected, a mask is variously called "mask," "masque," "veil," among others. Particularly, the term "veil" is inextricably linked to witchcraft and religion in Western civilizations. More recently, mask wearing in Western societies associates with violent rebellion and religious dissents (Eli, 2020; Zhang, 2021). For example, the veil has been conceived as "a weapon and a tool of resistance" (Fanon, 1965: 54–55). Talal Asad (2003, 2006) powerfully argue how the veiled Muslim women were perceived as a threat to the secular states such as France and were later banned of veiling themselves in public (see also Abu-Lughod, 2013). In this case, the idea of the masking as concealment reveals the contested nature of contemporary secularism which "proposes new concept of 'religion,' 'ethics,' 'politics,' and new imperatives associated with them" (Scott, 2007, as cited in Eli, 2020: 5). In essence, the politics of masking demonstrates the politics of religious freedom, racial equality and women's rights (Zhang, 2021).

Under the double social pressure of wearing a mask, people should not wear it if they are healthy. As a result, "there is no point for the general public to wear masks," "healthy people do not need to wear masks," "only those who already have symptoms need to wear masks," and "masks are included in the prescription drugs" became the main topics of the Federal Ministry of Health (Germany), Centers for Disease Control and Prevention (United States), and the Department of Health at the beginning of the outbreak. The official institutions in these Western countries also hold the same viewpoint, which to some extent reinforces the public perception that masks should be used in limited special situations.

6 The Solidary Society Vs. The Free One: Mask Wearing as Compliance or Confinement

The perceptions of the body are never isolated from people's mapping of their social relations (Betsch et al., 2020). Mask wearing is not only about one's management of their individual body, but also, if not more importantly, about one's negotiation of their status and relations in the larger society. As Douglas argues (1986, 1992), risk response is a result of a society's custom-built social structure and moral system. Debates about mask-wearing policy-making expose the power relations between individual citizens and state institutions which guide and regulate people's choices and affiliations. In Chinese and western societies we surveyed, mask wearing is contradictorily understood as compliance or confinement, reflecting people's different ways of socializing and group-making.

The Chinese believe that wearing a mask protects themselves and others around them. Deep in mind the Chinese interviewees shared the moral imperative of "not causing troubles to others." In China, it is believed that wearing a mask protects against infection. Both healthy and sick people should wear masks during an epidemic (Kahan, 2008). Wearing a mask can have a positive psychological effect because it reinforces the sense of distance and reminds people to be hygienic.

More importantly, the compliance of mask wearing is regarded not as a policy imposition rather than a consensus of care shared among relatives, friends, and even strangers. As Interviewee 11, a social worker, said:

> I advised my mom many times about wearing a mask, but the old man thought it was a hassle, so she didn't wear it. Until she read the WeChat Official Account, which said that not wearing an N95 mask would make her susceptible to the new coronavirus. Then my mom said, 'I didn't expect this virus to be so scary, for the health of the whole family, I must wear a mask.'
>
> (Interviewee 11 on November 12, 2020)

Especially for the Chinese interviewees co-inhabiting with their parents and children, mask wearing is not only for their own health but also for that of their family members. Many admitted to us that even if they or their parents did not feel comfortable with or even necessary of wearing a mask, they would wear one for the safety of other family members. Interviewee 2 told us a story of her family:

> I had to block the door with my own body when my father insisted going out without wearing a mask. I told him that for the sake of his granddaughter, he had to wear a mask, or he had no chance of leaving the apartment. He gave up and wore a mask whenever he got out.

Almost all the interviewees following the three-generational residence pattern shared with us similar stories of "fighting" for mask wearing among family members. It has a lot to do with the living condition of contemporary urban young Chinese families where most of them had to live with the parents in the only apartment they could afford in the big cities. The parents, however, are invited by the children to move in to help taking care of the grandchildren. Due to the different living habits, the three generations have to deal with a number of conflicts on a daily basis. Mask wearing was but one of the most urgent issues. Such conflicts are often not resolved but compromised in the name of caring, especially the care for the grandchildren. Disputes of scientific knowledge or individual choices could give way to the well-being and continuance of the family.

The pandemic's high mortality and contagiousness rates made the Chinese to believe that their health also depended upon others' active collaboration, and vice versa. Interviewee 14:

> I feel that everyone should wear a mask. I wear a mask. Others who do not wear a mask will make the number of infections grow. If everyone is infected, then I am also in danger.
>
> (Interviewee 14 on January 21, 2022, during the tense situation of the epidemic in Xi'an)

We also found the following call for collaboration on the Xinhuanet, the official news webpage of the Chinese governments:

> The Hong Kong community also needs to wake up to the fact that the biggest enemy at the moment is the virus. All people must be of one mind and take ownership of Hong Kong as a home and actively support and cooperate with the fight against the epidemic. "All for one, one for all", only with the concerted efforts of the whole community can Hong Kong truly overcome the difficult times.
>
> (Xinhuanet calls on the Hong Kong community to fight the epidemic, on February 19, 2022, source: http://www.news.cn/gangao/2022-02/18/c_1128392561.htm)

The Fear of infection is a common concern among the Chinese interviewees. Wearing a mask is a necessary but easy thing to do to prevent infection. As no one wanted to catch the coronavirus, people were united in the mask wearing. Mask wearing has become a symbol of common destiny. The masked faces together promoted a sense of group identity. Mask wearing has become a new representation of Chinese solidarity in the context of COVID-19.

On the contrary, the Western interviewees believed that wearing a mask or not is everyone's freedom. Each person was and could only be responsible for their own body. When interviewed about her mask wearing decision during Christmas travel in December, 2020, the Interviewee 18 said:

No one said they were infected because they didn't wear a mask. The government didn't say that either. Society just says it's better to wear a mask to prevent getting the virus. I just don't think I can breathe with a mask on. And if you wear glasses in winter, you can't see well with a mask. It's inconvenient.

(Interviewee 18 on December 22, 2020)

To many people like Judy in America, masks function as confinement (Shepherd, 2020). Mask wearing is interpreted as governmental or social compulsion and regulation at the cost of individual freedom and choices, which is highly suspected or challenged in the liberalist countries.

Fundamental to the debate of mask wearing as compliance or confinement is the contrasted idea of what makes a good citizen. The concept of individuality and collectivism revolves around the social relations (Lu et al., 2021). In the Chinese context, being responsible is one of the key sociocultural indicators of good (*shan*), benevolence (*ren*), and righteousness (*yi*). One has indispensable responsibility for being good and helpful to their family and lineage as regulated in the principal of filial piety. One also has indispensable responsibility for being good and loyal to the larger society as exemplified by the great literati and high-minded people. A good citizen is one that shoulders and dares to shoulder lots of responsibilities. As Interviewee 15, a Chinese college student in Xi'an during the lockdown in January 2022 told us:

Didn't the CDC say it? Mainstream media all remind residents of wearing masks to protect ourselves and our society. Everyone has a responsibility to wear a mask. We are not just about individuals anymore. Everyone's safety is connected to the fight against the outbreak. By wearing a mask, we're devoting ourselves to our country.

(Interviewee 15, in Xi'an on January 17, 2022, when the epidemic broke out there again)

Consequently, the Chinese view of collectivism is about relations and interdependence where individual obligations create group well-being. Solidarity is achieved as one being responsible for one's own and the collective health and safety (Betsch et al., 2020). Therefore, most Chinese were worried about being criticized or even attacked for not wearing a mask in public. Driven by the sense of responsibility and pressured by external responsibility, most Chinese have strictly adhered to the mask-wearing convention, thus providing a good guarantee for the overall health safety of the Chinese people. In an online conversation with our interviewee 16, she said:

Everyone on the street wears a mask ...... If I don't wear a mask, it will look like I'm weird, right? Once I went to the market downstairs to buy groceries without a mask, the shopkeeper asked me why I forgot to bring a mask. Although I didn't want to wear a mask when he asked me, I would answer 'I will remember to wear a mask next time.' I would still feel guilty in my heart and have a double-minded sense for not following the rules.

(Interviewee 16 June 8, 2021)

Compliance to practice that promotes the greater good at the expense of one's own well-being is thus an act of righteousness and heroism rather than submissiveness or cowardice. Such a view of collectivism also impacts how the Chinese respond to governmental policy and regulations. In China, the Covid-19 prevention and control measures such as the lockdowns, quarantines and travel restrictions have required extensive individual dedication and even sacrifice for the collective good. Even at present when the new Covid-19 variant Omicron is scientifically proved to be less deadly, Chinese people still observe very strict mask-wearing and lockdown measures. As some research points out, Chinese political system and its omnipresent institutional power which "excel at mass mobilization" may be able to make the citizens to comply with such anti-pandemic measures (Ang, 2020; see also He et al., 2020; Yan et al., 2020). It is also reasonable to expect that not all Chinese citizens would voluntarily follow such measures. A thorough discussion of how the centralized power can lead to collective response is beyond the scope of the chapter. However, we propose that cultural orientation is as critical a factor as institutional organization in determining anti-pandemic response (see also Yan et al., 2020).

We argue that many Chinese chose to obey such regulations to a large extent because of the care for the collective—their families, friends, neighbors and strangers—among whom one finds their value and esteem. Wearing masks outside is both self-protective and altruistic. The individual practice is believed to be able to shape the destiny of the community. The mask culture has promoted a social sentiment of solidarity and shared responsibility, which facilitates the maintenance of social order. As one of our friends commented, "I am not afraid of the Omicron at all. But if my family, my neighbors, my colleagues and people like you who just talk to me for a few minutes catch it because of my carelessness (referring not to wear a mask in public), I will feel deeply guilty. It is the 'social death' of which I am afraid."

In the primary sources we surveyed, the Western individualist culture emphasizes independence and individual rights, prioritizes personal values, and deeply trusts self-rule and self-control. It rejects any form of interference or obstruction of individual will by authority, religion, social rules or any other external factors. For the Westerners interviewed, rules or regulations against one's free will may damage one's citizenship and the rights associated with it. A good citizen is one that is responsible for one's well-being and vigilant of the imposition of or intrusion from especially the powerful others. A Spanish college student explained to us how he considered mask-wearing as an individual choice:

> When I was growing up, my parents didn't care about my dress code. Dad said we'd better to wear a mask to prevent getting infected by the virus. But if I didn't wear a mask, Dad wouldn't blame me. However, I guess your parents would have blamed you for not wearing a mask, right?
>
> (Interviewee 20 on February 1, 2022)

Mask-wearing as a confinement thus endangers the individualists' rights and power of private choice and independence. Non-mask wearers were not as much against the knowledge of mask-wearing as resisting official regulation or orders not of their own decisions. In this sense, the non-mask wearers in the Western counties surveyed were also trying to be good citizens, guarding their rights and individuality against the encroachment of the state power or others' imposition. Thus, in Western countries, official calls for people to wear masks are often made in non-coercive terms such as "suggestions" and "appeals." People have a choice of whether to respond to the appeals or not. The people's decisions once again reflect how they imagine their status in society and how social order and values can be properly maintained.

Yet unfortunately the debates of mask wearing entail not so much mutual understanding as racial and cultural stereotypes and discrimination against the Chinese home and abroad (Ma & Zhan, 2022; Ren & Feagin, 2021; Zhang, 2021). While individual freedom can be acknowledged as the cornerstone of Western civilization, it is not and never the excuse for sociocultural criticism or alienation. According to our interviews, the Chinese interviewees shared the attitude of trying to understand the Western measures of mask wearing despite feeling very sorry for them being less protective. On the other hand, the studies on the Chinese immigrants' situation in America particularly have identified many cases of anti-Asian hate crimes or discrimination (Ma & Zhan, 2022; Ren & Feagin, 2021; Zhang & You, 2022). Much of the harm is caused by the problematic imagination of the post-colonial world tainted by orientalist and Eurocentric stereotypes (Zhang, 2021). In our argument, we have strived to demonstrate that the different standards of collectivism and good citizenship need to be recognized and sensitively analyzed, instead of being utilized as a weapon of attack.

7 Conclusion

This chapter examines the sociocultural incentives that decisively direct people's attitudes and practice towards mask-wearing during the Covid-19 pandemic. Comparatively studying the Chinese and Euro-American people's different mask-wearing behaviors, we have shown that mask-wearing was contradictorily regarded as either a form of protection and of social responsibility, or a representation of concealment and of individual rights. Such ideas are centrally shaped by the two groups' perceptions of individuality and collectivism. We have found that the two groups of people have tried to protect themselves and societies in respective ways. They both have strived to be good citizens according to their specific cultural norms and moral codes. In this sense, neither group should be regarded as being "crazy" or "silly" as some media representation showed. Instead, more research needs to be devoted to advising policy-making in a more culturally sensitive manner with respect to people's particular sociocultural classification of risk response.

The ongoing Covid-19 pandemic will surely motivate policy makers to reconsider the effectiveness of many public pandemic prevention and control regulations including mask wearing. The politics of masking not only speaks to the contested process of scientific knowledge-making, but also showcases the far-reaching impacts of religious containment, orientalist stereotypes, racial discrimination and cultural alienation. Admittedly, it is extremely difficult to advocate unprecedented and sometimes uninviting information and policies during crisis situations as the public lacked the familiar experiences or cultural knowledge to apprehend them. We suggest that culturally-sensitive education, communication and compassion can facilitate fast implementation and public collaboration of crisis intervention policies. Consensus making based on particular ethical values, interpersonal relations and historical consciousness would be the key to promote mutual understanding and collective action.

Our writing also has its limitations. As the Covid-19 is still mutating and the pandemic is still in effect, different nation's anti-pandemic measures as well as people's perspectives of the Covid-19 continue to change. The policy changes are important cultural factors contributing to the different behaviors of people wearing masks in the Western countries and China. Our interviews have been held in four periods (see Table 1). However, more detailed analyses are needed to further compare and contrast the change of policy and perspectives of the four different periods. Factors such as age and gender can also have an impact on mask-wearing behaviors and need to be analyzed. Among the many factors, this chapter only focuses on cultural motivations of mask wearing in response to government policy while other institutional and economic aspects need to be further studied. In addition, due to the lockdown policy in China during the epidemic, we had to rely on online interviews for data gathering which are limited in scale and participant selection. Future studies with more participants of different sociocultural backgrounds need to be conducted to present a more comprehensive analysis of how sociocultural perceptions of risk shape risk response in policy and practice.

Fundings This research is supported in part by UIC Research Grant (R202048), Guangdong University Innovation and Enhancement Project "Synergy between cultural mechanism and risk governance from the perspective of urban-rural comparison" (R5202004), and The Joint Research Project of Guangdong Philosophy and Social Science Foundation "Coordination mechanism between 'risk adaptation' and social development in the post-epidemic era" (R202116).

References

Ali, I. (2021). Rituals of containment: Many pandemics, body politics, and social dramas during COVID-19 in Pakistan. *Frontiers in Sociology, 6*, 160–188. https://doi.org/10.3389/fsoc.2021.648149

Ang, Y. Y. (2020). When COVID-19 meets centralized, personalized power. *Nature Human Behaviour, 4*(5), 445–447.

Asad, T. (2003). *The Formation of the secular: Christianity, Islam, Modernity*. Retrieved January 12, 2020, From http://books.google.co.ma/books

Asad, T. (2006). *Reflection on La laïcité and the Public Sphere*. Retrieved January 12 2020, From https://www.ssrc.org

Betsch, C., Korn, L., Sprengholz, P., Felgendreff, L., Eitze, S., Schmid, P., & Böhm, R. (2020). Social and behavioral consequences of mask policies during the COVID-19 pandemic. *Proceedings of the National Academy of Sciences of the United States of America, 117*(36), 21851–21853. https://doi.org/10.1073/pnas.2011674117

Carlisle, M. (2020). Trump goes without mask for public tour of Michigan factory, says he 'didn't want to give the press the pleasure' of seeing him wearing one. *Time*. https://time.com/5840833/trump-michigan-ford-plant-tour-mask/

Crook, S. (1999). Ordering risks. In D. Lupton (Ed.), *Risk and sociocultural theory: New directions and perspectives* (pp. 160–185). Cambridge University Press.

Dai, Y., Li, Y., Cheng, C., Zhao, H., & Meng, T. (2021). Government-led or public-led? Chinese policy agenda setting during the COVID-19 pandemic. *Journal of Comparative Policy Analysis: Research and Practice, 23*(3), 157–175. https://doi.org/10.1080/13876988.2021.1878887

Douglas, M. (1966). *Purity and danger: An analysis of the concepts of pollution and taboo.* Routledge.

Douglas, M. (1986). *How institutions think.* Syracuse University Press.

Douglas, M. (1992). *Risk and blame: Essays in cultural theory.* Routledge.

Douglas, M., & Wildavsky, A. (1982). *Risk and culture.* University of California Press.

Dynel, M. (2021). COVID-19 memes going viral: On the multiple multimodal voices behind face masks. *Discourse and Society, 32*(2), 175–195. https://doi.org/10.1177/0957926520970385

Eli, T. B. (2020). The anthropology of the face mask: Rethinking the history of face covering controversies, bans and COVID-19 context. *Journal of Xi'an University of Architecture & Technology, 12*(V), 741–751.

Elliott, B., & Conneller, C. (2020). Masks in context: Representation, emergence, motility and self. *World Archaeology, 52*(5), 655–666. https://doi.org/10.1080/00438243.2020.1996770

Fanon, F. (1965). *A Dying Colonialism.* Grove Press.

Finkelstein, J., Donohue, J. K., Goldenberg, A., Baumgartner, J., Farmer, J., Zannettou, S., & Blackburn, J. (2020). *COVID-19, Conspiracy and contagious sedition: A case study on the militia-sphere.* Rutgers University.

Gerstein, A. (2021). Bodies and boundaries during the COVID-19 Pandemic. *Midwest Journal of Undergraduate Research, 12*, 160–188.

Gostin, L. O., Cohen, I. G., & Koplan, J. P. (2020). Universal masking in the United States: The role of mandates, health education, and the CDC. *JAMA, 324*(9), 837–838.

He, J., Shi, Y., & Liu, H. (2020). Crisis governance, Chinese style: Distinctive features of China's response to the Covid-19 pandemic. *Policy Design and Practice, 3*(3), 242–258. https://doi.org/10.1080/25741292.2020.1799911

https://www.usatoday.com/story/news/politics/2020/04/03/coronavirus-trump-says-cdc-recommending-voluntary-use-face-masks/2938705001/

Ingraham, L. (2020). Left fearmongering over COVID despite new data [transcript]. Fox News. Available online at: https://www.foxnews.com/transcript/leftfearmongering-over-covid-despite-new-data, (accessed 4 May 2021).

Kahan, D. M. (2008). Cultural cognition as a conception of the cultural theory of risk. In S. Roeser (Ed.), *Handbook of risk theory* (pp. 8–20). Harvard Law School Program on Risk Regulation Research Paper.

Kemmelmeier, M., & Jami, W. A. (2021). Mask wearing as cultural behavior: An investigation across 45 U.S. States during the COVID-19 Pandemic. *Frontiers in Psychology, 12.* https://doi.org/10.3389/fpsyg.2021.648692

Lu, J. G., Jin, P., & English, A. S. (2021). Collectivism predicts mask use during COVID-19. *Proceedings of the National Academy of Sciences, 118*(23). https://doi.org/10.1073/pnas.2021793118

Lupton, D. (1999). *Risk.* Routledge.

Lynteris, C. (2018). Plague masks: The visual emergence of anti-epidemic personal protection equipment. *Medical Anthropology, 37*(6), 442–457.

Ma, Y., & Zhan, N. (2022). To mask or not to mask amid the COVID-19 pandemic: How Chinese students in America experience and cope with stigma. *Chinese Sociological Review, 54*(1), 1–26. https://doi.org/10.1080/21620555.2020.1833712

O'Connell, F. (2020). *How democrats will weaponize coronavirus to beat Trump.* Washington examiner. https://www.washingtonexaminer.com/opinion/how-democrats-will-weaponize-coronavirus-to-beat-trump

Polo, M., & Komroff, M. (2002). *The travels of Marco Polo: [The Venetian].* Liveright Pub. Corp.

Ren, J., & Feagin, J. (2021). Face mask symbolism in anti-Asian hate crimes. *Ethnic and Racial Studies, 44*(5), 746–758. https://doi.org/10.1080/01419870.2020.1826553

Rippl, S. (2002). Cultural theory and risk perception: A proposal for a better measurement. *Journal of Risk Research, 5*(2), 147–165. https://doi.org/10.1080/13669870110042598

Shepherd, K. (2020). Tensions Over Restrictions Spark Violence and Defiance Among Protesters as Trump Pushes States to Reopen. Available online at: https://www.washingtonpost.com/nation/2020/05/13/protestviolence-coronavirus/, (accessed 4 May 2021).

Siegler, K. (2020). Across America, Frustrated Protesters Rally To Reopen The Economy. *National Public Radio*. Available online at: https://www.npr.org/2020/04/18/837776218/across-america-frustrated-protesters-rally-toreopen-the-economy, (accessed 4 May 2021).

Thomson, S., & Ip, E. C. (2020). COVID-19 emergency measures and the impending authoritarian pandemic. *Journal of Law and the Biosciences*, *7*, 1–33. https://doi.org/10.1093/jlb/lsaa064

The UK Government. (2020). *Self-isolating after returning to the UK: Your employment rights.* https://docs.qq.com/scenario/link.html?url=https%3A%2F%2Fwww.gov.uk%2Fgovernment%2Fpublications%2Fcovid-19-stay-at-home-guidance%2Fstay-at-home-guidance-for-households-with-possible-coronavirus-covid-19-infection%3B&pid=300000000$erdXDLfEcSfA&cid=144115225141547486

Tsang, P. M., & Prost, A. (2021). Boundaries of solidarity: A meta-ethnography of mask use during past epidemics to inform SARS-CoV-2 suppression. *BMJ Global Health, 6*(1), e004068.

World Health Organization Regional Office for Europe. (2020). *Statement–Older people are at highest risk from COVID-19, but all must act to prevent community spread.* https://www.euro.who.int/en/health-topics/health-emergencies/coronavirus-covid-19/statements/statement-older-people-are-at-highest-risk-from-covid-19,-but-all-must-act-to-prevent-community-spread

Wu, N., & Jackson, D. (2020). Trump: CDC recommends voluntary use of face masks for public to stem spread of coronavirus. *USA Today*.

Yan, B., Zhang, X., Wu, L., Zhu, H., & Chen, B. (2020). Why do countries respond differently to COVID-19? A comparative study of Sweden, China, France, and Japan. *Comparative Management & Learning Techniques, 50*(6–7), 762–769.

Zhang, M. (2021). Writing against "mask culture": Orientalism and COVID-19 responses in the West. *Anthropologica, 63*(1), 1–14. https://doi.org/10.18357/anthropologica6312021327

Zhang, Q., & You, Z. (2022). Fear not the want of armor, for mine is also yours to wear: Trust and Community Cultivation for Risk Response of a Chinese Immigrant Group in the United States. In Kuah Khun Eng (ed.), *Covid-19 Responses of Local Communities—Exploring Trust in the Context of Risk and Fear*, (pp. 155–169). Routledge.

Qiaoyun Zhang received her Ph.D. in Anthropology from Tulane University (USA) and was research fellow at the International Institute for Asian Studies (the Netherlands) and associate researcher at the School of Anthropology and Museum Ethnography of the University of Oxford (the UK). She has published more than 10 articles in leading SSCI, A&HCI, and CSSCI journals including *Journal of Contemporary China, Journal of American Folklore, and Asian Ethnology.* Currently, she is conducting research on cultural change in post-pandemic era. Zhang's research focuses on cultural recovery and preservation after disasters, and engages with topics of risk and culture, state-ethnic relations in China, and intangible cultural heritage safeguarding.

Yushan Wang majors in Media and Communication Studies (MCOM) at Beijing Normal University-Hong Kong Baptist University United International College (BNU-HKBU UIC). Wang's research interests include social identity studies, educational culture, cultural memory, and Chinese popular culture and globalization.

Exploring How Social Media Influences COVID-19 Vaccination Intention in China: The Mediating Effect of Perceived Risk, Negative Emotion, and Subjective Norms

Ye Tao, Chuanli Xia, Hong Zhao, and Ning Liu

1 Introduction

In the post-pandemic era, vaccination seems to be an effective strategy to curb the global spread of COVID-19 (Alagoz et al., 2021). However, according to the Updated data from Our World in Data, Chinese official news revealed that as of April 2021, the COVID-19 vaccination rate in China was only 4%, far lower than that in the UK and US. Public concerns and hesitations about COVID-19 vaccines are the main reasons for the low level of vaccination (Zhao et al., 2021). For example, while some people were worried about fake or faulty COVID-19 vaccines (Lin et al., 2020), others were concerned about the vaccine's side effects. Some groups also believed vaccination offered less protection than other defense measures (Zhao et al., 2021). Therefore, increasing the public's vaccination intention is essential for promoting vaccination behavior (Rakhmadhani et al., 2021).

Previous research showed that individuals' psychological mechanisms of vaccine behavior are particularly critical to vaccination (Mo et al., 2021). For example, the health belief model could be applied to examining the effects of health information circulated on social media, suggesting that consumption of health information on social media could influence vaccination intentions through self-efficacy, susceptibility, and perceived severity (Im Kampe et al., 2020).

Despite much scholarly discussion on vaccination behavior by relying on some health communication models, such as the health belief model, most previous studies have focussed on the perception of the threat of diseases and belief in vaccine efficacy (Chen et al., 2021). In the past few decades, the influence of media use

Y. Tao (✉) · C. Xia · H. Zhao · N. Liu
Beijing Normal University-Hong Kong Baptist University, United International College, Zhuhai, China
e-mail: yetao@uic.edu.cn

© The Author(s), under exclusive license to Springer Nature Singapore Pte Ltd. 2023

S. X. B. Zhao et al. (eds.), *Comparative Studies on Pandemic Control Policies and the Resilience of Society*, https://doi.org/10.1007/978-981-19-9993-2_13

on health behavior intentions and attitudes has begun to receive attention (Vahedi et al., 2018). The rapid development and wide use of information technology have created an open environment in which people are more inclined to obtain health information on social media (Puri et al., 2020) Exposure to differing health information may lead to different perceptions influencing intentions and behavior. Using a theoretical framework helps to integrate empirical findings and elucidate change processes and mechanisms of action of effective communication and other intervention strategies. A helpful framework for this purpose is the COM-B model summarising factors necessary for behavior to change across behavioral domains. Following this line of research, Crawshaw et al. (2021) argues that there are three psychological components—Capability (C), Opportunity (O), and Motivation (M), which influence behaviors based on the COM-B model. Because of the interacting nature of these components, one may increase, for example, motivation by increasing capability (e.g., knowledge and skills) and opportunity (e.g., access to resources and social influence) (Rubinstein et al., 2015). This study proposes a mediation model to examine how social media use influences individual psychological factors and the individuals' vaccination intention.

Risk perception, which refers to the judgments people make when asked to describe and evaluate hazardous activities and technologies, plays a crucial role in how people respond to health emergencies and influences public policy and individual behavior (Slovic, 1987). In addition, the information provided by using social media can influence people's emotions and subjective norms, which in turn impact the public's protective behavior (Liu et al., 2020).

In summary, this study delineates a detailed model that links social media use to vaccination intentions through mediating variables.

2 Literature Review

2.1 Social Media Use and COVID-19 Vaccination Intention

A series of studies have shown that during the COVID-19 pandemic, social media use may substantially impact people's attitudes and willingness to undergo COVID-19 vaccination (Qin et al., 2022). Social media are Internet-based, multi-dimensional, and popular channels of mass personal communication facilitating perceptions and interactions among users, and they also derive value primarily from user-generated content (Carr & Hayes, 2015). The social media analyzed in this research include the social networking services (such as Microblog, WeChat, and QQ), short video clients (such as TikTok and Pear Video), and online discussion forums.

Unlike traditional media, social media allow rapid global creation and sharing of content without excessive supervision, and users can choose what they want to read themselves (Puri et al., 2020). Many public health organizations use social media to carry out health education and advocacy for the general public (Tang et al., 2018).

For example, social media played an important role in informing the public during outbreaks such as H1N1 in 2009 and Ebola in 2014 (Sooknanan & Comissiong, 2020).

As the information on social media is primarily user-generated, it may cause problems such as information overload, inaccurate information, dissemination of unexamined and poor-quality information, and even the spread of outright misinformation and conspiracy theories (Marcelin et al., 2021). These issues may play a role in the current vaccine debate on social media and contribute to vaccine hesitancy (Larson et al., 2011). This skewed environment, combined with the speed with which information spreads on social media, affects vaccine acceptance, an effect proven to be very volatile in response to vaccine-related events (Puri et al., 2020). If misinformation prevails, it can negatively impact public awareness of COVID-19 vaccination knowledge (Asmundson & Taylor, 2020).

In contrast, some studies have found that anti-vaccine messages on social media are insufficient to directly reduce vaccination intentions (Mckinley & Lauby, 2021). Many studies have shown that the diversity of social media information and the interaction of users' norms of behavior leads to increased willingness for vaccination (Thaker, 2020). For instance, contrary to popular belief, young people who use social media are mostly positive about COVID-19 vaccination and want to protect their family and relatives by getting vaccinated. They consider vaccination a civic duty and have strong confidence in vaccine safety (Benis et al., 2021b). Social media use to deliver health information has become a strategy to strengthen the population's willingness to be vaccinated (Benis et al., 2021a). These findings suggest that we can use social media platforms to disseminate vaccination information to increase vaccination intentions (Ratanasiripong et al., 2018).

In summary, social media use can influence the reception of information about vaccine health communication. The dissemination of information on social media can influence vaccination willingness in multiple ways, impacting people's COVID-19 vaccination willingness to different extents. Therefore, our study investigates the relationship between social media use and COVID-19 vaccination willingness by drawing on previous research insights. To explore whether social media influences individuals' vaccination intentions, we posed a research question:

RQ1 Is social media use related to vaccination intentions?

2.2 *Perceived Risk of Vaccination*

Risk perception is generally believed to originate from social psychology and is also studied in communication. Theories on health behavior take risk perception as the core concept and study how it changes an individual's health attitude and behavior, such as the Health Belief Model (Rosenstock et al., 1988). Risk perception generally refers to people's attitudes and subjective judgment towards risk, influenced by personal psychological cognition and society. Many studies regard risk communication as an essential factor affecting public response behavior (Morrow, 2009).

Social media has become a new channel and tool for people to adopt coping behaviors. During public health crises, many people turn to the Internet for information that can positively impact their responses to such crises (Han & Cheng, 2020), thus reducing the perceived degree of risk. Though some studies have shown that public perceived risk is a mediating variable of the relationship between individual characteristics and public behavior, this important influencing factor has not received enough attention (Yıldırım & Güler, 2020). Therefore, it is necessary to consider the perceived risk of COVID-19 vaccination as an intermediary variable of vaccination willingness. Combined with the influence of knowledge level on vaccination intention in the COM-B model (Crawshaw et al., 2021), Awareness Skills will positively impact vaccination intentions. When people perceive the risk as low, they are more likely to have a positive attitude.

Based on the above discussion, we propose the following hypotheses:

H1 Social media use is negatively correlated with the perceived risk of vaccination.

H2 There is a negative correlation between the perceived risk of vaccination and vaccination intentions.

2.3 Negative Emotion

Negative emotion is a form of emotional expression largely dependent on the spread of positive or negative information posted by others on social media. Emotion is an indispensable part of the user experience on social media. According to the BetterrHealth Channel, the emotions that can become negative are hate, anger, jealousy, and sadness. Users express their emotion on social media, react to the posted content, and use emojis to communicate (Zenk et al., 2020). Previous research has shown that social media often provides negative coverage and publicity about COVID-19 vaccines, giving rise to negative emotions among the public. Therefore, social media use increases negative emotion, leading to emotional contagion and lousy behavior like cyberbullying and burning (Luo et al., 2021). Much of social media behavior is influenced by negative bias. In psychology, negative bias refers to humans recalling more unpleasant memories than positive ones. Adapting it to social media means that people are more likely to identify and be affected by negative information shared on social media (Hyvönen et al., 2014). The spread of vaccine misinformation on social media harms people's vaccine beliefs and behaviors (Featherstone & Zhang, 2020).

Negative emotion have different effects on vaccination intentions. On the one hand, lower negative emotions can promote trust in vaccines. Social media users with less negative emotion had greater confidence in the efficacy and benefits of the vaccine. They were more amenable to being vaccinated against COVID-19, registering their support and approval for the vaccine (Thaker, 2021).

In contrast, high levels of negative emotions threaten trust in vaccines. When social media users have high negative emotion, one direct cognitive manifestation is skepticism and distrust of vaccines. Once people post negative news about vaccines on social media, making other users believe in it, users become dissatisfied with the vaccine, and their willingness to be vaccinated against COVID-19 decreases (Chou et al., 2020).

Overall, the above discussion describes how negative emotion affect the relationship between social media use and vaccination intentions.

Therefore, we propose the following hypotheses:

H3 Frequency of social media use is positively correlated with negative emotion.

H4 Negative emotion are negatively correlated with COVID-19 vaccination intentions.

2.4 Subjective Norms

Subjective norms come into play when an important person or group of people approve and support a particular behavior. They are determined by other people's perceived social pressure for others to act in a certain way and their motivation to follow other people's views. In previous studies, subjective norms were shown to have a weaker influence on the formation of intentions than attitudes (Zenk et al., 2020). In addition, Krueger and colleagues (2000) show that subjective norms have nothing to do with individuals' intention to establish their enterprises. The authors, therefore, call for further research and improvement of the measures used. One of the most frequently cited weaknesses of the planned behavior theory is the weak relationship between subjective norms and intentions (Ajzen, 2011). As explained by Ajzen, author of the theory of planned behavior, the intent is heavily influenced by individual factors (such as attitude and perceived behavioral control). Armitage and Conner (2001) criticized the narrow concept of subjective norm variable, which leads to a weak correlation between subjective norm and intention. According to Rivis and Sheeran (2017), the confirmed correlation between descriptive norms and intentions suggests the possibility of the variable having predictive power, which provides a strong impetus for further research in this area. Descriptive norms refer to the actual activities and behaviors performed by others. Contrarily, social norms refer to others' perceptions of how an individual should behave. We believe that both variables (descriptive and social norms) are part of subjective norms (Armitage & Conner, 2001).

As interpersonal relationships arise and society develops, subjective norms emerge to meet social needs. Subjective norms take various forms, from descriptive to social norms. In recent years, the influence of subjective norms in different societies and countries has been closely related to the development of social media networks. Key findings suggest that subjective norms strongly predict intention-seeking (Vanderpool et al., 2020). Hence, social media use can influence perceptions of the likelihood of people around them being vaccinated against COVID-19 and can influence users' intentions about their COVID-19 vaccination behavior. These studies suggest that social network use may influence health behavior by changing people's understanding of subjective norms (Yong & Jordan, 2017).

In the social context of the COVID-19 outbreak, more perceived social norms about vaccination were associated with COVID-19 vaccination intentions, i.e., subjective norms influenced vaccination intentions. Subjective norms and perceived behavior control the intention to be vaccinated. These findings suggest that public health efforts to encourage vaccination should focus on subjective norms, such as those provided by professional organizations (Askelson et al., 2010). In multiple models, subjective norms were indirectly correlated with intentions through self-efficacy and attitudes ($p < 0.05$). Specifically, greater vaccine support from friends, parents, and doctors was associated with greater vaccine self-efficacy, which in turn was associated with vaccination willingness. In addition, the belief that parents and doctors are more supportive of vaccines is associated with more favorable attitudes towards vaccines, which are associated with increased willingness to be vaccinated. The results suggest potential targets for future interventions to promote COVID-19 vaccination. Therefore, subjective norms play a meaningful role in the willingness to receive COVID-19 vaccines, though the extent of the correlation is unclear.

In summary, we discussed the relationship between subjective norms and social media use and subjective norms and willingness to be vaccinated against COVID-19.

Therefore, we propose a research hypothesis and a research question:

Q2: Is Social media use associated with subjective norms?

H5: Subjective norms are positively correlated with COVID-19 vaccination intentions (Fig. 1).

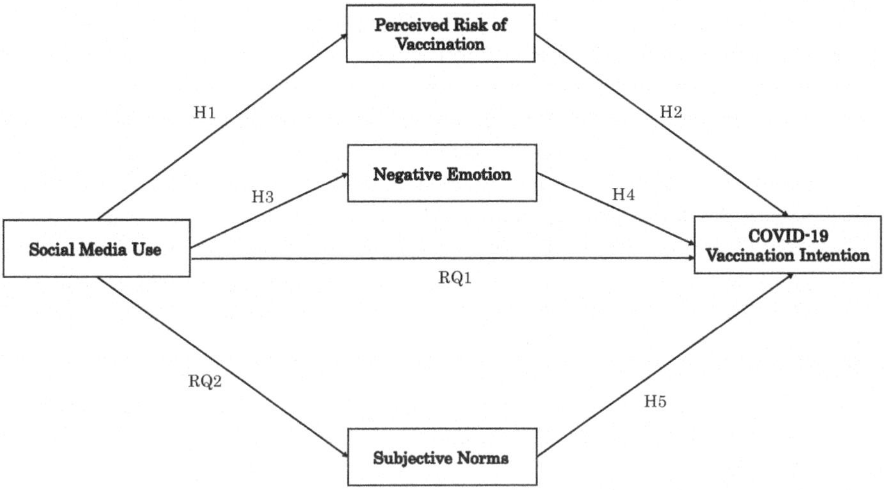

Fig. 1 A theoretical model of the mediating effect of social media use on COVID-19 vaccination intentions

3 Method

3.1 Data

Hummingbird Survey Company in China collected data from a nationwide online survey conducted between April 1 and April 6, 2021, in mainland China (excluding Hong Kong, Macao, and Taiwan). According to the sixth census data, this study used the quota sampling method to recruit respondents from 31 provinces and cities when the COVID-19 vaccine began to be popularised nationwide. In total, 3460 questionnaires were distributed and 1200 valid, completed questionnaires were received from Chinese residents aged 20 and above. As the age segmentation area of the China Census website is heavy, the segmentation involving adults starts at the age of 20, so this study selects people aged 20 and above.

3.2 Measurement

Social media use. This survey asks respondents to choose usage frequencies relating to the following three media (1 = none, 5 = very frequent): In terms of user usage, we have divided social media into the following two categories. 1. Social Networking Services (such as Microblog, WeChat, and QQ) (M = 3.13, SD = 1.14), 2. Short Video Client (such as Douyin, Pear Video, etc.) (M = 2.88, SD = 1.15), 3. online discussion forums ($M = 2.09$, SD = 1.17). The average frequency of social media use was ($\alpha = 0.58$, $M = 2.70$, SD = 0.85).

Perceived risk of vaccination. The risk perception indicated by the survey participants was evaluated using the scale developed by Morton and Duck (2001), which comprises three items. All the questions were measured through a Likert-type scale with a seven-point answer, ranging from nothing (1) too much (7). The internal consistency of the indicator was high ($\alpha = 0.6$).

Negative emotion. Overall negative sentiment was measured by asking participants how negative they felt about COVID-19 vaccination (1 = strongly disagree, 7 = strongly agree). The average of the five answers was used to measure negative emotion ($\alpha = 0.797$, $M = 2.39$, SD = 0.99).

Subjective norms. A subjective specification consists of five indicators, which are the percentage of respondents who predict that colleagues or classmates, relatives, friends, online friends, and people they know in daily life are likely to be vaccinated for COVID-19 ($\alpha = 0.88$, $M = 57.12$, SD = 17.60).

COVID-19 vaccination intention. The willingness to be vaccinated against COVID-19 was measured on a five-degree scale (1 = very reluctant, 5 = very willing) ($M = 4.39$, SD = 0.94).

Control variables. We collected control variables for gender, age, education, monthly household income, whether close relatives or friends were infected with COVID-19, and traditional media use that might have influenced the results. About 51.19 percent of the respondents were male. Respondents ranged in age from 20 to 69 ($M = 40.80$, SD = 12.90). We used the data from the sixth census as a quota sample for an age group of four years. About 24.4% of the respondents had obtained a bachelor's degree or above. According to the measurement of the fixed ratio, the monthly household income was divided into eight ranges, from 1000–3000 Yuan or less to 50,001 yuan or more ($M = 5.40$, SD = 1.39). About 2.3% of respondents had a family member or friend who had been infected with COVID-19 ($M = 0.02$, SD = 0.15). Since this study mainly explores social media use in the context of the new media network, the use of traditional media is taken as the control variable. The use of traditional media consists of three indicators: the frequency of use of newspapers/magazines, television, and radio (1 = none, 5 = very frequent) ($M =$ 57.12, SD = 17.60).

4 Result

In this study, SPSS Statistics software was used for data analysis, and regression Model 4 in Process was used to test the hypothesis. The control variables were input. First, the independent variable was social media use, the dependent variable was COVID-19 vaccination intention, and the mediator variables were the perceived risk of vaccination, negative emotions, and subjective norms.

In Research Question 1, after controlling for traditional media use and demographic characteristic variables, multiple linear regression was performed for social media use and willingness to be vaccinated against COVID-19. The results are shown in Table 1. The frequency of social media use positively affected the willingness to be vaccinated against COVID-19 ($b = 0.21$, $p < 0.001$). In study hypothesis 1, it was posited that social media use was inversely associated with the perceived risk of vaccination. As shown in Table 1, social media use had a negative effect on the perceived risk of vaccination ($b = -0.05$, $p < 0.001$). Therefore, Hypothesis 1 is valid. In hypothesis 2, it was assumed that the level of the perceived risk of vaccination was related to the willingness to undergo COVID-19 vaccination. As shown in Table 2, the perceived risk of vaccination was negatively correlated with the willingness to receive the COVID-19 vaccine ($b = -0.85, p < 0.001$). Therefore, hypothesis 2 is valid.

Hypothesis 3 assumed that social media use was positively correlated with negative emotion. Contrary to expectations, the results showed that social media use was inversely correlated with negative emotion ($b = -0.12$, $p < 0.05$), disproving hypothesis 3. In hypothesis 4, it was assumed that negative emotions were inversely associated with the willingness to receive the COVID-19 vaccine. As shown in Table 2, negative emotions had a negative effect on the willingness to receive the COVID-19 vaccine ($b = -0.07, p < 0.05$). Therefore, hypothesis 4 is valid. In hypothesis 5, it was assumed that subjective norms were positively correlated with the willingness to receive the COVID-19 vaccine. As shown in Table 2, subjective norms did have

Table 1 Demographic characteristics of the respondents ($N = 1200$)

Demographic variables		*N*	%
Gender	Male	614	51.2
	Female	586	48.8
Age	20–29 years	294	24.5
	30–39 years	277	23.1
	40–49 years	296	24.7
	50–59 years	205	17.0
	60–69 years	128	10.7
Education	Primary school or below	30	2.5
	Secondary school	114	9.5
	High school	169	14.1
	Technical secondary/vocational high school	228	19.0
	College	366	30.5
	University or above	293	24.4
Monthly household income	<5001 yuan	57	4.7
	5001–10,000 yuan	477	39.7
	10,001–20000 yuan	592	49.4
	20,001–50,000 yuan	72	6.0
	>50,001 yuan	2	0.2

Table 2 Direct and indirect effects within endogenous model ($N = 1200$)

	Explanatory variables			
	Social media use	Perceived risk of vaccination	Negative emotion	Subjective norms
Perceived risk of vaccination	−0.05***(0.00)	–	–	–
Negative emotion	−0.12**(0.04)	–	–	–
Subjective norms	4.71***(0.61)	–	–	–
COVID-19 vaccination intentions				
Direct effect (Total)	0.02***(0.03)	−0.85***(0.21)	−0.07**(0.02)	0.03***(0.00)
Indirect effect	0.19***(0.02)	–	–	–
Total effect	0.21***(0.03)	−0.85***(0.21)	−0.07**(0.02)	0.03***(0.00)

Note All entries are standardized estimates, after controlling for gender, age, household income, educational background, whether a family member or friend has had COVID-19, and traditional media use, Standard error is in parentheses.
**$p < 0.01$; ***$p < 0.001$

a positive effect on the willingness to receive the COVID-19 vaccine ($b = 0.03$, $p <$ 0.001), thus validating hypothesis 5.

Research Question 2 explored the relationship between social media use and subjective norms. The results in Table 2 show a positive correlation between social media use and subjective norms ($b = 4.71$, $p < 0.001$).

To sum up, first, social media use was positively related to COVID-19 vaccination intention. Second, the mediation analysis revealed that perceived risk of vaccination, negative emotion, and subjective norms could mediate the relationship between social media use and vaccination intention. Specifically, social media use led to a less perceived risk of vaccination, increasing people's intention to get vaccinated. Similarly, social media use reduced people's negative emotion, which strengthened their vaccination intention. In contrast, social media users were more likely to perceive a high vaccination rate among the public, making them more willing to vaccinate (Fig. 2).

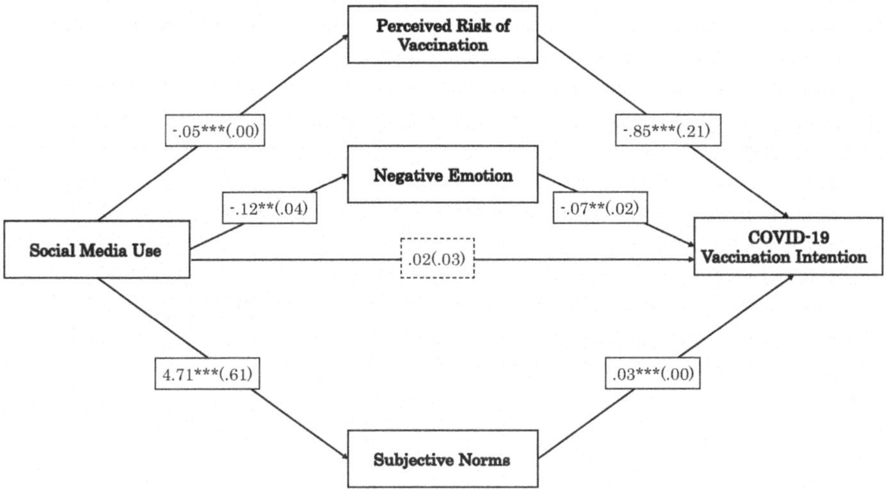

Fig. 2 Predicting COVID-19 vaccination intention. $^{**}p < 0.01$; $^{***}p < 0.001$

5 Discussion

This study explored the impact of social media use on COVID-19 vaccination intentions to enrich the existing literature on the effects of social media. Several noteworthy findings discussed in this study need to be elaborated.

Firstly, accessing health information from social media can improve health literacy and promote health behaviors (Huo et al., 2018). Therefore, regular exposure to positive and correct health information on social media can help users develop positive attitudes and perceptions about health behaviors, which can increase vaccination intentions.

Second, this study found that social media use to obtain health information was negatively correlated with the perceived risk of vaccination, in tune with previous research results. The results show that social media content influences vaccine-related knowledge, attitudes, and behaviors (Lee et al., 2021).

Thirdly, we found that social media use is negatively correlated with negative emotion. The higher the frequency of social media use may lead to fewer negative emotion. The finding contradicts earlier research that suggested that social media use leads to increased negative emotions and decreased vaccination intentions (Featherstone & Zhang, 2020). Social media, such as WeChat and Weibo, are strongly connected with acquaintances and friends, publishing much positive information about vaccination. For example, users often claim no adverse reaction to vaccination after vaccination. On the one hand, people need to create a good social image on social platforms and tend to post optimistic and positive information. On the other hand, people's desire to share using social media can lead to a vaccination bandwagon and create a "bandwagon" phenomenon. Many publicity platforms combine vaccination with a sense of social responsibility, making people consider vaccination an act of civic value, which leads to increased willingness to be vaccinated against COVID-19. This behavior reduced the negative feelings of social media users who had been hesitant about vaccines and increased their vaccine confidence. Our study considers the relationship between social media and negative emotions from a new perspective. On the other hand, in China's social media environment, there are filtering mechanisms such as public opinion censorship, ensuring that comments about vaccination posted on social media are primarily positive, reducing negative feelings about vaccines.

Fourth, we found that negative emotions were inversely proportional to the willingness to be vaccinated, i.e., the more negative emotion, the less the willingness to be vaccinated (Thaker, 2021). In recent years, many scholars believe that the cognitive difference caused by the frequency of media use impacts behavior and willingness. According to the health beliefs model, perceived benefits are one of the most critical prerequisites for healthy behavior. Social media exposure to negative information about vaccination increased people's negative feelings about vaccination. In this case, getting information about vaccinations from social media does not necessarily increase the willingness to get vaccinated. However, another scenario is that since people are constantly exposed to positive news and the benefits of the COVID-19 vaccine on social media, they develop fewer negative emotion and are more likely to desire vaccination against COVID-19. In addition, the reason we chose negative emotion over positive ones is that positive emotions do not play a mediating role in regression analysis.

In addition, like previous studies on subjective norms, we found that social media use can be influenced by subjective norms, leading to different vaccination intentions (Glanz et al., 2017). In the relationship between social media and subjective norms, we can see that previous studies have focussed on the influence of subjective norms on the frequency of social media use. These results show that the greater the influence of subjective norms, the higher the frequency of social media use. Unlike earlier scholars, we chose subjective norms as the mediating variable and the frequency of social media use as the independent variable. With a new research question and verifications of the results, we find that the higher the frequency of social media use, the greater is the influence of subjective norms.

The theoretical and practical significance of this study lies in that, in the context of the new media, where people are used to getting health information through social and official media, we aim to find links and influence mechanisms between media use and the relationship between vaccination intentions, to promote COVID-19 vaccination.

It should also be noted that we used official media as a control variable. It is unclear whether official media can influence vaccination intentions through the three aforementioned mediating variables, which points to a possible direction for further research.

In summary, social media is an essential carrier of vaccine health information, which can lead to differences in COVID-19 vaccination intentions. Social media strengthens the will of others to be vaccinated against COVID-19 by spreading vaccine mobilization and positive messages.

This study has several limitations that can be further addressed by future research. Firstly, the survey was conducted online, which limits the reach of individuals without access to the Internet.

Secondly, the type of questionnaire must rely on a simplified structure to avoid subject fatigue. For example, our assessment of social norms is based on a single question: the probability that someone close to you has been vaccinated against COVID-19. It is not measured by any more criteria and does not comprise multiple factors.

Finally, this study examined the impact of social media use on the individual level of willingness to receive the COVID-19 vaccine. The impact of social media on users can vary from country to country, depending on different political, cultural, and social structures in different social contexts. Such an issue should be considered when extending our findings to other countries and regions.

References

Ajzen, I. (2011). The theory of planned behavior: Reactions and reflections. *Psychology and Health, 26*(9), 1113–1127. https://doi.org/10.1080/08870446.2011.613995

Alagoz, O., Sethi, A. K., Patterson, B. W., Churpek, M., Alhanaee, G., Scaria, E., & Safdar, N. (2021). The impact of vaccination to control COVID-19 burden in the United States: A simulation modeling approach. *PLoS One, 16*, 1–28. https://doi.org/10.1371/journal.pone.0254456

Armitage, C. J., & Conner, M. (2001). Efficacy of the theory of planned behaviour: A meta-analytic review. *British Journal of Social Psychology, 40*, 471. https://pdfs.semanticscholar.org/ab83/87ae29317c651a129efc7a2297bc5f348a95.pdf

Askelson, N. M., Campo, S., Lowe, J. B., Smith, S., Dennis, L. K., & Andsager, J. (2010). Using the theory of planned behavior to predict mothers' intentions to vaccinate their daughters against HPV. *Journal of School Nursing, 26*(3), 194–202. https://doi.org/10.1177/1059840510366022

Asmundson, G. J. G., & Taylor, S. (2020). Chronophobia: Fear and the 2019-nCoV outbreak. *Journal of Anxiety Disorders, 70*(February). https://doi.org/10.1016/j.janxdis.2020.102196

Benis, A., Khodos, A., Ran, S., Levner, E., & Ashkenazi, S. (2021a). Social media engagement and influenza vaccination during the COVID-19 pandemic: Cross-sectional survey study. *Journal of Medical Internet Research, 23*(3), 1–15. https://doi.org/10.2196/25977

Benis, A., Seidmann, A., & Ashkenazi, S. (2021b). Reasons for taking the COVID-19 vaccine by US social media users. *Vaccines, 9*(4), 1–17. https://doi.org/10.3390/vaccines9040315

Carr, C. T., & Hayes, R. A. (2015). Social media: Defining, developing, and divining. *Atlantic Journal of Communication, 23*(1), 46–65. https://doi.org/10.1080/15456870.2015.972282

Chen, H., Li, X., Gao, J., Liu, X., Mao, Y., Wang, R., Zheng, P., Xiao, Q., Jia, Y., Fu, H., & Dai, J. (2021). Health belief model perspective on the control of COVID-19 vaccine hesitancy and the promotion of vaccination in china: Web-based cross-sectional study. *Journal of Medical Internet Research, 23*(9). https://doi.org/10.2196/29329

Chou, W. Y. S., Gaysynsky, A., & Cappella, J. N. (2020). Where we go from here: Health misinformation on social media. *American Journal of Public Health, 110*, S273–S275. https://doi.org/10.2105/AJPH.2020.305905

Crawshaw, J., Konnyu, K., Castillo, G., van Allen, Z., Grimshaw, J., & Presseau, J. (2021). *Factors affecting healthcare worker COVID-19 vaccination acceptance and uptake among the general public: A living behavioral science evidence synthesis*. McMaster Health Forum. https://www.researchgate.net/profile/Jacob-Crawshaw/publication/353044581_HCW_Vaccination_Living_Behavioural_Science_Evidence_Synthesis_v2_May_18/links/60e5daee1c28af345850c5c5/HCW-Vaccination-Living-Behavioural-Science-Evidence-Synthesis-v2-May-18.pdf

Featherstone, J. D., & Zhang, J. (2020). Feeling angry: The effects of vaccine misinformation and refutational messages on negative emotions and vaccination attitude. *Journal of Health Communication, 25*(9), 692–702. https://doi.org/10.1080/10810730.2020.1838671

Glanz, J. M., Wagner, N. M., Narwaney, K. J., Kraus, C. R., Shoup, J. A., Xu, S., O'Leary, S. T., Omer, S. B., Gleason, K. S., & Daley, M. F. (2017). Web-based social media intervention to increase vaccine acceptance: A randomized controlled trial. *Pediatrics, 140*(6). https://doi.org/10.1542/peds.2017-1117

Han, R., & Cheng, Y. (2020). The influence of norm perception on pro-environmental behavior: A comparison between the moderating roles of traditional media and social media. *International Journal of Environmental Research and Public Health, 17*(19), 1–20. https://doi.org/10.3390/ijerph17197164

Huo, C., Zhang, M., & Ma, F. (2018). Factors influencing people's health knowledge adoption in social media: The mediating effect of trust and the moderating effect of health threat. *Library Hi Tech, 36*(1), 129–151. https://doi.org/10.1108/LHT-04-2017-0074

Hyvönen, E., Alonen, M., Ikkala, E., & Mäkelä, E. (2014). Life stories as event-based linked data: Case semantic national biography. *CEUR Workshop Proceedings, 1272*, 1–4.

Im Kampe, E. O., Lehfeld, A. S., Buda, S., Buchholz, U., & Haas, W. (2020). Surveillance of COVID-19 school outbreaks, Germany, March to August 2020. *Eurosurveillance, 25*(38). https://doi.org/10.2807/1560-7917.ES.2020.25.38.2001645

Krueger, N. F., Reilly, M. D., & Carsrud, A. L. (2000). Competing models of entrepreneurial intentions. *Journal of Business Venturing, 15*(5), 411–432. https://doi.org/10.1016/S0883-9026(98)00033-0

Larson, H. J., Cooper, L. Z., Eskola, J., Katz, S. L., & Ratzan, S. (2011). Addressing the vaccine confidence gap. *The Lancet, 378*(9790), 526–535. https://doi.org/10.1016/S0140-6736(11)60678-8

Lee, S., Yamamoto, M., & Tandoc, E. C. (2021). Why people who know less think they know about COVID-19: Evidence from US and Singapore. *Journalism and Mass Communication Quarterly*. https://doi.org/10.1177/10776990211049460

Lin, Y., Hu, Z., Zhao, Q., Alias, H., Danaee, M., & Wong, L. P. (2020). Understanding COVID-19 vaccine demand and hesitancy: A nationwide online survey in China. *PLoS Neglected Tropical Diseases, 14*(12), e0008961. https://doi.org/10.1371/journal.pntd.0008961

Liu, L., Xie, J., Li, K., & Ji, S. (2020). Exploring how media influence preventive behavior and excessive preventive intention during the COVID-19 pandemic in China. *International Journal of Environmental Research and Public Health, 17*(21), 1–27. https://doi.org/10.3390/ijerph17217990

Luo, S., Xin, M., Wang, S., Zhao, J., Zhang, G., Li, L., Li, L., & Lau, J. T. F. (2021, May). Behavioral intention of receiving COVID-19 vaccination, social media exposures, and peer discussions in China. *Epidemiology and Infection*. https://doi.org/10.1017/S0950268821000947

Marcelin, J. R., Cortés-Penfield, N., Del Rio, C., Desai, A., Echenique, I., Granwehr, B., Lawal, F., Kuriakose, K., Lee, D. H., Malinis, M., Ruidera, D., Siddiqui, J., Spec, A., & Swartz, T. H. (2021). How the field of infectious diseases can leverage digital strategy and social media use during a pandemic. *Open Forum Infectious Diseases, 8*(2), 1–6. https://doi.org/10.1093/ofid/ofab027

Mckinley, C. J., & Lauby, F. (2021). Anti-vaccine beliefs and COVID-19 information seeking on social media: Examining processes influencing COVID-19 beliefs and preventative actions. *International Journal of Communication, 15*, 4252–4274. http://ijoc.org.

Mo, P. K. H., Luo, S., Wang, S., Zhao, J., Zhang, G., Li, L., Li, L., Xie, L., & Lau, J. T. F. (2021). Intention to receive the COVID-19 vaccination in china: Application of the diffusion

of innovations theory and the moderating role of openness to experience. *Vaccines, 9*(2), 1–15. https://doi.org/10.3390/vaccines9020129
Morrow, B. H. (2009, July). *Risk behavior and risk communication: synthesis and expert interviews* (pp. 1–52). Final Report for the NOAA Coastal Services Center.
Morton, T. A., & Duck, J. M. (2001). Communication and health beliefs: Mass and interpersonal influences on perceptions of risk to self and others. *Communication Research, 28*(5), 602–626. https://doi.org/10.1177/009365001028005002
Puri, N., Coomes, E. A., Haghbayan, H., & Gunaratne, K. (2020). Social media and vaccine hesitancy: New updates for the era of COVID-19 and globalized infectious diseases. *Human Vaccines and Immunotherapeutics, 16*(11), 2586–2593. https://doi.org/10.1080/21645515.2020.1780846
Qin, N., Shi, S., Duan, Y., Ma, G., Li, X., Shen, Z., Zhang, S., Luo, A., & Zhong, Z. (2022). Social media use, eHealth literacy, knowledge, attitudes, and practices toward COVID-19 vaccination among chinese college students in the phase of regular epidemic prevention and control: A cross-sectional survey. *Frontiers in Public Health, 9*(January). https://doi.org/10.3389/fpubh.2021.754904
Rakhmadhani, I., Yulida, E., Fahrina, & Jaelani, A. K. (2021). COVID-19 vaccine hesitancy on society and related legal regulations. *Proceedings of the International Conference on Environmental and Energy Policy (ICEEP 2021), 583*, 281–285. https://doi.org/10.2991/assehr.k.211014.061
Ratanasiripong, N. T., Sri-Umporn, S., Kathalae, D., Hanklang, S., & Ratanasiripong, P. (2018). Human papillomavirus (HPV) vaccination and factors related to intention to obtain the vaccine among young college women in Thailand. *Journal of Health Research, 32*(2), 142–151. https://doi.org/10.1108/JHR-01-2018-018
Rivis, A., & Sheeran, P. (2017). Descriptive norms as an additional predictor in the theory of planned behavior: A meta-analysis. *Planned Behavior: THe Relationship between Human Thought and Action, 43–62*,. https://doi.org/10.4324/9781315126449-4
Rosenstock, I. M., Strecher, V. J., Becker, M. H. (1988). Social learning theory and the health belief model. *Corporate Ownership and Control, 13*(2), 393–401. https://doi.org/10.22495/cocv13i3c2p11
Rubinstein, H., Marcu, A., Yardley, L., & Michie, S. (2015). Public preferences for vaccination and antiviral medicines under different pandemic flu outbreak scenarios. *BMC Public Health, 15*(1), 1–13. https://doi.org/10.1186/s12889-015-1541-8
Slovic, P. (1987). Risk perception. *Science, 236*(4799), 280–285.
Sooknanan, J., & Comissiong, D. M. G. (2020). Trending on social media: Integrating social media into infectious disease dynamics. *Bulletin of Mathematical Biology, 82*(7), 1–11. https://doi.org/10.1007/s11538-020-00757-4
Tang, L., Bie, B., Park, S. E., & Zhi, D. (2018). Social media and outbreaks of emerging infectious diseases: A systematic review of literature. *American Journal of Infection Control, 46*(9), 962–972. https://doi.org/10.1016/j.ajic.2018.02.010
Thaker, J. (2020). *Planning for a COVID-19 vaccination campaign: The role of social norms, trust, knowledge, and vaccine attitudes.*
Thaker, J. (2021). The persistence of vaccine hesitancy: COVID-19 vaccination intention in New Zealand. *Journal of Health Communication, 26*(2), 104–111. https://doi.org/10.1080/10810730.2021.1899346
Vahedi, Z., Sibalis, A., & Sutherland, J. E. (2018). Are media literacy interventions effective at changing attitudes and intentions towards risky health behaviors in adolescents? A Meta-Analytic Review. *Journal of Adolescence, 67*(July), 140–152. https://doi.org/10.1016/j.adolescence.2018.06.007
Vanderpool, R. C., Gaysynsky, A., & Chou, W. Y. S. (2020). Using a global pandemic as a teachable moment to promote vaccine literacy and build resilience to misinformation. *American Journal of Public Health, 110*, S284–S285. https://doi.org/10.2105/AJPH.2020.305906

Yıldırım, M., & Güler, A. (2020, January). Positivity explains how COVID-19 perceived risk increases death distress and reduces happiness. *Personality and Individual Differences, 168*, 110347.

Yong, S. D., & Jordan, A. H. (2017). The influence of social networking photos on social norms and sexual health behaviors. *Clinical Infectious Diseases*.

Zenk, L., Steiner, G., Cunha, M. P., Laubichler, M. D., Bertau, M., Kainz, M. J., Jäger, C., & Schernhammer, E. S. (2020). Fast response to superspreading: Uncertainty and complexity in the context of COVID-19. *International Journal of Environmental Research and Public Health, 17*(21), 1–13. https://doi.org/10.3390/ijerph17217884

Zhao, Y.-M., Liu, L., Sun, J., Yan, W., Yuan, K., Zheng, Y.-B., Lu, Z.-A., Liu, L., Ni, S.-Y., Su, S.-Z., Zhu, X.-M., Zeng, N., Gong, Y.-M., Wu, P., Ran, M.-S., Leng, Y., Shi, J., Shi, L., Lu, L., & Bao, Y.-P. (2021). Public willingness and determinants of COVID-19 vaccination at the initial stage of mass vaccination in China. *Vaccines, 9*(10), 1172. https://doi.org/10.3390/vaccines9101172

Ye Tao received her MA degree in communication studies from the Hong Kong Baptist University. She is currently working as an assistant instructor at Beijing Normal University-Hong Kong Baptist University United International College in the Faculty of Humanities and social sciences. Her research interests include health communication, and political communication.

Chuanli Xia is Assistant Professor in the Faculty of Humanities and Social Sciences at BNU-HKBU United International College. His research interests include political communication, public opinion, and Hong Kong studies.

Hong Zhao earned his master's degree in communication from the Hong Kong Baptist University after getting a bachelor's degree in journalism and communication from the Northeast Normal University. Zhao's main research interests include misleading communication, gender war, and cross-cultural communication.

Ning Liu got a master's degree in communication at BNU-HKBU United International College with the first-class scholarship. Her research interests are international communication and media studies, with the quantitative article "Image of Female Victim in News Report" newly published. She served as a research assistant at school and a reporter of People's Daily.

Media Discourses and China's Social Mobilization at the Early Crisis Stage of the COVID-19 Pandemic

Siu Han Chan, Ruijie Guo, and Xinyu Huang

1 Introduction

China, as the epicenter of the initial outbreak of the COVID-19 virus in late 2019 and early 2020, managed to contain the domestic spread at a very low level by the end of April of the same year. In early 2020, many observers predicted the massive COVID-19 outbreak to be China's "Chernobyl Moment," a calamitous juncture that may compromise the legitimacy or even threaten the survival of the Chinese state as a whole. They believed China's national conditions in early 2020 would be similar to the Soviet Union in the mid-1980s (Shih, 2020; Topaloff, 2020). However, the pandemic consolidated Chinese society at the early stage. The Chinese population also exhibited a high level of social and political trust, an essential lubricant of social life and smooth implementation of policies, especially at the time of crisis (Putnam, 2000; Van der Meer, 2017). Such observation is empirically corroborated by *Edelman Trust Barometer* released in early 2022. China was well-trusted by its nationals and ranked second in 2021 and first in 2022 on the global trust barometer. Most democratic European states, Japan, and the US recorded overall distrust among their peoples (Edelman, 2022). Among the four major institutions surveyed, the Chinese people instilled trust in their government more than in the business sector, media, and NGOs in descending order (Edelman, 2022). Chinese people's confidence in the government explains why compulsory itinerary tracking, lockdown, quarantine, or frequent COVID test requests can be imposed in large parts of China without much resistance. There was even an upsurge of nationalist sentiment after the alleged "Chernobyl Moment" (Zhang, 2020a, 2020b).

S. H. Chan (✉) · R. Guo · X. Huang
Beijing Normal University-Hong Kong Baptist University,
United International College, Zhuhai, China
e-mail: siuhanchan@uic.edu.cn

© The Author(s), under exclusive license to Springer Nature Singapore Pte Ltd. 2023
S. X. B. Zhao et al. (eds.), *Comparative Studies on Pandemic Control Policies and the Resilience of Society*, https://doi.org/10.1007/978-981-19-9993-2_14

As Arjun Appadurai astutely observes, the state of exception of COVID-19 connotes less about the usurpation of extraordinary powers by sovereign states as many worry. It is about "the concession that all national sovereigns are weak...that they cannot face this crisis without the help of society at large...No ruler in his right mind has denied that without a full-blooded *social mobilization*, there will be no survival this time...The social has been rediscovered by the state" (Appadurai, 2020, p. 222; authors' italics). Successful social mobilization from the top depends on people's trust and confidence in the government. They are willing to suspend (part of) their individualities and throw themselves into campaigns or directives for the greater good that may not always coincide with their best personal interests. In the process, the role of media is crucial. Media not only report an incident, but assume a more positive and activistic orientation in their representation. Sociologists frame such a process as *mediatiz*ation, where media enter and become a constituent part of the incident with the power to shape its course of action and development (Esser & Strömbäck, 2014; Hjarvard, 2013). In Simon Cottle's succinct words, media "performatively enacting [incidents]...constituting them on the public stage, and visualizing and narrating them in ways that can demand recognition and response" (Cottle, 2012, p. 260). In other words, media contribute significantly to the discursive formation of the cultural and political definition of the meaning of an incident, e.g., a minor disturbance or a major crisis of the social system, and whether actions should be taken and what lines of activities are considered legitimate and acceptable by the people.

Discourses are, as Foucault defines, "practices that systematically form the objects of which they speak" (Foucault, 2002, p. 53). Discourse organizes knowledge and perception, and structures social relations by way of collective subscription to the discursive logic of a discourse, which essentially excludes competing discursive formation. Discourse reinforces and limits the circulation of particular social meanings and interpretations and (re)produces social order. The potential rhetorical force of media discourse at both the cognitive and affective levels makes its positioning immensely important for constituency-building and social mobilization in society, especially in critical moments of the social system, like the early stage of the coronavirus outbreak. Approbative discursive orientations in the mainstream media can assuage social tension caused by implementing controversial and extraordinary measures. Yet, discourse is, at the same time, the strategic breaking point for resisting the dominant stream of meanings, instituting alternative definitions and symbols, and rebalancing social power.

Taking the double-edged nature of discourse into consideration, we examine how China successfully weathered the pandemic-bred crisis and restored the social system to a relatively (new) normal state of being intramurally in four months, by effectively cementing social trust and mobilizing the Chinese citizens. The performative enactment of the COVID-19 pandemic in state-sponsored media discourses during the initial period of the outbreak from January to April 2020, the *crisis* stage of Chinese society, is the primary focus of this paper. State-sponsored media refer primarily to state-owned media that are the official channels of the Chinese state, including *People's Daily, Xinhua News, Guangming Daily, Chinese Central Television (CCTV)*

News, and *Hubei Daily*. State-sponsored media also refer to media outlets closely align with the official stances, but are not directly owned by the state, such as *Southern Metropolis* and *Global Times*. State-sponsored media use different platforms, both the more conventional media outlets such as newspapers, television, and social media alike, to reach the Chinese public. Given the unique structural position of media in relation to the ruling party guided by the Marxist-Leninist perspective, state-sponsored media in China are instrumental in driving the leitmotifs of society. Such leitmotifs are sanctioned by the central government and the Chinese Communist Party.

Chinese society has, in principle, returned to the "normal state of being" by the end of April 2020. Since then, the national spread of the coronavirus has been effectively controlled. Pandemic prevention has become normalized and routinized at the societal level. The release of the official statement "Normalized COVID-19 Prevention and Control Requirements" in May 2020 also marked the termination of the pandemic "crisis." However, it does not mean that Chinese society is done with COVID-19 once and for all (Caiyun et al., 2021). The meaning of "crisis" has to be clarified here as its ubiquity in everyday and academic discourses has almost dwindled it into a flimsy idea. Sociologically speaking, crisis designates the sudden disruption of the normal state of being, often transitional, of the seemingly normal development of a social system. A crisis has three major dimensions: *threat*, *urgency*, and *uncertainty*. Hence a rigorously defined crisis refers to an incident that poses imminent threats to commonly held values, like life, safety, health, or the integrity of the social system, with largely unknown consequences and impacts, which actions must be taken immediately (Boin, 2005).

The three criteria make it evident that the coronavirus gradually ceases to be a crisis threatening the survival of the social system when we have more knowledge and means to deal with it, and the threats it brings become relatively manageable and predictable. Starting from May 2020, Chinese society has no longer been embroiled in the crisis phase of the pandemic. Intramurally, everyday life returns to "normal," and economic activities resume, with Polymerase Chain Reaction (PCR) tests, health code, and itinerary tracking as daily regulars. "Zero-tolerance policy" is adopted. Based on strict border control, guerilla tactics are deployed to curb the sporadic outbreaks caused by imported cases until December 2022.

Chinese people's high level of trust in the Chinese state and general public compliance with government's anti-pandemic directives indicate that the society was much better connected and deployed by the state for the common good and anti-pandemic campaign in China than elsewhere (Appadurai, 2020). This paper investigates how the state-sponsored media impart cultural meaning, collective sentiment, and empathy that strike a chord with Chinese people at the national level through carefully poised discourses, fostering despatialized solidarity and social mobilization indispensable to the success of rapid crisis termination. Three major discursive formations deployed by the state-sponsored media are militant discourse, altruistic discourse, and authoritative discourse.

Since the beginning, the Chinese state has resorted to war imagery to build up the tension and urgency of an extraordinary time that warrants decisive decision-making and extreme measures. The construction of militant discourse will be revealed. In addition, the state-sponsored media spare little effort in reporting the suffering of patients who caught coronavirus and the altruistic acts and stories of selfless medical personnel, patients, and other civilians. Such discursive strategy has been proven successful in arousing empathy, tolerance, and the suppression of the self in thinking and action. The temporary belittlement of selves makes voluntarism, collective orientation, and social mobilization much more viable. The third primary discourse is the authoritative address, in which *Zhong Nanshan*, known for being professional and courageous, was tactfully established in the state-sponsored media as an authoritative figure. As one of the informal official spokespersons, Zhong helped convince the public about official information's credibility and the inevitability of extreme measures.

Despite state-sponsored media usually dictates the mainstream discursive directions in China, the field of cultural meaning construction is not airtight. It would be misleading to characterize the overall media environment in China today straightforwardly as top–down direct control. The government's vigilance and widespread censorship are the reality. The formation of alternative meanings and discourses, as well as critical engagement with the official stances, are still possible, especially using social media. Social media refer to the online space where citizens are the central nodes of information construction, dissemination, and exchange. The Chinese state is generally lenient about criticisms unless such criticisms cross specific sensitive (political) thresholds, though it must be admitted that the thresholds are not always clear. In addition to state-sponsored discourses, we also witness the discursive formation on social media that explicitly contests the official interpretation of Dr. *Li Wenliang* and some of his colleagues as rumormongers, compelling the government to change its verdict during the pandemic crisis period. It may be seen as the remarkable triumph of the voices of the Chinese people, even though the government's concession must be contextualized and should not be taken as a norm. Social media has become an important space for Chinese citizens to negotiate their social participation and meaning construction.

2 The Militant Discourse: "We Are at War with COVID"

Resorting to militant discourse is the most salient discursive strategy seen in the Chinese state-sponsored media during the crisis stage of the COVID-19 pandemic. China is not alone in constructing the anti-pandemic campaign as a battle against an aggressive virus. Many major democratic countries, including France, America, Britain, and others, also have recourse to war-related imageries to mobilize their people, especially during the early outbreak (Malm, 2020). War is undoubtedly the time that the state can summon the full support of society. However, China is perhaps the country that upholds the war narrative most persistently. Framing the pandemic crisis as a war-like situation and positioning the virus as a microbiological enemy of the whole nation helped the Chinese government successfully solicit the support of the Chinese society for most of the rather demanding directives imposed during the crisis period.

The critical terminology *Zhanyi* (战疫), punning on the homo-phonic term *Zhanyi* (战役), meaning battle, was coined as early as January 25, 2020, to designate the anti-pandemic campaign as a wartime collective project. "*Zhanyi*" has a two-fold meaning. When the two characters "Zhan" and "Yi" come together as a single noun, *Zhanyi*, it gives the pandemic a distinctive identity of war. The war imagery brings together a specific set of motifs, and organizes them in a particular manner to create an intense perception of urgency, decisiveness, and the vision of collective responsibility and national solidarity. The term "*Zhanyi*" also has an activistic connotation when the two characters are read separately. In that sense, the character "Zhan" is an action verb meaning "to battle with," and the second word, "Yi," refers to the pandemic. So "*Zhanyi*" can mean both an imperative and advocacy "to battle the pandemic." By establishing *Zhanyi* as the catchword, the pandemic has been successfully framed as the "enemy" of Chinese society.

Naming is not only a process of giving a name to a happening; it is about recalling a familiar configuration of concepts and affection associated with a particular word (Lukin, 2013). Tapping the COVID-19 crisis into the modern national historical memory of wars and suffering serves to evoke the Chinese people's national consciousness and resonance:

> The contemporary history of China is a history of humiliation, great fall, and rise from the sick man of East Asia to the grand dragon of the East...The current battle against the pandemic is a crisis and a significant test for the Chinese nation. No risk or challenge can stop the Chinese people and country from marching forward. (Wang, 2020, para. 4)

Connecting war narrative with major public crisis has long been one of China's propaganda traditions. The construction of a war narrative in the *Wenchuan earthquake* in 2008 has proven to be effective in mobilizing the Chinese population for disaster relief effort (Zhang, 2015).

Setting the pandemic into the frame of existing national memory of wars provides a convergent focus for Chinese society to arrive at a collective understanding of an emerging situation full of uncertainties. A shared definition of the situation saves the politics of meaning that may delegitimize the state and compromise its capacity for mobilization in an emergency. The idea of war, to put in de Saussure's words, is like "[t]he center of a constellation…the point of convergence of an indefinite number of co-ordinated terms" (de Saussure et al., 2011, p. 126). The semantics associated with war in China includes culturally received notions such as "responsibility," "sacrifice," "courage," "resolute," and "urgency." It also consists of lexically denotative terminologies like "battle," "command," "the bugle call," and "army." All these related meanings were actively invoked and used to raise the collective consciousness by state-sponsored media, like *CCTV News*, *People's Daily*, and *Xinhua News*. Instructions and orders in warring circumstances command wholesale cooperation and compliance from society. Repetitive militant terms conveying the sensation of collectivism and urgency saturate the public space for a limited period, shaping and steering public opinion and sentiment through a homogenizing process as in other crises (Zhang, 2015, p. 128).

On the one hand, war-related imageries prepare Chinese society to accept the societal shift from *normality* to an *emergency*. The latter is an extraordinary state of being for the social system, in which the societal priority is given over to pandemic-related government dictates above other normal expectations. On the other hand, they work to psychologically equip Chinese citizens to respond to the top-down rally call of the government like a resolute "army." That includes accepting drastic measures like total lockdown and severe limitation on individual freedom and mobility that affected more than half of the Chinese population during the early crisis phase of the pandemic in 2020 (Zhong & Mozur, 2020). To borrow the words of the World Health Organization (WHO), "adherence to these containment measures has only been possible due to the deep commitment of the Chinese people to collective action in the face of this common threat" (WHO, 2020, p.17). War narrative makes such commitment possible in China.

Framing the coronavirus as a national enemy in *Zhanyi* reminds the Chinese that "we" as individuals and community members should take action to cope with the crisis together. Fighting against the microbiological enemy becomes "our" target. The militant consciousness sustained by *Zhanyi* chimes well with the entrenched altruistic value of collectivism in Chinese society, which the state-sponsored media constantly reinforce in the pandemic crisis.

3 The Altruistic Discourse: Praising Selfless and Sacrificial Behaviors

In addition to the militant discourse, state-sponsored media outlets were devoted to reporting patients' suffering and the altruistic acts of medical personnel in Wuhan, the pandemic's epicenter. Visualizing people's misery and praising selfless and sacrificial behaviors strike public consciousness and create a feeling of proximal affinity, which further intensifies the sense of shared responsibility and strengthens social solidarity at the national level (Zhang, 2015). The "Fangcang" column launched by the *CCTV* "News Investigation" program takes patients receiving treatment in *fangcang* hospitals, makeshift cabins used for housing and treating COVID patients, as the main subject, and reports from the perspective of patients in the form of Vlog. The living scene of suffering and human greatness in the news is powerful in articulating collective sentiment on the part of the audience.

Fu Dan, a patient hospitalized in a *fangcang* hospital, told a reporter Zhu Xidong (2020) on February 15 that panic often gripped her when she first arrived. However, she overcame her panic when she witnessed a self-motivated patient without a proper protective suit volunteering quietly to sort out the waste in front of the *fangcang* hospital. Fu Dan's Vlog[1] in the *fangcang* hospital inspired other patients to make similar contributions, like helping with supplies distribution, maintaining order and discipline, and dealing with garbage disposal (Zhu, 2020). Some Vlogs also showed harmonious scenes among doctors, nurses, and patients dancing together and supporting each other. Through this and similar moving stories, the audience shared the suffering and anxiety of the patients. Still, they felt the growing vitality and optimism in the *fangcang* hospital simultaneously. Carefully selected personalized stories and emotive accounts encourage identification and empathy with the plight of patients (Cottle, 2012).

Emotional reactions such as pity and empathy instigated by moral spectatorship are indispensable elements in the mobilization of solidarity and emotional engagement on the part of the audience (Lahusen, 2020). A tie between the media and the public is formed while the audience unconsciously transforms into subjects involved in the issues (Hartley, 2002). It involves a cognitive transition from a conception of affliction from elsewhere to some emotion-arousing "life miseries" (Zhang, 2015, p. 139). Collective empathy united the whole Chinese nation, which became the critical emotional foundation for collective mobilization.

Altruistic acts were widely reported in the state-sponsored media at the early stage of the pandemic. In the broadest sense, altruism means promoting the interests of others, which implies benevolent and sympathetic feelings. Altruistic orientations are an important foundation of collective lives. If everyone were egoistic all along, social life would not be possible. Individuals must identify with something larger than themselves and, at times, take action with goals that are externally directed to the groups they participate in (Durkheim, 2005). Pure egoism and altruism represent two poles of a continuum. Different societies have various overall balances of the two value orientations, and different societal situations may shift the balance too. It is fair to say that Chinese society, which was historically more collectivist and emphasized the virtue of self-sacrifice under the influence of Confucianism and Daoism, is more disposed to altruism, especially in a crisis (Lee et al., 2008).

Unsurprisingly, the Chinese mainstream media draw attention to the altruism and collectivism of the medical workers and volunteers in their reports during the COVID-19 outbreak. The altruistic discourse always praises individuals and acts that deny the self in favor of collective goals or well-being. Taking the reports on the voluntary medical teams who ran to the frontline released by *People's Daily* and *Xinhua News* as an example, the most used words and phrases are "resolute," "without hesitation," "fear no sacrifice," "united as one in fighting coronavirus" and "sacrifice for the sake of others." The familiar figure Zhang Dingyu, whom the *Xinhua News* crafted sacrificial spirit on February 1, had gone viral on the Internet

[1] In the Vlog, Fu Dan was shown broadcasting reminders to fellow patients through loudspeakers: "The remaining water in mineral water bottles should be poured into a bucket, and then the plastic bottles should be thrown into the dry trash can. Please separate dry and wet garbage. This can raise garbage disposal efficiency and better maintain our *fangcang* hospital's environment." From the video Fu recorded, we also witnessed the scene of mutual support among the patients. They chatted and cheered each other up when someone had a mood swing.

(Qian et al., 2020). As the director of Jinyintan Hospital[2] in the critical moment, Zhang possessed the essential quality of a good leader—calmness, determination, and organizational ability. However, upon these, what distinguishes him is his disregard for his medical condition as an Amyotrophic Lateral Sclerosis (ALS) patient, who suffers from a declining state of muscle wasting, difficulty swallowing, and even respiratory failure but insists on working in the frontline. In Durkheim's words, Zhang has actively renounced his self-interest (Durkheim, 2005). His colleague Wang Xianfei told the reporter on January 29 that he worried about the director's health condition and admired his restless dedication, "He had not allowed himself a day off since December 29." Zhang remained in his position even after his wife, a medical worker at another hospital, was diagnosed with COVID-19 on January 20, 2020. After his wife was admitted to the hospital, Zhang visited her late at night and stayed for only half an hour. "I did not say too much since both of us were exhausted," Zhang regretted, "I did not shoulder the responsibility of a husband though I was credited as a competent doctor. I cannot help but feel guilty for my wife." On the horns of a dilemma of the personal and the collective, he prioritized the latter, even if that means he was failing his family.

An editorial published on February 15 by *People's Daily* (Chen, 2020) explicitly venerated the devoted figure of Wu Longan, a voluntary medical doctor at Wuhan Jiangxia Miaoshan Community Health Service Center, who hid the truth from his 80-year-old parents. He said:

> When I heard about the severe epidemic in Wuhan, my hometown, I could not sleep a wink or eat well. I immediately contacted the epidemic prevention headquarters in Jiangxia District and learned there was a shortage of medical staff. I volunteered to back up the frontier on February 10 by taking up the duty as a doctor. I did not inform my parents since if I did so, they would be worried about the dangerous situation I was involved in. (para. 20)

Just like in the case of Zhang Dingyu, personal feelings and family needs were cast aside for the greater need of the nation as a whole.

The infusion of selfless stories and spirits in society may well be the reason for a sudden influx of volunteering medical care personnel into the pandemic-ridden city of Wuhan. These medical volunteers were very well aware that this was a tough battle, and they would likely die from contracting the then unknown virus. It is worth noting that the first medical team arrived in Hubei on January 24, just one day after Wuhan closed down all its outbound routes. Ten national emergency medical teams were dispatched to assist Hubei province on February 4. According to *Xinhua News*, the larger batch of medical personnel arrived on February 9, with a total of more than 6,000 medical personnel (Yuyuan, 2020). While the municipal states outside Hubei may have contributed to the mobilization, the high spirit of the doctors and nurses widely seen in media accounts revealed their enthusiasm and pride in being part of the rescue team. The demanding working condition and long hours of the medical personnel were receiving significant attention. According to a report by *People's Daily* on February 7, many medical staff wore "adult diapers" when they were on

[2] Jinyintan Hospital is situated in Wuhan and specializes in curing infectious diseases. It admitted many COVID-19 patients.

duty to minimize the waste of protective clothing caused by toilet use and save time for rescuing more patients (Zhu & Yang, 2020).

State-sponsored media also charted altruism among volunteers at the community level. *Xinhua News* released a report on February 23 in which a volunteer call was issued throughout Wuhan, and over ten thousand people answered the call immediately (Liao & Wang, 2020). All the volunteers interviewed said they did not hesitate to join the volunteer team after noticing the call. Themes like "working without complaints" and "bearing hardship" were recurring in news reports. Such kind of selfless spirit was vividly reflected in the dedication of the volunteer barber depicted by *People's Daily* on February 19, "*Qumo* had been cutting hair for several days without rest. Her hands were bruised, and the cuts had not healed yet. Even though she was supposed to take a rest some time, as soon as she heard that Huoshenshan Hospital needed our assistance, she responded to the call without hesitation" (Chen & Jin, 2020).

The more that citizens' cultural values were anchored on collectivism, the more likely they would be affected by the altruism reported. More importantly, such altruistic behaviors portrayed in the state-sponsored media profoundly impact both the medical personnel and volunteers at the grassroots level. In all the cases discussed above, individuals almost abnegate themselves and, if necessary, their lives to serve the collective, the greater good. Altruistic discourse hence promotes the spirit of mutual help, minimizes psychological barriers to work in otherwise unbearable conditions, and reduces the friction of social mobilization indispensable for the rapid termination of crisis in Chinese society.

4 The Authoritative Discourse: The Staging of an Anti-pandemic Icon

Living under the crisis phase of the coronavirus outbreak, the public has greater demand for unified and authoritative information with "truth-claims," furnished by institutions with social legitimacy, than on average (Foucault, 2002; Mills, 2008, p. 61). Figures well-trusted by the public would be essential in convincing the population to accept the official definition of the situation and the necessity of specific harsh measures. An 84-year-old epidemiologist *Zhong Nanshan* has been such an authority figure in China's anti-pandemic campaign. Extensive media coverage in the mainstream media successfully projects Zhong Nanshan, a renowned respiratory disease expert and currently the head of a High-level Expert Team at China's *National Health Commission* (NHC), as a responsible, bold, and resolute icon to be looked up to and trusted.

Zhong was pushed to the center stage in the media space during the COVID-19 outbreak primarily because of his professional prestige as a veteran and globally recognized epidemiologist known for his involvement in the 2002–2003 Severe Acute Respiratory Syndrome (SARS) epidemic that took 325 lives in mainland China.

Zhong was among the first people who were aware of the spread of SARS in late December 2002 and early 2003 and worked to contain the epidemic relatively quickly. Zhong's contribution to combating SARS earned him the excellent presidency of the *Chinese Medical Association* from 2005 to 2009. The state-sponsored media quickly chose Zhong to be the key spokesperson in the fight against the COVID-19 pandemic. Zhong broke the news about the communicability of the new coronavirus to the public in a live interview on *CCTV* on January 20, a fact that was formerly denied by the *Wuhan Center of Disease Prevention and Control*, and confirmed that 14 medical personnel had been infected (CCTV News, 2020; Yuan, 2020). Zhong's appearance on a state-owned television channel can hardly be an accident. Zhong is deemed suitable to publicize sensitive information that contradicts former official statements. Zhong's interview is a bugle call of China's *Zhanyi*, a battle against the pandemic. Almost immediately after the interview, China stepped up its preventive measures. On January 23, the city of Wuhan was locked down. Zhong Nanshan's definitive diagnosis of the epidemic became the basis for the collective recognition of the impending crisis and the imposition of drastic measures.

Besides his professional knowledge and accomplishments, Zhang also established himself as an upright respiratory epidemiologist who is tenacious with truth and unyielding to political pressure (Ye, 2020). When he was involved in the SARS prevention program in 2003, he blurted his doubts openly about the Chinese government's public assurance that the spread of the virus was under control in Beijing and chlamydia was to be blamed. Revealing truth in public that contravenes official proclamation is a potentially risky political undertaking in China, especially in an emergency. While dissents are generally tolerated in China, high-profile challenges to the government's stance may be penalized with house arrest, administrative detention, or convicted with the "Picking quarrels and provoking troubles" law. Zhong was rather forthcoming when he publicly told an inconvenient truth at that sensitive time and refused to treat patients with the government's recommendations. Eventually, Zhong was proven right, and his forewarning helped save many lives. Since 2003, "the spirit of Nanshan" has been widely promoted in China, and he becomes an iconic scientist who speaks the truth (Ye, 2020, p. 5).

Once again, in 2020, Zhong proved himself to be a righteous person. Four days after Dr. Li Wenliang's death, Zhong, in an interview with Reuters, bravely expressed his admiration and pride for Dr. Li and called him "the hero of China." At that time, the Wuhan government was still hesitant to redress the wrongs done to the whistleblower. Zhong's interview was reposted on the Voice of America (VOA)'s Chinese website. Many Chinese netizens praised Zhong in the comment section. For instance, @Joe Sun (2020) posted on February 12 that "Zhong is the backbone of China since he has outstanding professional skills, courage to take responsibility, and bravery to praise and criticize." @Dongdong yang (2020) posted on February 12 that "Zhong and Dr. Li and millions of others who dare to disclose the truth deserve the same pride and praise. They are the best warriors of China."

Being revered by many, Zhong is a very suitable candidate to play the leading, authoritative figure in society when the public direly needed authority and practical guidance during the early crisis period of the COVID outbreak. Zhong and his team used an established scientific model to predict that the case of infection in China to be peaked in mid to late February, and the number would level off before going down (Lin, 2020). The prediction, which was later proven accurate in the main, provided some critical orientations for the larger society and eased the highly charged emotions among the public during a greatly uncertain period. Moreover, Zhong helped prevent the further spread of the virus when he recommended citizens to stay home and suspend unnecessary travel plans during the seven-day Spring Festival Holiday in 2020 (Zhong & Mozur, 2020). Zhong had been instrumental in psychologically preparing the Chinese citizens to comply with highly demanding measures like the stay-home order, which later became mandatory in many of China. *New York Times* estimates that as many as 776 million Chinese, primarily urban dwellers, experienced some form of lockdown during the first year of the pandemic (Zhong & Mozur, 2020). The words of this respected and trusted figure helped make the harsh anti-pandemic measures appear to be more tolerable and better accepted by the Chinese citizens as strictly necessary.

Zhong also appears as a person who devotes to his career and patients in public. According to a reporter who has followed and interviewed Zhong Nanshan for many years, the happiest thing in his life has always been treating patients and saving people's lives personally (Jiang & He, 2020). Most people in their eighties would have already retired home, but Zhong still went to the heartland of the pandemic. On January 18, 2020, a photo of this weary 84-year-old doctor dozing on a high-speed train heading to Wuhan went viral in China's cyberspace. The post had generated a buzz among Chinese netizens as Zhong had earlier called on the public not to travel to Wuhan to avoid catching the disease. Zhong once said: "Hospitals are like battlefields. As a soldier, it is my duty to rush to the frontline," which resonates with the war imaginary of the pandemic (Department of Science & Technology of Guangdong Province, 2020). Zhong also appeased public emotion and attempted to boost national morale: "Wuhan has been a heroic city. With the support of the whole country, Wuhan will survive the challenge [posed by COVID-19 pandemic]" (Xiao & Xu, 2020).

At the early crisis stage of the outbreak of COVID-19, Zhong had been staged as a key icon in China's battle against the epidemic by the state-sponsored media. In recognition of his outstanding professional contribution and to further consolidate Zhong Nanshan's position as an authority at both medical and political levels, on September 8, Zhong received the Medal of the Republic, which is the highest state honor, and the national title "the People's Hero," from Chinese President *Xi Jinping*,[3]

[3] Zhang Nanshan is for sure far from an impeccable figure. He was embroiled in several scandalous accusations. He was accused of involving in brand endorsement in the context of a press conference for Yili Company in exchange for a 5 million RMB donation to his research team without disclosing the potential conflict of interests (RFI, 2020, May 12). Zhong Nanshan is also a prominent spokesperson for the medicine *Lianhua Qingwen* capsule and advocates it to be included in the national treatment plan for COVID-19. The drugs are highly sought after at some points, and the share price of the manufacturing company rises exponentially. Zhong is closely related to the company and has overlapping interests with its shareholders (*Epoch Times*, 2022, January 13).

in addition to the titles and honors he got in the last 20 years, including "one of the most inspiring figures since the establishment of new China," "revolutionary pioneer," "the most beautiful fighter," and so on (Department of Science & Technology of Guangdong Province, 2020). Branding Zhong as a courageous and respected man who works within the establishment in the state-sponsored media helps foster an image of openness and transparency at the top echelons in Chinese people's minds. The presence of an authoritative figure, whose words are well-received by the public, unifies individuals' perceptions of the crisis, strengthens the collective morale, and makes unpleasant imperatives more tolerable.

5 Social Media as the Emerging Space for Negotiating Social Participation

As aforementioned, the three discourses seen in the state-sponsored media facilitate the implementation of government policies and campaigns in the early crisis phase of the pandemic. However, the influence of spontaneous bottom-up responses from the public on social media during the same period should not be ignored. According to Gao's (2020) statistics, social media and various new platforms have become the main channels of information dissemination, including official information related to the COVID-19 pandemic in 2020. WeChat group and WeChat Moments, followed by television, are the two channels the public acquired most pandemic-related knowledge. Weibo posts and online news portals were widely used too. As illustrated before, it is fair to suggest that the dominant discursive orientations in Chinese society are driven primarily by the state through state-sponsored media outlets. Nonetheless, social media has emerged as a space between the Chinese government and society through which Chinese citizens give feedback to the state, voice their dissents and negotiate their social participation under a generally restrictive political environment.

On social media platforms, users engage in the exchange of arguments and feel themselves to be involving in some sort of collective interpretative work about what is happening around them (Trenz et al., 2020). On the one hand, Chinese citizens show a certain level of query, criticism, and disrespect to the party and the mainstream ideology online. The posts and comments on social media are hence the barometers of public attitude and opinions, which from time to time function to warn the government about collective sentiment and social tension on significant issues and help shift its policies toward a milder and more humane direction (Tai, 2006). On the other hand, limited information access and close surveillance of people's voices on potentially significant events or sensitive times are essential for the Chinese government to maintain social order and stability. Chinese people are not unfamiliar with this and are usually not too reactive to such limitations. However, in some critical situations,

However, the circulation of related information is limited within China. The authors reckon that it may have to do with some form of political maneuver or that Chinese people do not want to deconsecrate a figure they belove.

like Dr. Li Wenliang, the avalanche of online posts, comments, and retweets with rather sensitive contents force the government to take people's feelings and demands seriously.

The 33-year-old doctor *Li Wenliang* was an ophthalmologist working in Wuhan Central Hospital. On December 30, 2019, Dr. Li learned from the test report sent by one of his epidemiologist colleagues that their hospital admitted seven patients suspected of catching a contagious virus similar to SARS. Dr. Li then shared the news with his fellow doctors in a work WeChat group, warned them to take precautions, and urged them not to disperse this information. Yet his friendly reminder still circulated wildly on the internet and was not appreciated by the hospital and local authorities (Green, 2020). Four days later, on January 3, he was summoned to the local Public Security Bureau and was accused of "spreading rumors about epidemic" and "severely disturbing the social order" (Liu, 2020b). Dr. Li was asked to sign a "confessional note" for his misbehavior, but no further punitive measure was issued. Before that, on January 1, *Wuhan Public Security Bureau* (Beijing News, 2020, January 1) publicly condemned eight persons for spreading rumors about the epidemic. It is unclear whether Dr. Li was among the condemned as names were not disclosed in the statement.

The local health authorities stated that an unknown infectious disease was circulating in Wuhan, but no sign indicated that this was a communicable disease. In other words, the disease was believed to be of zoonotic origin only. A few days later, on January 9, the NHC expert evaluation team officially announced that a new coronavirus was the cause of the "pneumonia" found in Wuhan (The State Council Information Office, 2020). The public soon realized it was the same virus forewarned by Dr. Li and others like Dr. Ai Fen, who were publicly reprimanded for spreading rumors. The first massive wave of online discontent for the Wuhan local government's inaction and information non-transparency emerged. Dr. Li and the eight doctors admonished were seen by netizens as "whistle-blowers" who warned the public of a new communicable disease instead of "rumormongers" unfairly accused by the authorities. The official condemnation of the "rumormongers" and intentional underestimation, if not covering up, was interpreted by netizens as politically motivated. Netizens proposed that the underestimation reflected the local government's attempt to reduce adverse reports around the time of the provincial meetings of the *National People's Congress* (NPC) and the *Chinese People's Political Consultative Conference* (CPPCC) in Hubei province in mid-January.

As soon as Dr. Li uploaded the confessional note on Weibo and confirmed on February 1 that he was tested positive for the new coronavirus, there was growing public attention on his condition. The attention peaked when he was transferred to the emergency ward on February 6 and passed away at 2:58 a.m., February 7, 2020. The news of his death sparked unprecedented disappointment and even anger among the Chinese netizens and triggered a huge wave of grieving posts on Wechat Moments and Weibo. Bitter comments implicating the Wuhan authorities were widely seen under *CCTV*'s Weibo post about the rescue of Dr. Li Wenliang on February 7, 2020. In the words of user @buchiqingcaide gemini (2020):

> Who will speak for the truth if one is punished for telling the truth? In times of great crisis, [you] say you serve and protect the people. But who is guarding the lives of the white-robed doctors and nurses? Ignorant people must be punished.

In the words of user @Anougme_Xing (2020):

> It's the anger of 1.4 billion people! Being the magistrate of this place, [what you are doing] is a betrayal to your people and your country... medical staff in the pandemic frontline are martyrs, but who are you? What did you do for this country? Shame on you!

Millions of citizens took the death of Dr. Li as an outlet for their suppressed emotions during the pandemic. Chinese people widely commemorated him as a "folk hero," an anchorage of their feelings through which they can survive the crisis together (You & Zhang, 2022). According to the statistic by Qingbo Big Data (2020), from February 6 to February 12, 510,590 posts related to Dr. Li's death were posted on numerous social media platforms, primarily on Weibo. Nearly half of the netizens expressed their deep sorrow about his death and attacked Wuhan authorities' inaction and cowardice to be the cause of Dr. Li's precocious death. Unlike many other online discussions that usually cool down after a few days, the netizens "refuse to forget Dr. Li" and use his last Weibo post as a "digital wailing wall," expressing their gratitude and appreciation for his courage in speaking up (Yuan & Taylor, 2020).

The unprecedented outpourings of public sentiments and discontent with the local Wuhan government was soon addressed at the national level. To the surprise of many, despite withdrawing some of the most sensitive posts and WeChat columns, the Chinese state did not block Dr. Li's Weibo account nor ban the search for news about Dr. Li altogether using a keyword filter. It must be noted that these moves are not uncommon in China to prevent commotion in the online space. In this particular instance, the Chinese government responded to people's demands, hoping to channel the collective melancholia triggered by Dr. Li's death.

Immediately after Dr. Li's death on February 7, 2020, the *National Oversight Commission* issued a public statement citing that it had decided to send an investigation team to Wuhan to thoroughly examine "the issues related to Dr. Li Wenliang that the public reflected about" (Ma & Mai, 2020). On March 19, the investigation team released a report detailing what happened to Dr. Li after his initial post about coronavirus on the work WeChat group. It then claimed that the issue of the reprimand was "improper," and the law enforcement procedures were not standardized. The team also suggested that the reprimand be withdrawn (National Oversight Office, 2020). On the same day, the Wuhan police formally exonerated Dr. Li and apologized to his family (Liu, 2020a). The Chinese government also awarded him the May 4 Medal on April 20 and honored him as a "martyr" of the country (Yang, 2020). The stigma of rumormonger was then officially removed, and Dr. Li is now remembered as a hero by the Chinese nation.

High-sounded statement issued at a national level, openly redressing formal verdicts or governmental actions triggered directly by people's online comments, is very rare in China. This episode can be interpreted as a significant victory for the netizens in China. They had successfully negotiated with the Chinese government on treating Dr. Li Wenliang through their persistent voices and compelled the latter to revoke its former verdict. More importantly, the central government also responded to the incident at the legal level. In Beijing and Shenzhen (Zhu, 2020), a legal clause, referred colloquially to by many as the "Li Wenliang clause" (Xiao, 2020), was instituted to explicitly protect the rights of whistle-blowers in relation to sudden events of potentially significant consequences.

In other words, whistle-blowing has now gained legal protection in China. Whistle-blowers are not punished even if the information is inaccurate, so long as no malicious intent is involved (Yuan & Taylor, 2020). It is expected that more cities or even the nation will legislate on the whistle-blowing issue in the future. Such legislation implies greater protection of civil rights and, indirectly, an official recognition of Dr. Li's rightful behaviors. These institutional improvements are even more consequential at the societal level than clearing the stigmatization of Dr. Li and others.

6 Conclusion

In the early crisis stage (January–April 2020) of the COVID-19 pandemic, the Chinese government successfully produced militant, altruistic, and authoritative discourses through state-sponsored media outlets. The three discourses have shaped the public understanding of the crisis and steered collective participation in the pandemic relief effort. The war rhetoric geared up the perception of an impending danger and urgency to act. Altruistic discourse helped promote selfless acts and identification with the collective, hence ready submission to collective action and top-down imperatives. Through the words and deeds of the authority figure Zhong Nanshan, the Chinese government was able to carry out some strict and disturbing anti-epidemic measures without much social resistance. The three discourses together performatively engendered the crisis script in Chinese society and prescribed the appropriate lines of actions to be taken by its citizens during the initial outbreak. The collectively accepted crisis script gives the Chinese state carte blanche to act and mobilize human and material resources.

We can compare the media discourses in China with those in European and American societies. Media of the latter countries often challenge the legitimacy of state action, and public discourses are somewhat ambivalent about drastic measures. During the early outbreak of the COVID-19 pandemic, such societies became more divided, and people were more likely to have pandemic fatigue and be unwilling to cooperate. With the help of media discourses, the Chinese state effectively mobilized society directly from the top, which is the key to the rapid control of the pandemic outbreak in China. Successful crisis management then becomes the merit of the Chinese government and contributes to reproducing the imagined national community.

Admittedly, China was doing well in containing the virus and terminating the crisis of the social system during the initial outbreak in 2020. China also manages to keep the death toll of coronavirus very low compared to most other countries, at least up to the point of the completion of this paper in June 2022. According to Our World in Data, China has reported around five thousand cumulative deaths by the end of April 2022 (Ritchie et al., 2022). In the United States, nearly one million people have already died from contracting COVID-19. However, the tone set by a successful crisis management scheme may have become a liability for the Chinese government to embrace a more scientific attitude and adjust to the emerging situation when the pandemic is evolving into an endemic, a state of intensified virality but much less fatality. The potential exponential surge in cases, the normal development of an endemic, would directly contradict the low-tolerance tone—the former successful formula. Even with a low fatality rate and high vaccine coverage, China is almost alone in adhering to a strict zero-COVID policy and maintains severe restrictions on inbound travel until early December of 2022. It is against the more extensive background that most countries begin to treat the coronavirus as endemic to society like influenza in the first half of 2022.

Before the official release of the circular on "Further Optimizing the COVID-19 Response," named "The New 10 Guidelines" colloquially, on December 7, 2022, China was still running regular and widespread PCR testing program (The State Council, 2022, December 8). If a case was found, the infected and his close contacts were evicted and sent to isolation facilities, followed by the lockdown of targeted areas. The lockdown was very extensive as the Omicron variants are highly contagious. As of November 29, 2022, there were 32457 high risk zones across the country with various extents of lockdown measures imposed (CN-HEAlTHCARE, 2022, November 30). Many Chinese people's daily lives have been greatly affected by the sporadic closure of all non-essential businesses and activities in the containment zone and the strict limitation of mobility in lockdown. Even though the strictest measures were implemented, China's zero-COVID policy was becoming less effective in dealing with the evolved variants with augmented virality. The social and economic cost of maintaining a zero-COVID policy has increased exponentially toward the end of 2022. As China emerges from the initial crisis state of 2020, in which the threat, sense of urgency, and uncertainty posed by the coronavirus are significantly mitigated, the three discursive strategies are also losing their sway over the Chinese population. Pandemic-related social mobilization becomes far more arduous than before.

Voices questioning the zero-COVID policy flourished on social media, and many local governments were criticized heavily for being too restrictive and impersonal in their prevention campaigns in 2022. The growing social disgruntlement eventually culminated in a series of small-scale social protests scattered in several major Chinese cities, including Beijing, Shanghai and Guangzhou, in late November. The short-lived social tumult was sparked directly by the deadly fire on November 24, 2022 that claimed ten lives in a residential area under lockdown in Urumqi, Xinjiang, where most of the residents were prohibited to leave home for more than three months. Strict lockdown was widely attributed to be the cause of unnecessary fatalities in the accident. The social tumult online and offline reflects Chinese citizens' general exasperation with the tardy zero-COVID policy and harsh prevention measures in China. It is generally believed that the abrupt turn of China away from the zero-tolerance policy and the extensive relaxation on COVID-prevention measures in early December 2022 relates closely to the series of November social protests, though Chinese government never officially acknowledges the connection. The voices of the people once again seem to be making progress in changing governmental decisions, like in the case of Dr. Li Wenliang. The success of Chinese citizens in negotiating their participation and influence with the state must still be carefully interpreted. The responsiveness of the Chinese state to Dr. Li Wenliang's incident and the drastic relaxation of pandemic-related restrictions should not be hastily taken as the liberalization of the Chinese state. The rapid and positive responses of the Chinese state in the two instances only suggest its sensitivity to people's overflowing emotions in unsettling time and its awareness of the adverse consequences if such collective sentiment is not channeled properly. Chinese government understands well that heavy-handed actions may work well in a normal situation in China, but not necessary so in the crisis phase of the social system or when people's overall grievances have reached certain threshold. So, in the case Dr. Li Wenliang, conceding to people's legitimate demands is indispensable for maintaining the high legitimacy of government and hence its ability for social mobilization in crisis. For the social protests triggered by the Urumqi's fire, the outburst of citizens' anger online and offline sends a clear message to the Chinese state, compelling it to redress its policy.

Chinese citizens are not compromising in making their voices heard by the nation, especially on social media. They are still together exploring the potential of the online space for their participation and influence. Due to the limitation of scope, the authors cannot discuss the topic further here. The ongoing negotiation between Chinese citizens and the Chinese state through social media is an exciting research topic for the future.

Acknowledgements This article is a pleasant and productive collaboration between teacher and students. Mr. Guo Ruijie and Ms. Huang Xinyu are outstanding undergraduate students at BNU-HKBU United International College. They have contributed to writing a decent part of the article. The authors would like to thank Ms. Shen Wumi for her help during the preparation stage of the article.

References

Anougme_Xing. (2020, February 7). *Zhe zhendeshi minfen! Laizi shisiyi renmin de fennu! Nimen zuowei renmin fumuguan* [It's really the outrage of Chinese, the anger from 1.4 billion people! Being the magistrate of this place...]. Weibo. https://m.weibo.cn/6780943890/4469143778558744

Appadurai, A. (2020). The COVID exception. *Social Anthropology, 28*(2), 221–222.

Beijing News. (2020, January 1). Sanbu wuhan feiyan bushi xiaoxi 8 ren bei jingfang yifa chachu [Eight people were investigated by the police for spreading false information about Wuhan pneumonia]. *Beijing News*. Retrieved April 28, 2022, from https://www.bjnews.com.cn/detail/157787169515286.html

Boin, A. (2005). Disaster research and future crises: Broadening the research agenda. *International Journal of Mass Emergencies and Disasters, 23*(3), 199–214.

Buchiqingcaide gemini. (2020, February 7). *Ruguo shuo zhenxiang yeyao zaoshou zhiyi, shui hai gan wei zhenxiang fasheng?* [If one is punished for telling the truth, then who will speak for the truth?]. Weibo. https://m.weibo.cn/status/4469129056594155?sudaref=m.weibo.cn

Caiyun, C., Ping, Z., Yongxiang, Z., & Bo, L. (2021). Effect of the "Normalized epidemic prevention and control requirements" on hospital-acquired and community-acquired infections in China. *BMC Infectious Diseases*. Retrieved April 28, 2022, from https://pubmed.ncbi.nlm.nih.gov/34814857/

Chen, Y. D., & Jin Y. M. (2020, February 19). Wuhan lifa zhiyuanzhe: "Liyi zhibang" de "tou" deng dashi [Wuhan barber volunteers: Helping the neighbors is the most important thing among other events]. *People's Daily*. Retrieved April 29, 2022, from http://society.people.com.cn/n1/2020/0219/c1008-31595221.html

Chen, Y. D. (2020, February 15). Yuane yiliao "bianwai zhanshi" men: Gudan dan bu gudan [Medical aid to Hubei "extra warriors": Brave but not lonely]. *People's Daily*. Retrieved April 28, 2022, from http://society.people.com.cn/n1/2020/0215/c1008-31588745.html

CN-HEAlTHCARE. (2022, November 30). *Quanguo gaofengxian diqu mingdan* [The list of the high-risk areas at the national level as of November 29]. Retrieved January 19, 2023, from https://www.cn-healthcare.com/articlewm/20221130/content-1475488.html

Cottle, S. (2012). Mediatized disasters in the global age: On the ritualization of catastrophe. In J. C. Alexander, R. Jacobs, & P. Smith (Eds.), *Oxford handbook of cultural sociology* (pp. 259–283). Oxford University Press.

de Saussure, F., Bally, C., Sechehaye, A., & Baskin, W. (2011). *Course in general linguistics*. Columbia University Press.

Department of Science and Technology of Guangdong Province. (2020). *Zhong Nanshan: Keyan jiyao dingtian, yeyao lidi* [Zhong Nanshan: Scientific research should "hold the sky and stand on the ground"]. Department of Science and Technology of Guangdong Province. Retrieved April 28, 2022, from http://gdstc.gd.gov.cn/hykxjjs/content/post_2913748.html

Dongdong, Y. (2020, February 12). *Danan laishi! Kenwei renmin tingshen erchu de ren, Zhong Nanshan huoshi li wenliang doushi yingxiong* [Great danger is here! Zhong Nanshan or Dr. Li and millions of others who dare to stand up for the people deserve the same pride and praise. They are the best warriors of China]. Youtube. https://www.youtube.com/watch?v=I0G_uCCdLhA

Durkheim, E. (2005). *Suicide: A study in sociology*. Routledge.

Edelman. (2022). *2021 Edelman Trust Barometer*. Edelman. Retrieved April 28, 2022, from https://www.edelman.com/trust/2021-trust-barometer

Epoch Times. (2022, January 13). Zhonggong shuaiguo? Zhong Nanshan qixia gongsi she chuanbo bingdu beicha [Chinese Communist Party shifting the blame? Zhong Nanshan's company was investigated for "spreading the virus"]. *Epoch Times*. Retrieved April 29, 2022, from https://www.epochtimes.com/b5/22/1/12/n13500438.htm/amp

Esser, F., & Strömbäck, J. (Eds.). (2014). *Mediatization of politics: Understanding the transformation of Western democracies*. Palgrave Macmillan.

Foucault, M. (2002). *The archaeology of knowledge*. Routledge.

Gao, Y. (2020). Yiqing houbanchang, liangdi chutai "chuishao ren" zhidu [In the latter stage of the epidemic, two cities introduced a "whistleblower" system]. *Legal Weekly*. Retrieved April 29, 2022, from http://m.legalweekly.cn/fzzg/2020-07/30/content_8262896.html

Gao, J., Zheng, P., Jia, Y., Chen, H., Mao, Y., Chen, S., Wang, Y., Fu, H., & Dai, J. (2020). Mental health problems and social media exposure during COVID-19 outbreak. *PLoS One, 15*(4). https://doi.org/10.1371/journal.pone.0231924

Green, A. (2020). Li Wenliang. *The Lancet, 395*(10225), 682. https://doi.org/10.1016/s0140-6736(20)30382-2

Hartley, J. (2002). *Communication, cultural and media studies: The key concepts*. Routledge.

Hjarvard, S. (2013). *The mediatization of culture and society*. Routledge.

Jiang, X. D., & He, L. P. (2020, September 9). Gongheguo xunzhang "huodezhe Zhong Nanshan--dayi jingcheng xie daai" [Zhong Nanshan, winner of the Medal of the Republic, a great and conscientious doctor who loves his patients]. *People's Daily*. Retrieved April 29, 2022, from http://cpc.people.com.cn/n1/2020/0909/c64387-31854373.html

Joe, S. (2020, February 12). *Zhe jiu Jiaozuo zhongguo jiliang, you jienng, youzeren, youdandang* [Zhong is the backbone of China since he has professional capability, courage to take responsibility, and bravery to praise and criticize]. Youtube. https://www.youtube.com/watch?v=I0G_uCCdLhA

Lahusen, C. (2020). *Citizens' solidarity in Europe's civic engagement and public discourse in times of crises*. Edward Elgar Publishing.

Lee, Y. T., Norasakkunkit, V., Liu, L., Zhang, J. X., & Zhou, M. J. (2008). Daoist/Taoist altruism and wateristic personality: East and West. *World Cultures eJournal, 16*(2), 1–20.

Liao, J., & Wang, Z. K. (2020, February 23). Wuhan: Quancheng fachu zhiyuanzhe zhaomuling baoming renshu guowan [Wuhan: More than 10,000 people sign up for the city's call for volunteers]. *Xinhua Net*. Retrieved April 29, 2022, from http://www.xinhuanet.com/politics/2020-02/23/c_1125615897.htm

Lin, G. L. (2020, February 14). Zhongguo zhongyao caijing xinwen yizhou huigu [A weekly glance of China's top financial news]. *Reuters*. Retrieved April 29, 2022, from https://www.reuters.com/article/weekly-glance-china-financial-news-0214-idCNKBS2080HY

Liu, M. Y. (2020a). Duihua "chuanyao" bei xunjie yisheng: wo shi zai tixing dajia zhuyi fangfan [Conversing with the doctor admonished for spreading "rumor": I am just reminding everyone to take precautions]. *Xinjing News*. Retrieved April 29, 2022, from https://m.bjnews.com.cn/detail/158044099415716.html

Liu, M. Y. (2020b, January 31). Duihua chuanyao bei xunjie yisheng: wo shizai tixing dajia zhuyi fangfan [Interview doctors who have been reprimanded for spreading rumors: I am trying to warn the public]. *Beijing News*. Retrieved April 28, 2022, from https://m.bjnews.com.cn/detail/158044099415716.html

Lukin, A. (2013). The meanings of "war." *Journal of Language and Politics, 12*(3), 424–444.

Ma, J., & Mai, J. (2020, February 14). Coronavirus doctor's death becomes a catalyst for "freedom of speech" demands in China. *South China Morning Post*. Retrieved April 29, 2022, from https://www.scmp.com/news/china/politics/article/3049606/coronavirus-doctors-death-becomes-catalyst-freedom-speech

Malm, A. (2020). *Corona, climate, chronic emergency: War communism in the twenty-first century*. Verso Books.

Mills, S. (2008). *Discourse*. Routledge.

National Oversight Commission. (2020). Circular on the investigation of Dr. Li Wenliang's situation reported by the masses. *Xinhua Net*. Retrieved April 29, 2022, from http://www.xinhuanet.com/2020-03/19/c_1125737457.htm

Putnam, R. (2000). *Bowling alone: The collapse and revival of American community*. Simon & Schuster.

Qian, T., Li, P. X., Hou, W. K., & Li, C. Z. (2020, February 1). Pin jiandong shengming yu yimo jingsu--ji yiqing "fengbaoyan" zhong wuhan jinyintan yiyuan yuanzhang zhang dingyu [Racing for life in horrid epidemic: Zhang Dingyu, president of Wuhan Jinyintan Hospital in the "eye of the storm"]. *Xinhua Net*. Retrieved April 28, 2022, from http://www.xinhuanet.com/politics/2020-02/01/c_1125517358.htm

Qingbo. (2020). *Data on Dr. Li Wenliang's death*. Qingbo Big Data. Retrieved April 30, 2022, from http://yuqing.gsdata.cn/vindex/eventDetail?id=2214

Ritchie, H., Mathieu, E., Rodés-Guirao, L., Appel, C., Giattino, C., Ortiz-Ospina, E., Hasell, J., Macdonald, B., Beltekian, D., & Roser, M. (2022, April 29). *China: Coronavirus pandemic country profile. Our world in data*. Retrieved April 29, 2022, from https://ourworldindata.org/coronavirus/country/china

Radio France Internationale (RFI). (2020, May 12). Zhong Nanshan zao Cui Yongyuan zhikong weifa guanggao gangmei jie duoqi sihuo zhiyi [Cui Yongyuan accused Zhong Nanshan of illegal advertising for Yili company. Hong Kong media point to a few illicit practices involved].

RFI. Retrieved April 29, 2022, from https://www.rfi.fr/cn/%E4%B8%AD%E5%9B%BD/20200512-%E9%92%9F%E5%8D%97%E5%B1%B1%E9%81%AD%E5%B4%94%E6%B0%B8%E5%85%83%E6%8C%87%E6%8E%A7%E8%BF%9D%E6%B3%95%E5%B9%BF%E5%91%8A-%E6%B8%AF%E5%AA%92%E6%8F%AD%E5%A4%9A%E8%B5%B7%E7%A7%81%E8%B4%A7%E8%B4%A8%E7%96%91

Shi, Y. C. (2020, January 20). Guojia weijianwei gaojibie zhuanjiazu jiu xinxiang guanzhuang bingdu feiyan da jizhe wen [National Health Commission Covid Response Expert Team press conference on COVID-19]. *CCTV News*. Retrieved April 29, 2022, from http://m.news.cctv.com/2020/01/20/ARTIF4Fl7LEu8TRqIsnde93B200120.shtml

Shih, G. (2020, January 29). In coronavirus outbreak, China's leaders scramble to avert a Chernobyl Moment. *The Washington Post*. Retrieved April 29, 2022, from www.washingtonpost.com/world/asia_pacific/in-coronavirus-outbreak-chinas-leaders-scramble-to-avert-a-chernobyl-moment/2020/01/29/bc4eb52a-4250-11ea-99c7-1dfd4241a2fe_story.html

Tai, P. F. (2006). Social polarisation: Comparing Singapore Hong Kong and Taipei. *Urban Studies, 43*(10), 1737–1756. https://doi.org/10.1080/00420980600838176

The State Council. (2022, December 8). *COVID-19 response further optimized with 10 new measures*. Retrieved January 19, 2023, from https://www.cn-healthcare.com/articlewm/20221130/content-1475488.html

The State Council Information Office of the People's Republic of China. (2020). *Fighting covid-19 China in action*. The State Council Information Office of the People's Republic of China. Retrieved April 29, 2022, from https://english.www.gov.cn/news/topnews/202006/07/content_WS5edc559ac6d066592a449030.html

Topaloff, L. K. (2020, March 18). Is COVID-19 China's "Chernobyl Moment"? *The Diplomat*. Retrieved April 28, 2022, from https://thediplomat.com/2020/03/is-COVID-19-chinas-chernobyl-moment/

Trenz, H. J., Heft, A., Vaughan, M., & Pfetsch, B. (2020). *Resilience of public spheres in a global health crisis*. Weizenbaum Institute for the Networked Society – The German Internet Institute.

Van der Meer, T. W. (2017). Political trust and the "crisis of democracy." *Oxford Research Encyclopedia of Politics*. Retrieved 7 May. 2022, from https://oxfordre.com/politics/view/https://doi.org/10.1093/acrefore/9780190228637.001.0001/acrefore-9780190228637-e-77.

Voice of America. (2020, February 11). *Zhong Nanshan: Li Wenliang shi yingxiong, wo wei ta gandao jiaoao; xinguan yiqing huo zai eryue zhong huo eryue di dadao gaofeng* [Zhong Nanshan: Li Wenliang is a hero, I am proud of him; the new coronavirus epidemic may reach its peak in mid or late February]. YouTube. Retrieved April 28, 2022, from https://youtu.be/I0G_uCCdLhA

Wang, S. M. (2020, April 10). Zhongguo zhanyi qiangda heli cong he erlai [Where does China's strong collective energy in fighting the epidemic come from?]. *People's Net*. Retrieved April 28, 2022, from http://theory.people.com.cn/n1/2020/0410/c40531-31669323.html

World Health Organization (WHO). (2020). *WHO-China joint mission on COVID-19 final report*. World Health Organization. Retrieved April 28, 2022, from https://www.who.int/docs/default-source/coronaviruse/who-china-joint-mission-on-COVID-19-final-report.pdf

Xiao, S. (2020, September 26). "Li Wenliang Clause" Beijing first, Shenzhen follows. *Radio France Internationale*. Retrieved April 29, 2022, from https://www.rfi.fr/cn/%E4%B8%AD%E5%9B%BD/20200926-%E6%9D%8E%E6%96%87%E4%BA%AE%E6%9D%A1%E6%AC%BE-%E5%8C%97%E4%BA%AC%E5%85%88%E8%A1%8C%E6%B7%B1%E5%9C%B3%E9%9A%8F%E5%90%8E

Xiao, S. S., & Xu, H.Y. (2020, November 10). Kangyi xianjin shiji | Zhong Nanshan: jiankang suoxi shengming xiangtuo [Enlightening story about anti-pandemic campaign: Zhong Nanshan: It is all about the health and survival of patients]. *Xinhua News*. Retrieved April 29, 2022, from http://www.xinhuanet.com/politics/2020-11/10/c_1126722849.htm

Yang, Q. (2020, March 19). Wuhan Police: Revoking the letter of admonition to Dr. Li Wenliang and solemnly apologizing to his family. *Yunnan News*. Retrieved April 29, 2022, from http://news.yunnan.cn/system/2020/03/19/030621885.shtml

Ye, Y. (2020). *Nihao, Zhong Nanhsan [Hello, Zhong Nanshan]*. Guangdong Education Press.

Yuyuan, T. T. (2020, February 14). Chiyuan! kan quanguo yihu renyuan ruhe juezhan hubei [Head for Wuhan! Let's see how the medical staff fight the epidemic in Hubei together]. *Xinhua News*. Retrieved April 28, 2022, from http://www.xhby.net/zt/zzccfkyq/yw/202002/t20200214_6511942.shtml

You, Z., & Zhang, Q. (2022). Graves in one's heart: Grassroots memorialization of Dr. Li Wenliang during the COVID-19 outbreak in China. *Journal of American Folklore, 135*(535), 3–25.

Yuan, L., & Taylor, R. (2020). How thousands in China gently mourn a coronavirus whistle-blower. *The New York Times*. Retrieved April 28, 2022, from https://www.nytimes.com/interactive/2020/04/13/technology/coronavirus-doctor-whistleblower-weibo.html

Yuan, Y. F. (2020, January 21). Weihe ganjue yiqing turan baofa? Guojia weijianwei gao jibie zhuanjiazu jieda [Why did the pandemic break out suddenly? National Health Commission High-level Expert Team here to answer the question]. *Guangming Daily*. Retrieved April 28, 2022, from https://mp.weixin.qq.com/s/sLbEdzWw68mKoa7OVnJ6rQ

Zhang, C. (2020a). Right-wing populism with Chinese characteristics? Identity, otherness and global imaginaries in debating world politics online. *European Journal of International Relations, 26*(1), 88–115.

Zhang, C. (2020b). Covid-19 in China: From "Chernobyl moment" to impetus for nationalism. *Made in China Journal, 5*(2), 162–165.

Zhang, W. M. (2015). *Media in China: Constructing "war narrative" in natural disaster coverage*. Unpublished doctoral dissertation, the University of Adelaide.

Zhong, R., & Mozur, P. (2020, February 15). To tame coronavirus, MAO-style social control blankets China. *The New York Times*. Retrieved April 28, 2022, from https://www.nytimes.com/2020/02/15/business/china-coronavirus-lockdown.html

Zhu, Y. (ed.). (2020). *Beijing Municipal Public Health Emergency Response Regulations*. Beijing Municipal People's Government Website. Retrieved April 29, 2022, from http://www.gov.cn/xinwen/2020-09/27/content_5547685.htm

Zhu, L. C., & Yang, Y. S. (2020, February 7). 90 hou zhongzheng yixue ke hushi Zhang Mingxuan: di yici gei ziji chuan zhiniaoku [Zhang Mingxuan, a post-90s intensive care nurse, said it was the first time she put on a diaper]. *People's Daily*. Retrieved April 28, 2022, from http://he.people.com.cn/n2/2020/0207/c192235-33774274.html

Zhu, X. D. (2020, February 15). Fangcang yiyuan nei bingyou bian "zhanyou," likai dimi yu jusang, "xiao huanxi" congwei xiaoshi [Patients in the makeshift cabin become comrades. Farewell to depression and frustration, and small joys never disappear]. *Changjiang Net*. Retrieved April 28, 2022, from http://news.cjn.cn/gnxw/202002/t3566891.htm

Zixue, T. (2006). *The Internet in China: Cyberspace and civil society*. Routledge.

Siu Han Chan is an Associate Professor in Sociology at BNU-HKBU United International College, China. She was trained in Hong Kong, UK, and Austria. She specializes in cultural sociology, historical sociology, sociology of knowledge, and social and political thought. Her research interests include (post)colonial Hong Kong culture and society, Chinese tradition and modernity, and the indigenization of Chinese sociology.

Ruijie Guo has recently graduated from the Programme of Literary Studies of English at BNU-HBKU United International College. Mr. Guo's research interests include public policy, sustainability science, and sociology.

Xinyu Huang is a final year undergraduate student majoring in Media and Communication Studies at BNU-HBKU United International College. Ms. Huang has keen interests and experiences in the media industries. She worked as intern at Southern Metropolis Daily and People's Daily. Her research interests include the political economy of media, new media studies, and sociology.

Netizen Sentiments and Demands in the Context of Normalized Epidemic Prevention and Control: A Comparative Study of Three Provincial Capitals in China

Qingyuan Zhao, Liang Nie, and Xiaying Xu

1 Introduction

Netizens are active on the internet and are present in online communities (Willya Achmad, 2021). While China successfully contained the nationwide outbreak of COVID-19 in early 2020, the recurrence of small-scale outbreaks across the country ("small-scale" here refers to outbreaks of less than 100 cases within one city) continues to preoccupy the authorities. The role of provincial and municipal governments is critical in preventing and managing further outbreaks, yet the unevenness of economic development across China is reflected in the varying capacities of governments to provide social welfare, public services, and medical resources. Accordingly, this study aimed to uncover whether uneven economic development was linked to differences in the emotions and demands made by urban "netizens" during the epidemic.

To achieve this aim, three Chinese provincial capitals (Guangzhou, Xi'an, and Harbin, whose levels of economic prosperity differ) were chosen and the demands and emotions that residents of the cities expressed via social media were analyzed. This revealed whether the sentiments of Internet users ("netizens") were related to economic development. The results may have implications for adjusting public policy to improve reactions to the pandemic.

Q. Zhao (✉) · L. Nie · X. Xu
United International College, Beijing Normal University-Hong Kong Baptist University, Zhuhai, China
e-mail: qingyuan1998@foxmail.com

© The Author(s), under exclusive license to Springer Nature Singapore Pte Ltd. 2023

S. X. B. Zhao et al. (eds.), *Comparative Studies on Pandemic Control Policies and the Resilience of Society*, https://doi.org/10.1007/978-981-19-9993-2_15

2 Review of the Literature

2.1 Level of Economic Development, Population Mindset, and the Pandemic

China's economy is marked by regional disparities: despite central government attempts to close development gaps between the country's eastern, central, western, and northeastern regions, these have continued to widen (Ren & Yang, 2010). China's east is the most developed region (Sun, 2021) in both absolute and per capita terms, based on various socioeconomic indicators (Yuan, 2020). However, it also contains the most uneven levels of intra-regional development, although these disparities are gradually being smoothed out by targeted government policy.

Writing in 2012, Yang et al. found that the rapid socioeconomic transformation of China since 1978 had presented various challenges, including imbalances in urban and rural development and a widening gap between the wealthy and the poor (Yang et al., 2012). Such developments would be anticipated to increase people's levels of stress, and indeed, five years later, Yang et al. (2017) found that residents of towns with lower GDPs reported more uncertainty, which would likely aggravate their negative emotions (Folkman, 2008). But how should negative emotions be understood? Drawing on earlier classifications (Ekman & Friesen, 1971; Izard, 2007; Shaver et al., 1987), Bollen et al. (2011) identified six emotions expressed by netizens on social media: nervousness, depression, anger, euphoria, confusion and fatigue, while anger, disgust, fear, sadness, surprise, optimism and happiness were pinpointed by Xu et al. (2008) for the Chinese context, later extended by Liu et al. (2015) into approval, fear, doubt, worry, opposition, anger, sadness, surprise, and emotionlessness. Finally, Zheng (2014) put forward the concept of positive, neutral, and negative emotional polarity in language, determined by comparing the number of positive and negative words in a text.

Recent research has investigated emotional reactions to the COVID-19 pandemic. Abd-Alrazaq et al. (2020) analyzed 2.8 million tweets whose emotional content included fear, anger, sadness, and joy. Lamsal (2021) also presented a Twitter data set comprising over 310 million specific COVID-19 English tweets and their emotional scores, allowing the spatial and temporal dimensions of the public pandemic discourse to be gauged more clearly. Liu et al. (2021) classified the emotions of netizens in China into optimism, happiness, anger, sadness, fear, disgust, and shock.

While the impact of economic disparities on the mindsets of people in China and the emotional expressions of Internet users in the context of the pandemic have been investigated separately, few studies have analyzed how differences in economic development were linked to the emotional expressions of Internet users in China during the COVID-19 period. The present chapter aims to fill this gap.

2.2 Economic Levels and People's Demands During the Pandemic

Gaps in regional economic development affect the quality of public services and the capacity of local administrations to govern, in turn shaping the demands of netizens.

Duan and Pang (2011) argued that differential economic development between eastern China and the central and western regions has contributed to the variable provision of public services, including education, medical and health care, employment, and social security. Gaps in regional welfare are reflected by differences in amounts of material wealth, the soundness of social security, and the quality of the living environment, which all point to obvious regional distinctions in China (Lu et al., 2012). Chu et al. (2021) showed that the financial capacities of urban governments significantly influenced efforts to prevent and control COVID-19 and were particularly important at later stages of the pandemic. For every unit increase in local government capacity, the researchers detected a 2.4% rise in the ratio of recovered COVID-19 cases per capita (Chu et al., 2021). Poorer economic conditions are likely to be reflected in increased online demands being made of politicians to perform their functions more effectively and for improved public services (Stoycheff et al., 2020).

The pandemic has put local government and public service provision to the test. The impact of the pandemic on netizens has occupied many researchers, such as Su et al. (2020), who analyzed content sourced from Weibo and Twitter before and after the lockdown to examine its impacts on the psychological state of Chinese and Italian people. The results showed that in Lombardy (Italy), people paid more attention to leisure during the lockdown, while in Wuhan (China), people focused on their social group, their religion, and their emotions. Lwin et al. (2020) highlighted a general fear of medical supply shortages, with the focus of anger shifting from foreign involvement to the authorities' approach to notifying people about lockdowns. When the pandemic began in China, Wang and Ying (2020) identified citizens' dissatisfaction with information resources (the speed of information disclosure), medical resources (care for health care personnel), and protective/living resources such as masks, medical alcohol, and vegetables.

Yin et al.'s M-SDI model (2020) is based on the amount of public discussion about a particular topic and can be used to predict trends in public opinion before the next major news event (as shown in Fig. 1). Han et al. (2020) used the KANO model (as shown in Fig. 2) in the context of COVID-19 to classify public demands for disclosure of public health emergencies into basic, expected, charismatic, and undifferentiated demands. Abd-Alrazaq et al.'s "infoveillance" study (2020) classified pandemic tweets into 12 topics and 4 categories: the origin (discuss the origin of COVID-19) of the virus, its sources (the causes leading to the transfer of COVID-19 to humans), its impact on people, countries, and the economy, and ways of mitigating the risk of infection. Yang et al. (2021) classified 20 topics by their content features and then divided the aspirations of the Chinese residents into 5 categories: medical and pandemic prevention, community management, education, traffic logistics, and economic measures. Their study showed how demands varied over time.

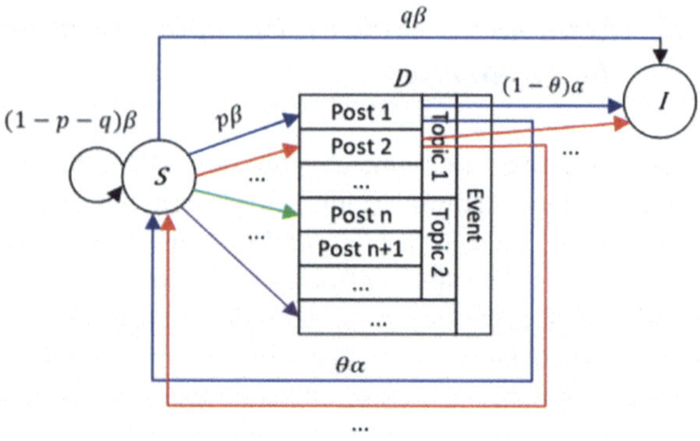

Fig. 1 Yin et al.'s M-SDI model (2020)

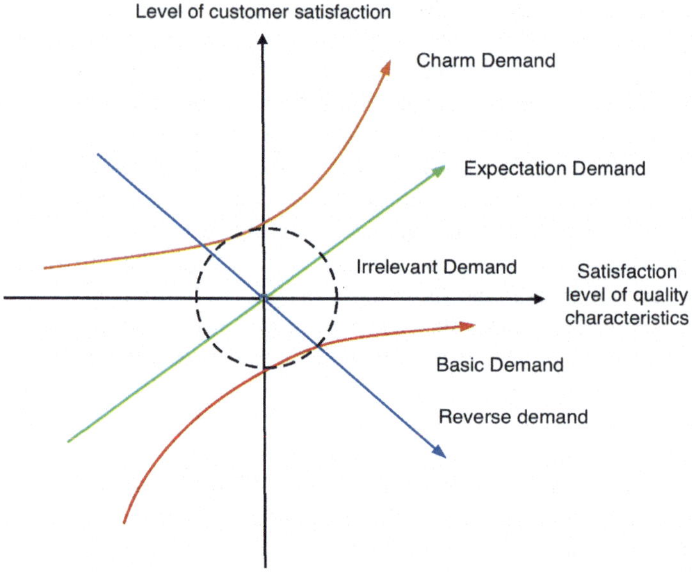

Fig. 2 The KANO model of Han et al. (2020)

Although the above-mentioned studies have explored the needs of residents during the pandemic, most of them have focused on the initial period of the pandemic. Besides, they have focused on the needs of a single country or have compared two nations, rather than examining the impacts of regional economic differences in the same country. To address this gap, the present chapter explores netizen demands in the context of later pandemic prevention and control measures in several mainland Chinese cities.

3 Data Sampling and Processing

3.1 Sampling

In this study, the cities of Guangzhou, Xi'an, and Harbin were chosen to explore the demands and emotions of netizens as expressed by their Weibo posts during the first three days of local COVID-19 outbreaks. The sampling process combined economic development factors, the size of each outbreak and the duration of the presence of Internet users' posts, which were used to select the final three cities for this study.

Economic Factors
All three cities are provincial capitals. Guangzhou is the capital of Guangdong Province (the leading province by GDP in 2021), Xi'an, the capital of Shaanxi Province (15th highest GDP in 2021), and Harbin, the capital of Heilongjiang Province (24th highest in 2021). Across China, Guangzhou ranked #4 in 2021 for GDP, Xi'an ranked #22, and Harbin ranked #44.

Factors in the Development of the Pandemic
All three cities were hit by small-scale COVID outbreaks in December 2021. On the selected dates, confirmed COVID-19 diagnoses were no more than 15 in any of the cities, all of which applied prevention and control measures to different degrees. During this time, their netizens actively commented on social media and their emotions could be extrapolated from these posts.

The Emotional Expression of Netizens: Timeline
In response to COVID-19 outbreaks, netizens tend to post their worries, expectations, and concerns about its development. The official media releases positive information promptly, which comforts the public and maintains its mood (Zhang & Gan, 2020). However, at such early stages of outbreaks, netizens are less influenced by epidemic prevention measures and public opinion, and their attitude towards the epidemic and its prevention and control is therefore reflected most faithfully at this point. For this reason, data collected early in the outbreak were chosen for this study.

3.2 Data Sampling Methods

Data were sampled from the Chinese social media site *Sina Weibo*, with data crawling software applied to crawl according to username, release time, and text content.

After a small-scale outbreak occurred in Guangzhou on December 3, 2021, data for the 3rd, 4th, and 5th of December were screened using the keyword "epidemic in Guangzhou" on Weibo. After removal of non-netizen content such as media reports, official publicity, marketing advertising, irrelevant content, etc.), a total of 5,624 items were collected for sentiment analysis.

In Xi'an, sporadic cases of COVID-19 began to appear on December 9, 2021 and were attracting attention and discussion from netizens three days later. Data were screened using the key phrase *Xi'an epidemic situation* for the December 12–15 period, from which 4718 pieces of data were extracted. In Harbin, cases began to emerge on December 1, so data between then and December 3 were screened using a similar phrase, *Harbin epidemic situation*, obtaining 4662 pieces of data after cleaning.

In total, 15,004 netizens' demands were obtained from the three cities. To optimize coding efficiency and compensate for limitations on manpower, half of the data sourced from the three databases were randomly sampled. These 7502 pieces of data comprised 2812 from Guangzhou, 2359 from Xi'an, and 2331 from Harbin.

3.3 Building the Coding Table

The coding was particularly influenced by two statements: first, Yang et al.'s (2021) identification of the indicators of medical and epidemic prevention, community management, school education, transportation and logistics, and economic measures; and second, the observation that the "Chinese people are increasingly concerned with emotions and community, while Italians are more concerned with leisure" (Su et al., 2020, p. 1). In Su's study, the netizens' demands were coded as *health care and epidemic prevention*, *work life*, *transport and travel*, *school education*, *leisure and recreation*, and *interpersonal communication*. Subsequently, at the micro-level, demands for health care and epidemic prevention were subdivided into demands for overall *epidemic prevention*, *government prevention and control*, and *health care resources*, following Wang and Ying (2020).

When coding for emotion, the nine categories proposed by Liu et al. (2015) were selected, to which *optimism* was added as a category of positive emotion that had emerged during the new epidemic. Due to the normalization of COVID-19, *surprise* was removed after a trial coding session, leaving the following categories: *approval*, *fear*, *doubt*, *anxiety*, *opposition*, *anger*, *sadness*, and *no emotion*.

3.4 Reliability Test

Before formal manual coding, SPSS was used to randomly select 15% of the total sample size (1125 posts), for use by two coders specializing in communication studies. After training, a trial coding run was performed for these data, and the reliability of the process was calculated using Krippendorff's alpha. After two failed coding rounds and further discussion, a consensus was reached. In the third coding session, the coding reliability for the *Emotional Classification, Event Classification, and Event Subcategory Classification* sections reached alpha values of 0.8544, 0.9159, and 0.9109, respectively—all within the acceptable range of reliability.

Table 1 Macro sentiment overview of social media content of Internet users in three cities

	Guangzhou (%)	Xi'an (%)	Harbin (%)	Average (%)
Positive	5.8	7.5	4.2	5.8
Negative	92.3	88.4	93.4	91.4
Neutral	1.9	4.1	2.4	2.8

4 Presentation of Results and Discussion of Data

4.1 Macro and Micro Presentations of Internet Users' Emotions in Different Cities

4.1.1 Macro Perspective: General Negative Emotions

This section focuses on the distribution of macro and micro emotions in netizens' posts concerning the COVID-19 outbreak. Table 1 shows that these were overwhelmingly negative, with negative emotions expressed in at least 85% of the posts and a mean of 91.4% across all three cities, with only 5.8% of all posts expressing a positive valence, on average.

Thus, the pandemic appears to have prompted widespread negative sentiments among Internet users, regardless of the city and the level of GDP, which supports the view of Lwin et al. (2020) that "negative sentiment during a pandemic has a global trend of continuous growth". However, we found that the negative online sentiment expressed during the pandemic was, in fact, extraordinarily high (>90% of posts). This may have been due to the persistence of the pandemic and the impact of prevention and control measures on people's lives, such as traffic control, home office, and travel restrictions.

4.1.2 Micro Perspective: Broad Anxiety and Evolving Disgust and Doubt

The differences between the three cities are apparent when considering negative valence in the posts. Table 2 shows the percentages of the nine major emotions expressed by the posts of Internet users in the three cities.

For users based in Guangzhou and Xi'an, *anxiety* was the most common emotion, accounting for 58.7% and 61.3% of posts respectively. This indicates the concern of people about how their normal life rhythms would be impacted, as demonstrated by the following excerpts from anonymous correspondents: "It is hard to go back to Guangzhou, please don't let me tomorrow. I can't get on the high-speed train out of Shenzhen North Station. I want to go to work on Monday." and "I'm crying. I'm not going to be persuaded by Chengdu to stay in Xi'an after the winter break, right?".

Table 2 Breakdown of the sentiment of Internet users' posts in the three cities

		Guangzhou (%)	Xi'an (%)	Harbin (%)
Positive	Approval	2.1	3.6	1.5
	Optimism	3.7	3.9	2.7
Negative	Doubt	1.5	1.5	19.2
	Anxiety	58.7	61.3	26.4
	Oppose	0	0.3	3.9
	Anger	11.1	5.8	8.6
	Distress	3.5	3.4	5.1
	Disgust	17.5	16.1	30.2
Neutral	Emotionlessness	1.9	4.1	2.4

Both posters were anxious about the impact of the outbreak on their everyday life activities such as travel, work, and schooling.

In Harbin, however, the prevailing emotions were different. Rather than anxiety, *disgust* predominated, comprising 30.2% of posts, as compared to a maximum of 20% in the other two cities. This may be because Harbin's netizens had been subjected to multiple pandemic control measures before the December outbreak and had lost patience, leading to widespread boredom: "Let's say there's no end to it, there's no end to it," one poster remarked.

Moreover, *doubt* was articulated by many of Harbin's netizens, appearing in 19.2% of posts while only expressed in 1.5% of posts from the other two cities. Harbin residents also questioned the prevention and control measures 12.8 times more frequently than inhabitants of Guangzhou and Xi'an. Harbin, with the lowest GDP ranking, was more opposed to the government's pandemic measures. One commenter remarked, "During this wave I do not want to publish a testimonial. I am convinced, the form is full, the effect is tragic, why are we not reflecting?" while another person stated, "Harbin's means of prevention and control is one-size-fits-all. Lock down stores, lock down schools, lock down everything. We are economically backward, so it's now more difficult."

4.2 Macro and Micro Representations of the Demands of Internet Users in Different Cities

This section of the study focuses on the three main types of macro demands that Internet users made in the three cities along with the corresponding six micro demand types, and whether these correlated with urban GDP levels. We first consolidated the six demand types (*health care and epidemic prevention*, *work life*, *transport and travel*, *school education*, *leisure and recreation*, and *interpersonal communication*) into three major categories: "anti-pandemic demands", "living demands",

Table 3 Proportions of macro demands made by Internet users

		Guangzhou (%)	Xi'an (%)	Harbin (%)	Average (%)
Anti-pandemic demands	Pandemic prevention and control	55.7	52.7	67.5	58.6
Living demands	Necessity	22	35	27.9	28.3
	Transportation				
	Education				
Psychological demands	Entertainment	22.3	12.4	4.6	13.1
	Interpersonal interaction				

Table 4 Proportions of micro demands made by Internet users

	Guangzhou (%)	Xi'an (%)	Harbin (%)	Average (%)
Pandemic prevention and control	55.7	52.7	67.5	58.6
Necessity	1.8	0.6	3.1	1.8
Transportation	13.6	27.9	14	18.5
Education	6.6	6.5	10.8	7.9
Entertainment	19	8.5	2.7	10
Interpersonal interaction	3.3	3.9	1.9	3

and "psychological demands". Among these, anti-pandemic demands contain one sub-category: "Pandemic prevention and control". Livelihood demands consist of "necessities", "transportation", and "education" while psychological demands correspond to "entertainment" and "interpersonal interaction". The sub-categorical data are listed in Table 3, and the sub-categories are detailed in Table 4.

4.2.1 Anti Pandemic

Macro Perspective: The Widespread Desire to End the Pandemic

Table 3 shows that pandemic prevention and control was the most significant demand, exceeding 50% in all three cities, and appearing in an average of 58.6% of posts (67.5% in Harbin), demonstrating the eagerness of Internet users to eradicate the pandemic.

Table 5 Subcategories of epidemic prevention and control demands

	Guangzhou (%)	Xi'an (%)	Harbin (%)
Overall prevention and control	90.2	84.4	62.5
Government prevention and control capacity	9.5	12.5	37.0
Medical services	0.3	3.1	0.3

Micro Perspective: Government Demands for Prevention and Control Capabilities Change According to Cities' GDP

To capture differences in pandemic prevention and control demands more effectively, pandemic prevention and control demands were subcategorized into *Overall prevention and control*, mainly for posts with no specific claims but which commented on the epidemic itself, *Government prevention and control capacity*, primarily for prevention and control demands, such as those related to the speed of information disclosure and other preventive measures, and *Medical services*, which mainly expressed demands for various medical resources, such as masks, disinfectants, and vaccines (Table 5).

Table 5 shows that the frequency of demands for governmental measures was inversely related to the cities' GDPs. In Guangzhou, the city with the highest GDP, fewer than 9.5% of Internet users' posts involved claims about the government's capacity to prevent and control the epidemic, increasing to 12.5% in Xi'an and 37% in Harbin, the least economically developed city. One netizen wondered, "Can the epidemic prevention and control be more serious?" Another commented:

> If there is any new situation, I hope we can achieve precise prevention and control, don't just lock down a unit and a building regardless of the problem, don't have anything to do with a nuclear acid, but also distinguish the specific situation, otherwise, there is no way to live.

As Chu et al. (2021, p. 1) stated, "urban governance capacity affects COVID-19 prevention and control, while effective and timely assistance in terms of financial and material resources plays a more important role in pandemic control." In addition, Duan and Pang (2011) argued that "differences in the capacity of basic public services due to differences in the level of economic development of cities all influence the government's ability to prevent and control", all of which can make difference to the demands of netizens in different regions.

Notably, only 1.23% of the quotes referred to demands for medical resources, indicating this was no longer a major issue at this point in the pandemic. Nonetheless, a few netizens objected to the lack of access to masks, disinfectants, vaccines, and timely medical services, due to the strict controls on such products overseen by the State Administration of Market Supervision and Administration (People's Daily Overseas Edition, 2021).

Table 6 Posts concerning living requirements (excluding outbreak prevention and control claims)

		Guangzhou (%)	Xi'an (%)	Harbin (%)
Life requirements	Necessity	49.5	73.7	85.7
	Transportation			
	Education			
Spiritual appeal	Entertainment	50.5	26.3	14.3
	Interpersonal interaction			

4.2.2 Living Requirements

Macro Perspective: Living Requirements Versus GDP Development

Table 3 shows that demands for epidemic prevention and control exceeded 50% of posts from all three cities (nearly 70% in Harbin). Since demands for epidemic prevention and control would obscure the proportion of posts focusing on basic living demands, they were removed from the data, as presented in Table 6.

Table 6 shows that the proportion of posts referencing living requirements after posts about pandemic measures had been removed were Guangzhou = 49.5%, Xi'an = 73.7%, and Harbin = 85.7%. The size of these proportions was inversely linked to the cities' GDPs. As the level of urban economic development declined, demands linked to basic living requirements increased. As Duan and Pang (2011) indicates, "the gap in economic development between the eastern and central and western regions of China has led to differences in basic public services such as education, health care, employment, and social security." Thus, the level of GDP growth appeared to be negatively correlated with demands related to basic needs posted by netizens during the outbreaks of COVID in the three cities.

Micro Perspective: Urgent Transport and Schooling Needs

Of the three basic living subcategories, *transportation* appeared most frequently in posts (18.5%). Demands for transport in Xi'an reached 27.9% of posts after the municipal government requested that "people in medium- and high-risk areas should not travel unless necessary" (China Government Online, 2021) after three new COVID cases were identified on December 15 (People.com, 2021). Measures included the closures of roads and residential communities and the suspension or reduction of public transport services. The response of many netizens was encapsulated in the hashtags *#Xi'an epidemic* and *#Don't make me stay here wanting to go home*, with the inconvenience to residents also apparent: "There is another epidemic in Xi'an, I want to know if there is any impact from Xi'an to Beijing ah? If there is any recent travel from Xi'an to Beijing Capital Airport, please give a suggestion."

The second most frequent demand was related to *education*. Official government guidelines state that "once the risk level of the novel coronavirus outbreak changes in the area where the school is located, the local requirements for prevention and control of the pandemic should be strictly implemented" (National Health Commission, 2021). Compliance of schools with these guidelines was expected, with the comments made by affected netizens from each urban area reflecting their hopes "that the university will be unblocked" or "that offline classes will be resumed as soon as possible", etc. One anonymous student remarked: "Guangzhou outbreak again … really begging it can be controlled properly! I have to keep going to the library to revise [as] it's almost exam time [and we] don't have any other cases" while another commenter, presumably an educator, observed: Although the college students in ss are asking if they can have an early holiday, we haven't finished our classes yet, an early holiday would really be the end. I just hope we do not close the school."

Least frequent were the posters' demands for *daily necessities*, with an average share of less than 2% of posts. Thanks to initiatives launched by the central government to protect livelihoods by "promoting production, ensuring the circulation of supplies, and implementing green channel policies" (People's Daily, 2020b), the supply of goods was not significantly affected.

4.2.3 Psychological Appeal

Macro Perspective: GDP-Related Psychological Demands

As Table 3 shows, psychological demands appeared relatively infrequently but were also correlated with the GDP of each city. Residents posting from wealthy Guangzhou made the most psychological demands related to leisure and interpersonal communication, at 22%. Because their basic needs were met, these netizens were more concerned with disruption to their leisure activities and social interactions; the content of posts typically centered on the wish to attend music festivals, to go out and have fun, or complaints about being separated from loved ones, etc., as these excerpts clarify: "The epidemic in Guangzhou is about to start again, right? It better not affect me going out to party with my housemates and watch live (musical performance by live singers)." and "I cannot see my boyfriend this week because of this shitty epidemic in Guangzhou. It is hard to have a normal meal together. I hope for world peace." Residents of Guangzhou, with its high GDP, were relatively unaffected due to the abundance of supplies and trust in the government's ability to prevent and control the epidemic. Instead, they were more concerned with the spiritual aspects of their lives. By contrast, Xi'an and Harbin's netizens rarely mentioned any psychological demands, which occurred in only 4.6% of the latter's posts. The implication here is that the inhabitants of these cities simply had no time to consider such needs due to economic necessity and administrative incompetence, both of which tended to feature more prominently in the posts of local netizens.

Table 7 shows the proportion of psychological demands on social media posts (Guangzhou = 22.3%, Xi'an = 12.4%, and Harbin = 4.6%), which was positively related to urban GDP. This supports Su et al.'s (2020) finding that the pandemic increased people's focus on leisure, emotion, and community increases. We found

Table 7 Psychological demands of Internet users in the three cities

		Guangzhou (%)	Xi'an (%)	Harbin (%)
Psychological demands	Entertainment	22.3	12.4	4.6
	Interpersonal interaction			

Table 8 Macro-level Internet users' attitudes towards the need for outbreak prevention and control

	Guangzhou (%)	Xi'an (%)	Harbin (%)	Average (%)
Positive	8.1	12.6	5.2	8.6
Negative	89.1	83.5	92.7	88.4
Neutral	2.8	3.9	2.1	2.9

that differential proportions of psychological demands among Internet users during the epidemic were linked to each city's GDP.

4.3 The Relationship Between Internet Users' Main Demands and Their Emotions

4.3.1 Urban GDP Development Levels and Internet Users' Attitudes Towards Government Preventive and Control Measures

Macro Perspective: The Demand for Epidemic Prevention and Control Surrounded by Negative Emotions

Table 8 illuminates the netizens' overall emotional orientations towards the local preventive and control measures applied in their cities. It shows that negative attitudes were discernable in over 80% of all posts about macro epidemic prevention and control, an average of 88.4%, demonstrating how widespread the negative valence was, as mentioned in Sect. 4.1 of this chapter.

Micro Perspective: Widespread Questioning and Opposition in Harbin

As shown above, social media content on macro-level outbreak prevention and control was classified into three micro categories of demands for (1) overall nonspecific prevention and control, (2) government prevention and control, and (3) medical resources referencing material needs such as masks and disinfectant. Due to their lack of detail, posts in the first category were excluded from analysis in this study, while very few posts belonged to the third category. Therefore, this section focuses on claims that refer to government control, as shown in Table 9.

Table 9 Micro-level netizens' attitudes towards the government's preventive and control measures

	Approval (%)	Optimism (%)	Doubt (%)	Anxiety (%)	Opposition (%)	Anger (%)	Distress (%)	Disgust (%)	Neutrality (%)
Guang-zhou	30.2	7.4	17.4	21.5	0.7	2.7	2.7	13.4	4
Xi'an	53.5	5.2	20.6	11.0	3.9	0.6	0	1.3	3.9
Harbin	4.7	0.4	70.9	1.3	14	2.7	0.7	5.0	0.4

Table 9 shows that in Guangzhou and Xi'an, 30.2% and 53.5% of posts expressed approval of the government's epidemic prevention and control measures, respectively, whereas merely 4.7% did so in Harbin, indicating that netizens in these wealthier cities were more confident of the government's epidemic prevention and control measures. Such confidence was articulated in the following remark:

> It is not easy for all health care workers, they face more risks than others head-on. Strengthen epidemic prevention and control; everyone cooperates with the epidemic prevention and control requirements to reduce the risk; let us all get through this together!

By contrast, only 4.7% of posts from the lower-GDP city of Harbin approved of the government's prevention and control measures, with 70.9% of posts questioning these as compared to 0.6% in Xi'an and 17.4% in Guangzhou, demonstrating the dissatisfaction Harbin residents expressed with their government and its lockdown measures. Posts containing directly oppositional sentiments were 0.7%, 3.9%, and 14% of the totals in Guangzhou, Xi'an, and Harbin, respectively. Regional economic differences, variable administrative capacities, grassroots mobilization, and social welfare protection translates into significant differences in local governments' ability to prevent and control COVID outbreaks. In turn, this prompted complaints and doubts from Internet users in different regions about these areas, including unfavorable comparisons with measures taken in other cities. One poster commented: "I've become insensitive to the repeated outbreaks. I hope the Harbin government will not adopt a broad-brush approach in dealing with the epidemic and listen to the people's voices" while another even more critical voice remarked, "Learn from other cities how to control it! You've got it easy; you can forget your responsibilities, but what about us, the people and the grassroots? Do we have no life and no work?"

The data above are indicative of the regional range of sentiment expressed towards the outbreak measures. Positive attitudes were more common in areas with higher GDP growth, such as Guangzhou and Xi'an, where the government's administrative capabilities are stronger, unlike Harbin, where complaints and dissatisfaction were commonly directed at the local government.

4.3.2 Internet Users' Attitudes Towards Basic Living Requirements

Macro Level: A Generally Negative Sentiment Regarding Basic Life Needs

The data shown in Table 10 were obtained by first removing data that referred to other needs, leaving the content of posts addressing basic life needs. The table shows that the emotional valence of posts on basic life needs during the pandemic was overwhelmingly negative, at over 95% in all three cities and reaching an average of 96.3%.

Table 10 Macro-level Internet users' attitudes towards access to basic needs

	Guangzhou (%)	Xi'an (%)	Harbin (%)	Average (%)
Positive	2.8	1.4	1.6	1.9
Negative	96.4	96.3	96.2	96.3
Neutral	0.8	2.3	2.2	1.7

Micro-Level: Worries Went Mainstream and Proliferated in Harbin

Among the three categories of daily necessities that were analyzed, that of basic supplies or materials (such as vegetables, rice and other household goods) was effectively met in all cities and was only mentioned in 1.8% of the netizens' posts. Therefore, we focus here on emotions related to basic living (such as the common transport needs of life) and educational needs (such as the need for a normal return to school).

Tables 11 and 12 show that the most common sentiment expressed in the three cities was *worry* in the two categories of *traffic* and *transportation* and *school education*. Some indicative comments are as follows: "Anyone taking a flight from Xi'an to Beijing today? What's the policy in Beijing! Anxiously waiting - I have an 8 pm ticket and am panicking!"; "Don't! Don't cancel IELTS!"; and "No way, no way! I won't be able to go home—even in my third year!"

Meanwhile, posts from Harbin were more negative than others about school education, but *anxiety* featured in less than 50% of these posts, compared to over 80% in other cities. Closer inspection demonstrates that the proportion of *boredom* posts linked to school education from Harbin was nearly 20% higher than for those of Guangzhou and Xi'an netizens. One parent expressed this emotion succinctly:

> It's so back and forth, it's only been less than two weeks of classes, and it's back online! I can't wait to know if my son's private school can refund half his tuition fees: brain ache, brain ache.

The frustration of another college student was apparent in these words:

> The school year has started, and now, the current situation of contemporary college students in Harbin: blocked for a month, unblocked for half a month and stopped again, unblocked again for two days, and blocked again. I regret coming here for college, and I'm going crazy.

Boredom, to a large extent, was caused by the enforcement of preventive pandemic measures. If an outbreak occurred, primary and secondary schools switched to online teaching instantly, with students attending class from home. Similarly, universities acted swiftly, shutting their schools to limit students' social contact. The repetition of such measures would likely provoke the expression of boredom among Internet users.

Table 11 Micro-sentiment of netizens' posts on traffic and mobility

	Approval (%)	Optimism (%)	Doubt (%)	Anxiety (%)	Opposition (%)	Anger (%)	Distress (%)	Disgust (%)	Neutrality (%)
Guangzhou	0	1.8	0.5	80.5	0	2.9	1.1	12.4	0.8
Xi'an	0.2	1.2	0.3	85.1	0	2.7	0.9	7.6	2
Harbin	0	1.8	0.3	75.2	0	5.8	4.6	10.7	1.5

Table 12 Micro-sentiment of Internet users' posts on school and education

	Approval (%)	Optimism (%)	Doubt (%)	Anxiety (%)	Opposition (%)	Anger (%)	Distress (%)	Disgust (%)	Neutrality (%)
Guangzhou	0	0	1.1	86.5	0	3.2	3.2	5.4	0.5
Xi'an	0.7	1.3	0.7	86.3	0	2	2	5.9	1.3
Harbin	0	0.4	8.7	47.2	0.8	9.5	6	25	2.4

5 Conclusion

Just as "news is frozen in history when it comes out" (Wang & Guo, 2020) so the social media content of Internet users during the pandemic provides an equally valuable record of what happened in each city at a particular point in time.

COVID-19 affected people both physically and mentally. The present analysis of emotions and demands in the social media content of Internet users in the three cities demonstrates the predominance of negative emotions, despite the continued availability of goods and medical supplies during the outbreak. These negative emotions, expressed in 91.4% of posts, saturated the demands made by the netizens and evidenced the huge psychological pressure that people experienced at this time. While it was crucial to eliminate cases of the pandemic, the effects of measures on the mental health of urban residents—covered in depth by the present study—must be considered carefully by planners; the need for "timely psychological guidance" (Su et al., 2021) should not be underestimated. The data presented in this chapter confirm that the government should ensure the timely provision of mental health support services, i.e. professional and accessible psychological counseling across all regions of China affected by the pandemic, and measures to control it.

In the meantime, the demands of netizens demonstrate the significant impacts that the lockdown measures had on their day-to-day lives. While demands to curb the spread of COVID-19 were undoubtedly important and accounted for a sizable proportion of content, netizens also demanded access to transportation, as witnessed in complaints about "overly strict travel restrictions", "extended closure measures", and "rigid control measures", all of which had a material impact on people's everyday lives, work, and income. Indeed, both the People's Daily and People's Daily Overseas Edition's accounts indicate that "the broad-brush expansion of substantive restrictions has seriously affected Residents' normal work and life and enterprises, (and) the resumption of work and production" (People's Daily, 2020a), and "the broad-brush approach to blocking roads is a manifestation of lazy governance" (Chivalry Island, 2022). The data presented in this chapter show that maintaining adequate general and medical supplies—as well as continuing to run transportation services—is essential for preventing and controlling COVID-19. Therefore, it is necessary for governments at all levels to explore the feasibility of more scientific and targeted measures in practice.

Another crucial contribution of this paper is its identification of the relationships between a city's level of economic development and the demands and sentiments of its Internet users. Previous studies have clarified that disparities in economic growth among different cities in China may impact the ability of administrations to govern effectively (Duan & Pang, 2011), particularly during the pandemic (Chu et al., 2021). In those cities whose GDPs are relatively high, the government's administrative capacity is accordingly higher, with better healthcare resources and social security available. This chapter has shown that, unlike their counterparts in lower-GDP urban areas, netizens in wealthier cities were less disapproving of government prevention and control measures, demanded such measures less frequently, and expressed fewer basic living requirements. However, their demand for psychological support was generally higher. Future studies should aim to investigate whether these findings apply to other cities. Further in-depth analysis may support local governments to develop more scientifically tailored prevention and control measures.

Appendix

Coding table for categories of macro and micro claims

Categories of demands	Categories of codes	Explanations and examples
Anti-pandemic	1 = Pandemic prevention and control	1.1 General control E.g., "Hope the pandemic will end soon" "We will overcome this situation" 1.2 Government capacity in controlling and preventing (information disclosure and other measures) E.g., "The controlling policy does not make sense" "The government's response was quick and timely" 1.3 Medical services (vaccinations and COVID testing) E.g., "I am wondering where I can get a COVID test when arrive in Guangzhou?" "Thanks for the vaccine"
Living demands	2 = Necessity	Including money (unemployment), water, food, and other necessities E.g., "If it goes on like this, I am going to be out of money and have nothing to eat" "I'm gonna lose my job"
	3 = Transportation	Traveling, taking transportations or being out E.g., "I really want to go back home from the university" "My travel plans fell through"
	4 = Education	Test arrangement, daily campus life and future education plan E.g., "Will the teacher licensing exams be postponed due to the pandemic?" "No more Zoom University"

(continued)

(continued)

Coding table for categories of macro and micro claims		
Categories of demands	Categories of codes	Explanations and examples
Spiritual appeal	5 = Entertainment	Activities that give people pleasure, such as watching performances and going to a party E.g., "Can't watch the concert again. Sad" "I was going to go out to dinner"
	6 = Interpersonal interaction	The need to interact with other people E.g. "Miss my boyfriend, long-distance relationships suffer under the pandemic" "Miss my parents and wanna go back home"

Macro and micro coding table of mood categories		
Categories of macro sentiment	Categories of micro-emotions	Explanations and examples
Positive	**1 = Approval** (of decisions made by the authority)	Satisfaction with the pandemic policies or efforts the government has made to control the temporary outbreak E.g. "The government's response was quick and timely" "Thanks for the staff members' work!"
	2 = Optimism (toward the pandemic)	Overt positive emotions toward the pandemic E.g. "Don't worry, the epidemic will finish soon." "We will overcome this situation"
Negative	**3 = Doubt** (on decisions made by the authority)	Question authorities react to control the pandemic E.g. "Why was this case not detected during routine COVID test?" "Why does the government reward the positive? Do they have money and no place to spend it?"

(continued)

(continued)

Macro and micro coding table of mood categories		
Categories of macro sentiment	Categories of micro-emotions	Explanations and examples
	4 = Anxiety (about the life/specific demand under pandemic)	Nervous or worried about something E.g. "I am not sure whether I can come back home successfully. I miss my hometown so much" "Can I still go to the concert tomorrow?"
	5 = Opposition (to decisions made by the authority)	Outright opposition/accusation of the authority during the pandemic E.g. "As for the control of pandemic, the government is responsible for a serious dereliction of duty" "There is no worse epidemic control than Harbin"
	6 = Anger (to the life/specific demand under pandemic)	Feel something happened due to the pandemic in a cruel or unacceptable way. Accompanying with irrational arguments E.g. "Fuck the shit COVID-19"
	7 = Distress (to the life/specific demand under pandemic)	A state of extreme sorrow, suffering or pain under the pandemic E.g. "I feel like the pandemic will never end" "We will never go back to normal life. Crying…"
	8 = Disgust (to the life/specific demand under pandemic)	Feeling of very strong dislike, especially for the unpredictable and constant outbreaks of the pandemic E.g. "The COVID haunts our lives, sick of it." "The pandemic is endless…Don't be so hard on me lol"
Emotionlessness	**9 = Emotionless**	No obvious emotion is found

References

Abd-Alrazaq, A., Alhuwail, D., Househ, M., Hamdi, M., & Shah, Z. (2020). Top concerns of Tweeters during the Covid-19 pandemic: Infoveillance study. *Journal of Medical Internet Research, 22*(4), e19016. https://doi.org/10.2196/19016

Achmad, W. (2021). Citizen and netizen society: The meaning of social change from a technology point of view. *Jurnal Mantik, 5*(3), 1564–1570.

Bollen, J., Mao, H., & Pepe, A. (2011). Modeling public mood and emotion: Twitter sentiment and socio-economic phenomena. *Proceedings of the International AAAI Conference on Web and Social Media, 5*(1), 450–453.

Chivalry Island. (2022, April 20). *(Dao shu shuo)Yi dao qie feng kong du lu, jiu shi lan zheng* [(Uncle Island says) A one-size-fits-all blockade of roads is lazy government]. https://weibo.com/ttarticle/p/show?id=2309404760437479834507&luicode=20000061&lfid=4760437478399153&featurecode=newtitle&sudaref=m.weibo.cn

China Government Network. (2021, December 16). *12 yue 15 ri yi qing fang kong zui xin xiao xi!* [December 15 epidemic prevention and control update!]. http://www.gov.cn/fuwu/2021-12/16/content_5661082.htm

Chu, Z., Cheng, M., & Song, M. (2021). What determines urban resilience against COVID-19: City size or governance capacity? *Sustainable Cities and Society, 75*, 103304. https://doi.org/10.1016/j.scs.2021.103304

Duan, Y., & Pang, J. (2011). Jiben gonggong fuwu gonggei shuiping yu quyu jingji fazhan de shizheng yanjiu [An empirical study on the level of basic public service provision and regional economic development]. *Productivity Research, 3*, 69–71.

Ekman, P., & Friesen, W. V. (1971). Constants across cultures in the face and emotion. *Journal of Personality and Social Psychology, 17*(2), 124–129. https://doi.org/10.1037/h0030377

Folkman, S. (2008). The case for positive emotions in the stress process. *Anxiety, Stress, & Coping, 21*(1), 3–14. https://doi.org/10.1080/10615800701740457

General Office of the National Health Commission General Office of the Ministry of Education. (2021, August 20). Guo jia wei sheng jian kang wei ban gong shi Jiao yu bu ban gong ting guan yu yin fa gao deng xue xiao、zhong xiao xue xiao he tuo you ji gou xin guan fei yan yi qing fang kong ji shu fang an (di si ban) de tong zhi —zhong hua ren min gong he guo jiao yu bu zheng fu men hu wang zhan [Notice of the General Office of the National Health and Wellness Commission, General Office of the Ministry of Education on the Issuance of the Technical Plan for Prevention and Control of the New Coronary Pneumonia Epidemic in Higher Education, Primary and Secondary Schools and Child Care Institutions (Fourth Edition)]. http://www.moe.gov.cn/jyb_xxgk/moe_1777/moe_1779/202108/t20210824_553683.html

Han, W., Chen, S., & Chen, A. (2020). Jiyu KANO moxing de tufa gonggong weisheng shijian xinxi gongkai de gongzhong xuqiu yanjiu [A study on public demand for information disclosure of public health emergencies based on KANO model]. *Information Studies: Theory & Application, 43*(5), 9–16.

Izard, C. E. (2007). Basic emotions, natural kinds, emotion schemas, and a new paradigm. *Perspectives on Psychological Science, 2*(3), 260–280. https://doi.org/10.1111/j.1745-6916.2007.00044.x

Lamsal, R. (2021). Design and analysis of a large-scale COVID-19 tweets dataset. *Applied Intelligence, 51*(5), 2790–2804. https://doi.org/10.1007/s10489-020-02029-z

Liu, C., Xie, C., & Wan, X. (2015). Weibo qingxu yu weibo chuanboli de guanxi yanjiu ——jiyu 24 qi gonggong shijian xiangguan weibo de shizheng fenxi [Research on the relationship between microblog sentiment and microblog communication power: An empirical analysis based on 24 microblogs related to public events]. *Journalism & Communication, 22*(9), 92–106, 128.

Liu, Z., Qin, Q., & Zhao, W. (2021). Weibo huanjing xia xinguan feiyan yiqing shijian dui wangmin qingxu de yingxiang fenxi [Analysis of the impact of the COVID-19 outbreak on netizens' emotions in the microblogging environment]. *Journal of Intelligence, 40*(2), 138–145.

Lu, J., Chen, Y., & Guo, Z. (2012). Shehui fuli, jingji zengzhang, yu quyu fazhan chayi—jiyu Zhongguo shengyu shuju de ouhe shizheng fenxi [Social welfare, economic growth and regional development differences: A coupled empirical analysis based on provincial data in China]. *Chinese Journal of Population Science, 3*(31–43), 111.

Lwin, M. O., Lu, J., Sheldenkar, A., Schulz, P. J., Shin, W., Gupta, R., & Yang, Y. (2020). Global sentiments surrounding the COVID-19 pandemic on Twitter: Analysis of Twitter trends. *JMIR Public Health and Surveillance, 6*(2), e19447. https://doi.org/10.2196/19447

People's Daily. (2020a, February 17). Dui "yi dao qie" de fang yi cuo shi shuo bu,guo ji fang kong xu ji shi jiu zheng [No to "one-size-fits-all" epidemic prevention and control measures, overkill must be corrected in time]. https://baijiahao.baidu.com/s?id=1658768289076130450&wfr=spider&for=pc

People's Daily. (2020b, July 16). Ren min ri bao: bao ji ben min sheng, rang ren min guo shang hao ri zi [People's Daily: Preserving basic livelihoods and enabling people to live a good life]. https://baijiahao.baidu.com/s?id=1672331806971578295&wfr=spider&for=pc

People.com. (2021, December 15). Xi an 3 que zheng bing li gui ji gong bu! Yi zhui zong guan kong mi qie jie chu zhe 237 ren [Trajectory of 3 confirmed cases in Xi'an announced! 237 close contacts have been traced and controlled]. http://sn.people.com.cn/BIG5/n2/2021/1215/c226647-35052136.html

People's Daily Overseas Edition. (2021, December 10). Kou zhao chan pin zhi liang xian zhu ti sheng gong ying chong zu_zhong guo wang [Mask product quality significantly improved with adequate supply]. http://fc.china.com.cn/2021-12/10/content_41819210.htm

Ren, J. J., & Yang, G. L. (2010). ZhongGuo quyu jingji fazhan chayi jiqi chengyin fenxi [Analysis of regional economic development differences and their causes in China]. *Economic Geography, 30*(5), 784–789.

Shaver, P., Schwartz, J., Kirson, D., & O'Connor, C. (1987). Emotion knowledge: Further exploration of a prototype approach. *Journal of Personality and Social Psychology, 52*(6), 1061–1086. https://doi.org/10.1037/0022-3514.52.6.1061

Stoycheff, E., Nisbet, E. C., & Epstein, D. (2020). Differential effects of capital-enhancing and recreational Internet use on citizens' demand for democracy. *Communication Research, 47*(7), 1034–1055. https://doi.org/10.1177/0093650216644645

Sun, R. (2021). Quanmian jiancheng xiaokang shehui yu quyu jingji junheng fazhan—jiyu xin gudian zhuyi quyu kongjian junheng lilun [Building a moderately prosperous society and balanced regional economic development—based on neoclassical regional spatial equilibrium theory]. *Inquiry into Economic Issues, 7*, 10–19.

Su, Y., Xue, J., Liu, X., Wu, P., Chen, J., Chen, C., Liu, T., Gong, W., & Zhu, T. (2020). Examining the impact of COVID-19 lockdown in Wuhan and Lombardy: A psycholinguistic analysis on Weibo and Twitter. *International Journal of Environmental Research and Public Health, 17*(12), 4552. https://doi.org/10.3390/ijerph17124552

Wang, D., & Guo, C. S. (2020). Zheng he kuang jia yu jie shi shui ping:hai nei wai bao zhi dui "yi dai yi lu" bao dao de dui bi fen xi [Framing and construal level: A comparative analysis of press coverage of China's "One Belt One Road" Initiative]. *Journalism and Communication Research, 27*(3), 5–20, 126.

Wang, J. X., & Ying, S. P. (2020). Renzhi, qingxu yu xingdong:Yiqing yingji xiangying xia de shehui xintai [Cognition, emotion, and action: Social mindset in emergency response to an epidemic]. *Exploration and Free Views, 4*(232–243), 291–292.

Xu, LH., Lin, H. F., Pan, Y., Ren, H., & Chen, J. M. (2008). Qinggan cihui benti de gouzao [The construction of emotion vocabulary ontology]. *Journal of the China Society for Scientific and Technical Information, 27*(2), 180–185.

Yang, J., Liu, Y., Qi, T., & He, S. (2021). Zhongda gonggong weisheng shijian zhong minzhong suqiu de zhuti wajue yu yanbian toushi [Thematic mining and evolution of public demands in major public health events]. *Library Tribune, 41*(4), 121–131.

Yang, T., Wu, D., Zhang, W., Cottrell, R. R., & Rockett, I. R. H. (2012). Comparative stress levels among residents in three Chinese Provincial Capitals, 2001 and 2008. *PLoS ONE, 7*(11), e48971. https://doi.org/10.1371/journal.pone.0048971

Yang, T., Yang, X. Y., Yu, L., Cottrell, R. R., & Jiang, S. (2017). Individual and regional association between socioeconomic status and uncertainty stress, and life stress: A representative nationwide study of China. *International Journal for Equity in Health, 16*(1), 118. https://doi.org/10.1186/s12939-017-0618-7

Yin, F., Lv, J., Zhang, X., Xia, X., Wu, J., Yin, F., Lv, J., Zhang, X., Xia, X., & Wu, J. (2020). COVID-19 information propagation dynamics in the Chinese Sina-microblog. *Mathematical Biosciences and Engineering, 17*(3), 2676–2692. https://doi.org/10.3934/mbe.2020146

Yuan, D. (2020). Zhongguo quyu jingji fazhan ji quyu jingji chayi yanjiu shuping [A review of regional economic development and regional economic differences in China]. *Market Modernization, 4*, 137–138.

Zhang, F., & Gan, H. (2020). Yiqing xinli shikong juli dui gongzhong qingxu de yingxiang yanjiu——jiyu xinguan feiyan yiqing qi weibo wenben mianban de jisuan fenxi [Research on the influence of psychological spatio-temporal distance on public sentiment during the epidemic: A computational analysis based on microblog text panel data during the epidemic period of COVID-19]. *Journalism, 6*, 39–49.

Zheng, Y. (2014). Jiyu qinggan cidian de zhongwen weibo qinggan fenxiyanjiu [Research on sentiment analysis of Chinese microblogs based on sentiment dictionary] [Master's thesis, Zhongshan University]. https://kns.cnki.net/kcms/detail/frame/list.aspx?dbcode=CMFD&filename=1014054398.nh&dbname=CMFD201501&RefType=1&vl=WThEOFn69NKOV5EWIyw955vYBxNVmZNmIuFOMS3keH-CraD_7nxK2xQNqUP9vaZM

Qingyuan Zhao with a master's degree in communication, was a research assistant at the Beijing Normal University-Hong Kong Baptist University United International College while writing this chapter. His research interests are audience research, social media research and public opinion research.

Liang Nie is a Media and Communication Study student from Beijing Normal University-Hong Kong Baptist University United International College (BNU-HKBU UIC). Her research interests include digital media and digital humanity.

Xiaying Xu received his PhD in Communication from the University of Macau in 2016. He currently serves as Associate Professor in the Public Relations and Advertising Programme, Department of Communication, Faculty of Humanities and Social Sciences, BNU-HKBU United International College. His research interests include cultural and creative industries, cultural and media studies.

COVID-19 and the Chinese Outbreak Narrative: Representations of Heroes, Victims and Minor Characters in Chinese Epidemic Films

Ziran Lu and Li Li

Long before the arrival of COVID-19, in both China and the West, the devastation wrought by epidemic outbreaks and the public and institutional responses to them have been the subject of both feature films and documentaries. COVID-19 is only the latest in a long line of epidemics that have prompted film-makers to dramatize the stories of those who suffer the consequences of infection, and those who struggle to protect the community. These stories have multiple functions: to engage and entertain the audience, to warn them, and to model public feelings about and responses to the actions of those institutional and individual agents who are caught up in the epidemic narrative. The critical literature on such narratives has focused on the outbreak narrative in the West (e.g. Wald, 2008). Comparatively little attention has been paid to Chinese dramatizations of epidemic narratives, either before or during the COVID-19 outbreak, which, at the time of writing, is still ongoing. The present chapter, then, addresses this absence by presenting a brief description of major Chinese epidemic movies of the past 70 years, followed by an analysis of recent feature films including *Chinese Doctors* (《中国医生》, 2021) that deals specifically with COVID-19.

A recurring theme of Chinese epidemic films is the necessity for heroic collective action against unseen enemies; the overt message, unsurprisingly, is that the audience should comply with government public health advice. The characterisations in the feature films tend to be two-dimensional: there is a focus on victims of low social status, while the healthcare workers, the military, and government and Communist Party officials are shown to be dedicated and altruistic in their work and in the sacrifices they make. The simplicity of the themes, the pseudo-realism of the form, and the stereotypical quality of the characterisation limit the effectiveness of the

Z. Lu · L. Li (✉)
Macao Polytechnic University, Macao SAR, China
e-mail: lili@mpu.edu.mo

© The Author(s), under exclusive license to Springer Nature Singapore Pte Ltd. 2023

S. X. B. Zhao et al. (eds.), *Comparative Studies on Pandemic Control Policies and the Resilience of Society*, https://doi.org/10.1007/978-981-19-9993-2_16

films as dramatic interrogations of the complexities of a crisis, however, they may contribute to their function as explicit public health messages during a national emergency.

1 Chinese Epidemic Films: A Brief Descriptive Overview

A brief survey of major Chinese epidemic narratives on film serves to show the recurrent themes of the emerging genre. The first film in China that can be called an epidemic narrative is a documentary entitled *Prevention of Bubonic Plague* (《预防鼠疫》) released in 1948, one year before the founding of the Republic of China, when the Communist Party of China (hereafter referred to as CPC) already occupied a dominant position in the long battle with its counterpart Kuomintang. This documentary was the first of many films to address the dangers of epidemic outbreaks. According to statistics provided by the *People's Daily*, the official newspaper of the Central Committee of the CPC, 'as of December of 2021, in the past five years, the total number of films made in China has exceeded over 4000.' Chinese epidemic movies only make up a small proportion of those produced. Even so, drawing on sources such as Douban films,[1] CNKI database[2] and Wikipedia, the present authors have identified no fewer than 56 films that can be described as epidemic narratives, released in mainland China[3] between 1948 and 2021. Among these 56 films, 21 are feature movies and 35 are documentaries. Among the 21 feature films, 11 were released theatrically in cinemas and the other 10 were presented online.

Based on the data collected, Chinese epidemic narratives on film can be summarized as follows:

1.1 Cinematic Representations Based on Real Epidemics

A number of films take as their subject matter actual epidemics, historical or contemporary, that have affected China. The epidemics concerned include plague, pestilence, smallpox, rape bacteriophage, bilharzia, AIDS, SARS, and COVID-19. COVID-19 is the subject of most of these films (26), with other epidemics covered being SARS (7 films), schistosomiasis (6 films) and AIDS (5 films). Other types of diseases featured less frequently include smallpox and the bubonic plague. It is interesting to note that some diseases, such as cholera and influenza, have not been the subject of film dramatizations in China; however, at least one movie, the prescient online film, *Virus of Doomsday Crisis* (《病毒之末日危机》), released in 2015, is exceptional in that it

[1] https://m.douban.com/movie/.

[2] https://www.cnki.net/.

[3] The epidemic films discussed in this chapter are limited to those made in mainland China, and exclude ones produced in HK, Macao and Taiwan.

did deal with an unnamed, imaginary virus, which can be taken as representative of a range of past or future disease carriers.

Of the relatively small but still substantial number of films in China that are based on epidemic narratives, then, COVID-19 features prominently; even so, the COVID-19 films are part of a longer tradition of Chinese movies that address issues that arise from actual, or in a few cases, imaginary, outbreaks of disease.

1.2 Temporal Distribution

The impact of COVID-19 on Chinese epidemic narratives can be seen more clearly in Fig. 1, which shows the output of epidemic films over the past 70 years, since *Prevention of Bubonic Plague* was released in 1948. Generally, the graph shows that the production of epidemic narratives in China was relatively steady, with a ten-year hiatus between 1965 and 1976, which almost coincides with the years of the Cultural Revolution (1966–1976). The peak in 2020 coincides with the outbreak of COVID-19, demonstrating that the new coronavirus has been a particular inspiration for Chinese film-makers, of both documentaries and features. However, as we shall see, the uptick in COVID-19 films also shows the broader availability of production resources to a wider range of film-makers and the possibility of release on social media as well as through traditional theatrical venues.

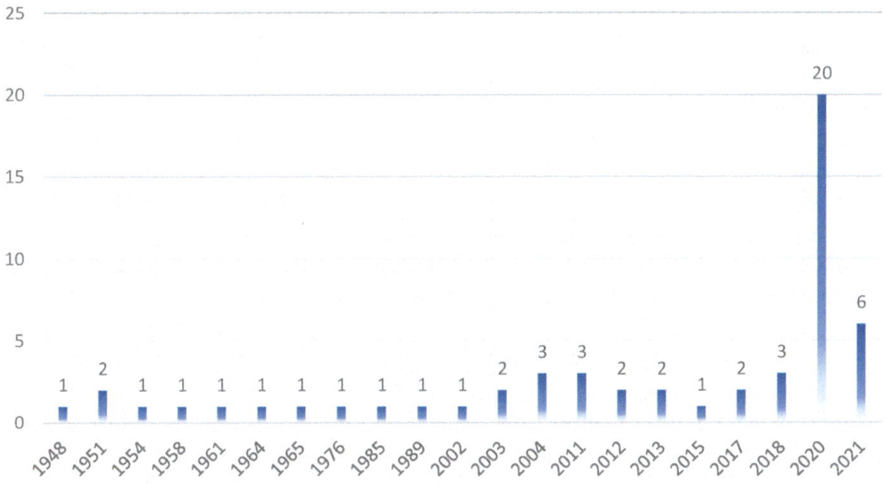

Fig. 1 Temporal distribution of Chinese epidemic films

1.3 Dominance of Documentaries

Documentaries account for more than 60% of all epidemic films made in mainland China on the subject of epidemics, totalling 35 out of 56 films. Again, we see, in the recent documentary treatments of COVID-19, a preference for formats more suited to social media than cinema theatres or even television. Among these documentaries, the majority are multiple-episode ones, with the shortest episode being only five minutes. A typical example is the 12-episode documentary *My Battle Against COVID-19 and Me* (《我和我的战疫》), released in 2020. The 33-episode documentary *Wuhan: My Diary of the Battle Against COVID-19* (《武汉:我的战 "疫"日记》, 2020) was made in the form of a VBLOG. The personal nature of these online documentaries, specifically, the perspective of the victim of the disease, still recalls the first-person point of view of earlier films, such as *Our Lives* (《我们的生活》, 2004), the first Chinese documentary about the daily experiences of people suffering from AIDS, which was filmed by the patients themselves. These documentaries offer the broader public an insight into a particular victim's experience of a disease, which tends to be characterised as an account, in the form of a diary or journal, of that individual's 'battle' against an unseen enemy.

1.4 Diversified Themes

While documentaries, particularly those now produced for social media, focus on the first-person experience of the epidemic, feature films have a wider range of narrative themes. It can be broadly argued that most western feature films featuring epidemics (such as the celebrated *Contagion,* or *Outbreak* or even the 'zombie' genre that includes *Night of the Living Dead*, and its sequels, as well as *29 Days,* and *World War Z*) fall into the category of disaster movies, sometimes inflected with horror; thus the disasters themselves, together with their devastating damage and resulting chaos, are, usually, depicted in great detail and sometimes with relish; whereas the focuses of the Chinese epidemic films are very different. The themes of Chinese epidemic feature films are diverse: politically sanctioned messaging relating to the healthcare advice, celebration of the CPC's role in caring for its citizens, manifestations of collective altruism and individual sacrifice, and the impact of the epidemic on stories about romantic love (see Lu, 2022, pp. 13–20). While there is a greater thematic diversity in the Chinese feature films than in the documentaries, certain recurrent patterns of characterisation and plot can still be observed.

2 Responses of Different Characters in Dealing with Epidemics

In this section, attention is paid to the responses of different characters in dealing with various epidemics involved in Chinese theatrical feature films that involve outbreak narratives.

2.1 Two Sets of Saviours: Governmental and Non-governmental Heroes

2.1.1 Heroic Governors

It need hardly be stated that China is a socialist state under the leadership of the CPC, which thus largely influences media production. In the 70-year history of Chinese epidemic movies, the central government, the local governments and/or their representatives thus always play a guiding and leading role, generally being portrayed as saviours who pluck infected people out of 'deep waters'.

Among the 11 feature films produced over the past 70 years, we focus in this section on *Dry Trees Meet Spring* and *Chinese Doctors* released respectively in 1961 and 2021. These two movies are particularly prominent in demonstrating the leading role of CPC, represented, respectively, by Chairman Mao Zedong (1893–1976) and President Xi Jinping. 'Dry trees meet spring' is a Chinese idiom, which literally means when springs comes, a dry tree will come back to its life; and it is also a metaphor for a dying patient or something regaining vitality. In this film, 'the dry trees' refer to the poor ordinary people, those who were suffering from schistosomiasis in particular, and 'spring' symbolizes Chairman Mao and the 'new' China under the leadership of the CPC. Taking the prevention and treatment of schistosomiasis as its starting point, and the life of Ku Meizi (苦妹子, literally 'Miserable/Suffering Girl') as a clue, the film *Dry Trees Meet Spring* is made to demonstrate Chairman Mao's untiring efforts, wise leadership and strong determination to prevent and eradicate the schistosomiasis that has been rampant throughout China, as well as to highlight the emphasis and protection of the people's lives by the CPC, in contrast with Kuomintang (Nationalist Party), the opposite party.

Thus in *Dry Trees Meet Spring*, the central and local governments are respectively represented by Chairman Mao, the county committee, and the Schistosomiasis Prevention and Control Station in the affected village; in *Profoundly Affecting* (《惊心动魄》, 2003), the governmental institutions are represented by the provincial governor, the Office for Dealing with SARS, and the Ministry of Railways of China; and in the COVID-19 movie, *Chinese Doctors* (《中国医生》, 2021), they are represented by President Xi Jinping, the National Health Commission, and the Wuhan Municipal Government. In *Dry Trees Meet Spring* and *Chinese Doctors*, the heads of state, Chairman Mao and President Xi, respectively, are not represented by actors, but their presence is conveyed positively within the films via dialogues, slogans, music, and media coverage.

For example, the dialogue between characters in the Chinese epidemic movies is overtly supportive of the CPC and its messaging. Having learned that she is suffering from schistosomiasis and that the disease has already developed into the late stage, Ku Meizi, the protagonist, says to Luo, the director of the Schistosomiasis Prevention and Treatment Center, 'I want to work, I want to live, I want socialism!' In a conversation with deputy director of the Center, Luo states, 'Human life is the most precious thing! The old China took away the lives of our people, and the new China wants to take them back! This is our responsibility!' At the end of the movie, after Ku Meizi is successfully treated, she expresses her gratitude to the deputy director of the Center, who responds, 'I didn't save you. It was not me, but the Party!' The examples of dialogues cited here directly indicate that it is the Party (CPC) who is the true saviour of Ku Meizi. She is thus characterised as a representative of those who have suffered from the disease and have been successfully cured, ultimately, by the CPC. The script might seem like unsophisticated and obvious propaganda, but it does indicate the importance to the government of the time of establishing its role as protector of its citizens, a role that it strives to maintain during the COVID-19 pandemic.

Apart from dialogues, in *Dry Trees Meet Spring* large banners displaying slogans also silently convey the role of the leadership of the CPC, represented by Chairman Mao, in its battle against the epidemic. The cloth banners bearing slogans appear twice in the movie, the first time being when the villagers learn that Chairman Mao has come to give some guidance and instructions to the work of disease prevention and control. A huge banner, which takes up almost the entire screen, urges the characters, and, indirectly, the viewing audience, to, 'Respond to Chairman Mao's Call, Strive to Eradicate Schistosomiasis!' (see Fig. 2). Not long afterwards, a tractor slowly drives onscreen, with a large banner, as wide as itself, commanding—'Go to the Front Line of Schistosomiasis Control'.

In addition, music plays a direct and important role in demonstrating the function of Chairman Mao as the saviour of the people. For example, when the director of the Treatment Center tells the villagers that Chairman Mao has come, in person, to guide the prevention and treatment of schistosomiasis and has just hold a meeting overnight, a song is heard, with the lyrics proclaiming—'The chrysanthemums reach out to the sun, the spring flowers are fragrant again in October, dead woods are full of green branches in spring, the kindness of Chairman Mao is long lasting' (see Fig. 3)

Following the tradition of *Dry Trees Meet Spring*, a later feature film that demonstrates the leadership of the CPC, this time during the COVID-19 pandemic, is *Chinese Doctors* released in 2021. The film is explicitly dedicated to the 100th anniversary of the founding of the CPC, and so its function to positively represent the leadership of CPC is self-evident. In this movie, the guidance and instructions of the CPC are largely presented by the abundant presence of banners onscreen, as well as by reports interspersed in the film, from traditional state television networks

Fig. 2 Slogan in *Dry Trees Meet Spring*

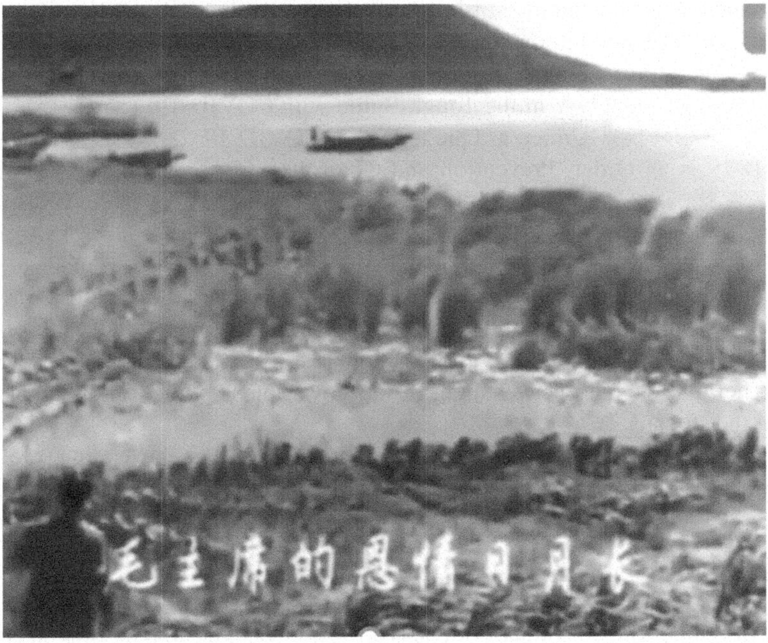

Fig. 3 Music in *Dry Trees Meet Spring*

such as CCTV (China Central Television)[4] and newer social media outlets such as *The Paper* (澎湃新闻).[5]

As noted above, banner slogans appear twice in *Dry Trees Meet Spring* as an important way to indicate the leadership of Chairman Mao in fighting the disease; however, the frequency of the use of banner slogans in *Chinese Doctors* is much higher. At 19:23 of the film, a close-up of a large red banner appears, for the first time, at the entrance of a hospital—'Staying True to Our Original Aspiration and Founding Mission' (不忘初心 牢记使命). The same banner reappears at 19:59. This slogan creates for its audience an immediate association with President Xi Jinping. On October 18, 2017, in the report of the 19th National Congress of the CPC, President Xi proposed that the party embark on an educational initiative with the goal of 'Staying True to Our Original Aspiration and Founding Mission', and stated that 'the original aspiration and founding mission of the Chinese Communists is to seek happiness for the Chinese people and to seek rejuvenation for the Chinese nation.' This educational initiative started in June, 2019 and lasted until early January, 2020. When Chinese audience see the slogan, 'Staying True to Our Original Aspiration and Founding Mission', it is natural for them to associate it with this CPC initiative and its proposer. The film thus explicitly reinforces the government's message and can be seen as part of the educational programme.

In the later episodes of the film, further huge red slogans are displayed in various scenes, such as at the entrance of hospitals, on the walls of all the hospital wards, on buses, on airplanes and on buses carrying supplies to support Wuhan, as well as at the entrance of residential buildings. The banners bear such slogans as: 'Spare no Efforts to Resolutely Win the Tough Battle with COVID-19' (集中力坚决打赢疫情防控这场硬仗), 'United as One to Defeat COVID-19' (众志成城 战胜疫情), 'COVID-19 is an Order; Prevention and Control is a Responsibility!' (疫情就是命令 防控就是责任), 'Resolutely Curb the Spread of COVID-19, and Resolutely Win the Battle' (坚决遏制疫情蔓延势头 坚决打赢疫情防控阻击战). As in the earlier movie, but to a much greater degree, the linguistic landscape is made to speak of the determination of the CPC in leading the public 'battle' against the invisible enemy. The characters in the film—and the audience—are made to move through an environment that urges solidarity in a resolute response to reduce the impact of the virus. The high degree of investment of the government in this role is abundantly clear in the slogans.

In *Chinese Doctors*, in addition to the use of such slogans, the leading role of the government is also portrayed through television footage interspersed throughout the film. These scenes include President Xi Jinping's instructions on the fight against the

---

[4] http://www.cctv.com. China Central Television (CCTV), established in 1958, is a Chinese state-owned broadcaster led by the Chinese Communist Party. With a network of over 50 channels broadcasting different programmes, CCTV is accessible to over one billion viewers in many different languages.

[5] http://www.thepaper.com. *The Paper*, launched in 2014 as an offshoot of the Shanghai United Media Group publication *Oriental Morning Post*, was founded as an attempt to capture the readership of mobile internet users.

COVID-19 virus, the visit of the officials of National Health Commission to Wuhan, and the announcements by the Wuhan Municipal Government.

Among the reports, the online outlet, *The Paper*, appears only once, with all of the rest coming from the more traditional media outlet, CCTV. The reports, together, make up a narrative of the entire process of the early, middle and late stages of the battle against COVID-19 in Wuhan. These reports, both on television and online, again focus on the sensible instructions and wise leadership of the CPC Central Committee and President Xi Jinping, as well as the positive role of the National Health Commission and the Wuhan Municipal Government in the battle against COVID-19. For example, the film shows a CCTV news report from January 20, 2020, in which President Xi gave important instructions on dealing with COVID-19, emphasizing that the safety and health of the Chinese people should be put first and that the spread of the disease should be resolutely curbed. As the film progresses, it is punctuated by news reports periodically giving President Xi's further instructions on the prevention and control of the virus, alongside domestic and international responses. The media coverage spreads from Wuhan to the global community, with a report from January 31, 2020 confirming that the outbreak of COVID-19 constituted a public health emergency of international concern.

While the use of *verité* reports from the Chinese official media (CCTV) allows the film to show President Xi himself as a guiding presence during the pandemic, the insertion of a report from the social media outlet, *The Paper*, focuses on a press conference given by the World Health Organization's Joint Expert Mission on COVID-19. At this event, a representative of this body of international experts expresses his thanks to the people of Wuhan, the city where a case of COVID-19 was first reported, 'We need to recognize the contribution of the people of Wuhan, the world owes them. When the pandemic is over, I hope we will have the opportunity to thank the people of Wuhan for their contribution during the process of the fight against the pandemic.' The use of social media in film acknowledges the role of the internet in contemporary China; the film demonstrates that this new medium can also be appropriated as a means of showing the CPC in a positive light, here by focusing on the gratitude of the international community for the leadership shown by the Chinese government and people.

To summarise, a large portion of the inserted media clips in *Chinese Doctors* focuses on the advice from the central government (President Xi Jinping's instructions appear three times during the film), the guidance from the National Health Commission, and relevant announcements from the Wuhan government. In addition to the domestic coverage, the social media report inserted in the film focuses on the recognition and praise of the World Health Organization with respect to China's fight against COVID-19. Together, the reports give a sense of documentary verisimilitude to the dramatized epidemic narrative, framing it through selective reports from the new and old media, and emphasizing the CPC's motto of 'people first and life first.' As with the earlier feature movie, the identity of the CPC as a political institution and its perceived relationship with its citizens are seen in terms of active paternal wisdom that is admired beyond China's borders. It is worth mentioning that the only

Chinese epidemic film among the 11 listed, in which the government or its representatives are not portrayed as saviours of the people or as effective leaders, is *The Fall of Ming* (《大明劫》). It is significant that this is a historical drama, and so the government is the imperial court of the Ming Dynasty (1368–1644), led by Emperor Chongzhen (1611–1644). This is a deeply allegorical movie in which the plague and other diseases result in the fall of the Ming Dynasty. The Emperor is shown to have little control over the events, imprisoning and then reluctantly releasing a general, Sun Chuanting, whose character is the locus of wisdom and heroism in the movie. Obviously, in this example, the historical genre distances the portrayal of a weak government and dissociates the portrayal from any relationship to current affairs in contemporary China. However, the narrative of the film also reinforces the sense that a Chinese government that fails effectively to tackle public health effectively will fall; thus, the film indirectly reinforces the message that the success of any governing institution in China is necessarily bound up with its ability to deliver effective public health policies.

2.2 Heroic Citizens: Doctors and Soldiers

In Chinese epidemic narratives, then, the political institutions frame a uniform, collective response to the crisis, which is communicated by slogans on screen, and inserted news bulletins. The response of the audience to this framing is modelled in dialogues, spoken by 'ordinary' characters in the films, that are enthusiastically supportive of the government position. However, between the government representatives and the ordinary citizens there are other 'heroic' groups that function to model appropriate collective responses to the crises, namely medical and military professionals. In the Western outbreak narrative, it is often the observant individual who is first alerted to the threat, but is ignored by those in authority, until his or her position is vindicated. In the more collective Chinese narrative, individuals are shown to be representative members of social categories that are selflessly dedicated to sacrificing themselves for the good of other people. *Chinese Doctors*, as the name suggests, purports to tell the story of medical staff in Jingyintan Hospital in Wuhan, which was the first institution to admit cases infected with COVID-19 in 2019 and from other hospitals in other parts of China. The medical staff includes doctors from Wuhan and volunteers from all over China, who become involved in the fight against the virus. The characters are partly based on real people, such as President Zhang, the director of Jinyingtang Hospital, Wen Ting, the Director of the Intensive Care Unit, and Dr. Xin, the Director of the Department of Anesthesiology, alongside purely fictional creations like Yang Xiaoyang, a new doctor who, at first, did not dare intubate a patient, and Tao Jun and Wu Chenguang, who represent doctors from other places of China who have come to Wuhan to provide voluntary medical assistance. Their function in the film is to show experts acknowledging the wisdom of the national health policy, despite the personal sacrifices they have to make in doing so, and to evoke admiration and sympathy from the viewing audience.

In addition to the health care workers, another important vehicle to embody collectivism in Chinese epidemic movies is the military. Among 11 the epidemic films listed in the present chapter, *Profoundly Affecting* is distinctive in its focus on the consciousness and commitment of the military in the face of the threat posed by an epidemic, in this case SARS. Its heroine is the fictional Yang Ping, whose role as a military doctor combines the virtues of both heroic categories. The film centres on a suspected case of SARS on a train; when Yang Ping learns of it, she volunteers to mobilize the soldiers to identify the infected passenger. She encounters no obstacle to her suggestion: in response to her request to mobilize the soldiers, the military captain declares, 'With the PLA (People's Liberation Army), we're not afraid any longer!' Later, when Yang Ping sees some passengers in uniform on the train, she urges them to follow her to the dining carriage, saying, 'Regard it as an emergency gathering!' Without any delay or objection, the soldiers choose to join her. Within a short period of time, the various types of military and quasi-military personnel on the train, including trainee army doctors, armed police officers, a film projectionist in the army, and naval personnel, quickly assemble into a fighting team and arrive at the dining carriage of the train, to be assigned the task of identifying the suspected SARS case. At a moment of crisis, the military, like the medical professionals, come together, without question or delay, to join the battle against the epidemic. Again, the function of such a representation is to evoke admiration and respect among the audience for a group of disparate professionals selflessly ignoring their own safety to follow official policy that privileges the greater good.

Chinese epidemic narratives on film, in short, are not in the business of undercutting authority or engaging in complex moral dilemmas. In a time of emergency, the Chinese government, led by the CPC, is represented as taking the lead in developing an effective policy, which is enacted by self-sacrificing professionals and enthusiastically adopted by 'ordinary' citizens, who function as role-models for the response of the viewing audience. The narrative conventions might be criticized by outsiders as simplistic messaging, which in some cases takes and reshapes history to support the government position, but it can also be argued that such films, in their privileging of individual self-sacrifice for the collective good, function as effective public health messages in times of emergency.

2.3 The Saved: The Weak and the Underprivileged

So far, we have focused largely on the heroic agents in Chinese epidemic narratives, namely the government, and the medical and military professionals. We now turn to the conventional representation of victims of viruses in Chinese epidemic movies. In these films, the infected or rescued are regularly portrayed as socially underprivileged or weak. In *Dry Trees Meet Spring*, the infected person is Ku Meizi, the child bride of Fang Dongge. In *Profoundly Affecting*, the first suspected case of a SARS infection is a migrant worker on a construction site. In fact, the first case of SARS in China was found on November 16, 2002, in Foshan city, Guangdong Province, where he was working a chef. In *Chinese Doctors*, the infected cases who are highlighted include a delivery worker, Jinzai, his pregnant wife, and an 82-year-old man with a high blood pressure and diabetes. The infected characters highlighted in the films are either, like Ku Meizi, Jinzai and his wife, and the migrant workers, at or near the bottom of the social ladder or are those with a low immunity or a higher risk of being infected, such as pregnant women and elderly patients with other underlying diseases.

Clearly, in the face of a devastating pandemic, no social category is immune. It is, therefore, legitimate to ask: why are the infected cases in Chinese epidemic narratives often portrayed as underprivileged or weak? The answer obviously lies in the film-makers' desire to impress upon the audience a collective response to a social emergency. The message is that the principle of the CPC and the Chinese governmental departments is to prioritise the lives of ordinary citizens, no matter how old they are or how sick they are with other medical conditions. The narrative function of these characters is not to evoke in the viewing audience admiration and respect but, rather, recognition, empathy and sentimental identification.

It can be no coincidence that the key representative of the 'ordinary' citizen is a lowly delivery worker, a symbol, if there ever was one, of the new China's consumer-based economic infrastructure. With the great popularity of online shopping in China, delivery has become an ever-growing and highly visible industry, and those who are responsible for delivering the parcels to online shoppers even get a special name, 'Kuaidi Xiaoge' (快递小哥), literally 'young delivery man', most of them being young males. From the viewpoint of disease transmission, delivery men have to come into everyday contact with many people and various packages from a variety of sources. Their working practices may indeed cause the virus to spread faster and wider; but from the viewpoint of life, both in movies and in the real world, when the general public is not allowed to go out during extended periods of lockdown, the delivery men have become not only 'life savers' for those who are trapped at home, but also an important part for fighting epidemics by ensuring the circulation of various materials needed. By making one of the victims of COVID-19 a delivery boy, *Chinese Doctors* highlights the social role played by this new class of worker, and both the dangers and the services to the community that they provide.

2.4 'Minor Characters': Care, Cooperation and Recognition

In addition to the two main categories of characters in epidemic movies, namely, the heroic saviours and the lowly saved, there are also numerous minor characters who play small but significant supporting roles in the films. These minor characters can help to develop the storyline and deepen the themes of the movies. Some scholars have explored the function of minor characters in Chinese films produced by certain directors (e.g. Liu, 2016; Lu, 2015) and movies with a certain subject matter (e.g. Xie, 2010); however, no studies have explored the function of minor characters in epidemic narratives.

In *Profoundly Affecting*, for example, a little girl played by Guan Xiaotong, a Chinese actress, first appears in the movie when she and her mother are pushed apart and separated by the crowd at a train station. A migrant worker who happens to be next to her helps to carry her to be reunited with her mother. Her second appearance is in the train compartment, where she sees again the migrant worker who has helped her. Later, the migrant worker begins to show symptoms of infection, and the little girl is forced to be quarantined in a separate compartment on the train. To while away the hours in quarantine, the girl folds paper cranes for the infected migrant worker. In China, of course, there are many legends about paper cranes; they are, in general, intended to express the good wishes of the maker and bring luck to the recipient. Towards the end of the film, the little girl walks out of the train with her mother, holding a string of paper cranes in her hand. Though she may have been infected because of the migrant worker, she still chooses to send her good wishes to him. The girl, therefore, becomes a symbol of innocent benevolence in her grateful attitude towards someone who has shown her kindness but who might also have endangered her life and caused her inconvenience.

Further minor characters who also function as symbols of selfless benevolence appear in *Chinese Doctors*. One character, Zhang Xiaofeng, is unable to be with her parents before their death; when she receives their belongings—a mobile phone, a modern token of everyday communication—from one of the volunteer doctors, her calm acceptance despite her immense sadness is a model of dignified restraint. Another minor character, who has lost his grandparents, and parents, is finally persuaded by President Zhang to donate to the hospital the remains of his beloved family members for autopsy. His personal grief is subjugated to a higher sense of duty. All these minor characters in the films are designed to be models of empathy, self-restraint and kind-heartedness at critical moments, in spite of their own pain and suffering.

However, there is an exception to the portrayal of minor characters as symbols of benevolence: in *The Fall of Ming* a young soldier is infected and then successfully cured by the hero-doctor, Wu Youke. Though the soldier could then have left the military camp, he chooses to stay an extra night, which has resulted in his being burned alive by the general, Sun Chuanting, the general, who fears that the plague might be further spread by infected patients. The tragedy of this minor character might evoke sympathy in the viewing audience but not because the character's actions and fate symbolise benevolence or self-restraint. The cruel fate of the young solder might prompt the audience to reflect on the barbarity of Sun Chanting's response to the plague and prompt criticism of the health policy of the period, a criticism which, again, is distanced by virtue of the film's historical setting.

There is one further, significant category of minor characters in Chinese epidemic films, namely foreigners. In Western epidemic films, foreigners are usually depicted as 'the Other', or in other words, the original sources of infection and/or those who need to be rescued from the ravages of the epidemic. Wald (2008) argues that the popular Anglo-American outbreak narrative usually presents civilised Western culture in opposition to supposedly primitive or exotic cultures. She observes that source of the outbreak is often a Third World or 'exotic' country: in the film *Outbreak*

(1995), for example, the source of the virus is an African monkey (Wald, 2008, p. 68). The film, *Contagion* (2011), made three years after Wald's discussion was published, largely follows the formula, too. The virus originates in China and 'patient zero', an air hostess, is infected in a casino in Macao, when she shakes hands with an unsanitary chef who has been preparing diseased pork. The Western epidemic movie conventionally portrays the disease as an external threat to the domestic polity.

However, in Chinese epidemic movies, foreigners usually serve as witnesses to the successful responses of the Chinese government's fight against the epidemics. For example, in *Profoundly Affecting*, two foreign lovers appear three times. The first shot of them shows them getting on the train with their train tickets. They appear again, later, in a carriage on the train, when a young female trainee doctor and a young male soldier go to check their body temperatures: they push open the carriage door and are embarrassed to find the two foreigners hugging and kissing. The last shot of them is when the train finally arrives at the destination after the suspected cases have been identified and transferred: the two lovers get off the train, with their masks, making victory gestures with their fingers. The two foreigners, then, do not represent a threat to anything but the sense of sexual propriety of the trainee doctor and soldier who discover them kissing in a public place. Their main function, indeed, is to serve as witnesses to the successful measures taken on the train to contain and deal with the epidemic outbreak. They indicate a desire in Chinese epidemic narratives for external recognition and approbation for the domestic measures taken.

This desire is also seen in *Chinese Doctors*. When WHO officials are passing by the empty streets of Wuhan city in a car, they wonder in amazement, 'How have you managed to get more than 10 million people to listen to the arrangements and be quarantined in their homes?' One of the health workers proudly responds, 'The Chinese are, by nature, united and altruistic.' At a subsequent international conference, the WHO official praises China's achievements and global contribution, stating that 'the world owes a thank-you to Wuhan, whose people have made a tremendous contribution to the world's anti-epidemic efforts.' The function of the foreign minor characters are, again, not to pose an outside threat to the health of the nation-state, as in the Western outbreak narrative, but to provide external vindication of domestic health policy, which is portrayed as an example to other countries.

3 What Is Under-Represented/Missing?

As we have seen in the discussion of the characterisation of heroes, victims and minor characters in the foregoing sections, Chinese epidemic films present simplified narratives of complex realities with the purpose of presenting to the audience models of collective action, self-restraint, and obedience to authority, with the purpose of evoking in the audience admiration for authority, sympathy for those who make sacrifices, and pity for ordinary victims. The reality of an epidemic outbreak is much more complicated than such narratives allow. Compared with what actually happens in an epidemic outbreak, what institutional and public responses are under-represented or missing in these movies?

3.1 Chaos and Helplessness

We have seen that Western epidemic narratives share tropes with disaster and horror movies: there are countless representations of chaos and helplessness both from the government and from the general public; at the same time, Western governmental departments usually serve as an obstacle, at least at an early stage, during the attempts of lone scientists to address the threat the epidemic poses. For example, in *Pandemic* (2007) and *Contagion* (2011), the efforts of the mayor and the governor in trying to cover up the pandemic at an early stage are depicted in great detail. By contrast, in Chinese epidemic movies, as we have seen, the government and/or its representatives always play a decisive and leading role in the fight against epidemics; the cooperation, collectivism and self-sacrifice of the general public are strongly emphasized.

It is in the nature of an epidemic outbreak for there to be chaos and a sense of helplessness. In practically every country of the world in the early stages of COVID-19, there were not enough facilities and spaces in hospitals for infected patients, and some patients could not obtain timely treatment, and some could not see their family members, or even failed to say a proper goodbye to their dying family members because of the quarantine measures. China was no exception to this rule. Not every professional acted heroically, and not every person who was infected or lost a family member acted with a sense of dutiful self-sacrifice and restraint. It can be argued that a more realistic sense of the complexities and varieties of response during an epidemic might find their representation on film. Uhrig (2005, p. 6) argues that 'cinema attendance can be both a personally expressive experience, good fun, and therapeutic at the same time.' A more nuanced and complex presentation of the dramas that arise from epidemics might provoke a cathartic and reflective response from audiences without necessarily undermining social order. As things stand, however, the Chinese epidemic narrative is securely bound to the goal of reinforcing social cohesion, at the expense of permitting a deeper exploration of the impact such epidemics have on individuals and social groups.

3.2 Representation of Media

Perhaps the most glaring omission in contemporary Chinese epidemic narratives is their near total neglect of social media. Until the outbreak of COVID-19, Chinese epidemic movies have acknowledged, to a certain degree, the development of different media in China. For example, in *Dry Trees Meet Spring*, the messages to be transmitted to the villagers are broadcast through the village radio; in *Profoundly*

Affecting, communication occurs mainly by telephone, and in *Chinese Doctors*, information is disseminated through television reports, CCTV in particular. Nevertheless, as discussed earlier in this chapter, the media represented in the films serve only to demonstrate the wise instructions and guidance of the CPC, as represented by Chairman Mao and President Xi, respectively, in *Dry Trees Meet Spring* and *Chinese Doctors*, or to show the international recognition of the efforts and achievements the Chinese government has made during its battles against the virus.

According to the '44th Statistical Report on Internet Development in China'[6] released in August 2021, internet users in China reached 854 million as of June 2019, with a popularity rate of 61.2%. There are various social media platforms in China, such as Wechat, QQ, Snapshot, and Weibo, and the number of active social media users in China reached 1.04 billion in 2020.[7] Since 2019 when the first cases of COVID-19 were reported in Wuhan, millions of internet users have posted on their social media platforms messages and videos concerning the disease, some of which are complaints, criticisms and requests for help. In such an era of social media, epidemic narratives might well portray the role of these messages in informing official responses to the epidemic. It is significant, then, that with the exception of the online news outlet, *The Paper*, which is modelled on the traditional printed press and television news format, no social media appears in any of these Chinese epidemic movies. In this respect the Chinese epidemic narrative lags behind reality, which, despite media censorship, allows for a greater dialogue with institutional authority than ever before. From WeChat posts to VLOGs and online documentaries, Chinese social media are also shaping public responses to epidemics; and future epidemic narratives might take this fact into consideration.

4 Conclusion

This chapter has sought to sketch the conventional nature of the 'Chinese epidemic narrative' on film, as it has developed over the past 70 years, culminating recently in a spate of feature movies dealing with COVID-19, as typified by *Chinese Doctors*. The recent films can be seen as conforming to a pattern established by earlier epidemic movies, and, to a certain degree, the pattern is not surprising. Generally speaking, the epidemic narratives are driven by the need to respond to a viral outbreak that threatens society as a whole. Into this scenario step two sets of saviours, represented in turn by the wise governors who pronounce strict but wise policies, and the heroic medical and military professionals who make personal sacrifices to enact them. They struggle to attend to those infected, whose lowly status, age or weakness does not disbar

[6] http://www.cac.gov.cn/2019zt/44/index.htm.

[7] https://tech.sina.com.cn/roll/2020-03-13/doc-iimxxstf8590329.shtml.

them from deserving official and professional support. Minor characters embody the virtues of self-discipline, social benevolence, and altruism. When foreigners are characterised, they are not a threat but a source of external vindication of domestic policy.

It might be easy to write Chinese epidemic narratives off as simplistic messaging; however, as Wald (2008) argues, all outbreak narratives articulate a partial view of a complex reality. She complains that the Western outbreak narrative is characterised by a conventional xenophobia largely absent from its Chinese counterpart, and underplays the value of systemic social responses to the causes of disease while overplaying lone individuals' heroic actions in the face of hostile institutions. The Chinese epidemic narrative arguably overplays systemic institutional responses to the advent of the epidemic while underplaying dissent, with the aim of reinforcing audience support for potentially unpopular policies. Of course, no epidemic narrative can wholly encapsulate the chaos, helplessness, despair or tragedy of an actual outbreak such as COVID-19. However, once the present emergency has passed and there is space to reflect on the lessons learned from it, we might hope for richer, more diverse dramatizations of one of the major health crises of the twenty-first century.

Appendix

附录: 国产疫病题材电影列表 (56 部, 1948–2021)
Annex: List of 56 Chinese epidemic movies (1948–2021)
This list presents the tradition in which the recent wave of COVID-19 films is situated. It is not always easy to translate the titles; for example, 'The Most Beautiful Walkers Against the Flow,' released theatrically, and the similarly titled 'The Most Beautiful Motion Against the Flow,' released online, both refer to doctors who travel to an infected area, while most people are going in the opposite direction. Even so, the titles give a sense of the sentiments that the movies attempt to convey.

1. 剧情片 (Feature films, 20)

院线电影 **(Theatrical films)**

1. 枯木逢春 (Dry Trees Meet Spring, 1961)
2. 惊心动魄 (Profoundly Affecting, 2003)
3. 最爱 (The Most Beloved, 2011)
4. 非典情人 (Lovers During SARS, 2012)
5. 大明劫 (The Fall of Ming, 2013)
6. 最美逆行者 (The Most Beautiful Walkers Against the Flow 2020)
7. 最美逆行 2 之国际救援 (The Most Beautiful Walkers Against the Flow 2, 2021)
8. 战疫时期的爱情 (Love During COVID-19, 2021)
9. 中国医生 (Chinese Doctors, 2021)
10. 战疫之疯狂的口罩 (Crazy Masks During COVID-19, 2021)

11. 穿越寒冬拥抱你 (Cross the Cold Winter to Give you a Hug, 2021)

网络电影 **(Online films)**

12. 病毒之末日危机 (Virus of Doomsday Crisis, 2015)
13. 机器之血 (Bleeding Steel, 2017)
14. 致命拯救 (Fatal Rescue, 2017)
15. 天网狼蛛 (The Spider and the Web, 2018)
16. 美少女危机 (Giohazard, 2018)
17. 花悸 (The Revenge of Plants, 2018)
18. 狄公灭鼠 (Killing Rats by Di Renjie, 2020)
19. 最美逆行 (The Most Beautiful Motion Against the Flow, 2020)
20. 王者联盟之神秘病毒 (Alliance Kings, 2020)
21. 战疫 2020 (Fighting against COVID-19 in 2020, 2020)

纪录片 **(Documentaries, 35)**

22. 预防鼠疫 (Prevention of Bubonic Plague, 1948)
23. 扑灭天花 (Eradication of Smallbox, 1951)
24. 预防传染病 (Prevention of Infectious Diseases, 1951)
25. 防治血吸虫病 (Prevention and Treatment of Schistosomiasis, 1954)
26. 日本血吸虫生活史(Life History of Schistosoma Japonicum, 1958)
27. 送瘟神 (Saying Goodbye to Plagues, 1965)
28. 疟疾与蚊子 (Malaria and Mosquitoes, 1964)
29. 防治油菜菌核病 (Prevention and Control of Rapeseed botrytis, 1976)
30. 防治血吸虫病 (Prevention and Treatment of Schistosomiasis, 1985)
31. 爱滋病 (AIDS, 1989)
32. 西站非典 (Beijing West Railway Station During SARS, 2003)
33. 非典时期的吃喝玩乐 (Daily Life During SARS, 2004)
34. 灾难时刻 (Disastrous Moments, 2004)
35. 我们的生活 (Our Lives, 2004)
36. 中国艾滋病实录 (AIDS in China, 2011)
37. 凉山艾滋病调查 (An Investigation on AIDS in Liangshan, 2011)
38. 预防艾滋病 (Prevention of AIDS, 2002)
39. 非典九年 (Nine Years after SARS, 2012)
40. 非典十年· 我的十年 (Ten Years after SARS, My Ten Years, 2013)
41. 中国医生战疫版(Chinese Doctors' Fighting Against COVID-19, 2020)
42. 武汉战疫纪 (Fighting Against COVID-19 in Wuhan, 2020)
43. 中国战疫纪 (Fighting Against COVID-19 in China, 2020)
44. 中国战疫录(Fighting Against COVID-19 in China, 2020)
45. 武汉: 我的战 "疫" 日记 (Wuhan: My Diary of the Battle Against COVID-19, 2020)
46. 逆行者: 江苏援湖北医疗队战疫纪实 (Fighting Against COVID-19: Medical Team from Jiangshu Province to Hubei Province, 2020)
47. 我们在战 "疫" (We are Fighting Against COIVD-19, 2020)
48. 我和我的战疫 (My Battle Against COVID-19 and Me, 2020)
49. 同心战 "疫" (United to Fight Against COVID-19, 2020)

50. 战“疫”故事 (Stories about Fighting Against COVID-19, 2020)
51. 战“疫” (2020) (Fighting Against COVID-19, 2020)
52. 乡村战疫记 (Fighting Against COVID-19 in Rural Areas, 2020)
53. 佛山战疫实录 (Fighting Against COVID-19 in Foshan City, 2020)
54. 2020, 中国战疫 (China's Fighting Against COVID-19 in 2020, 2020)
55. 我的战“疫” (2020) (My Fighting Against COVID-19, 2020)
56. 战疫启示录 (Record of Fighting Against COVID-19, 2021).

References

Liu, Q. (2016). *A study on minor characters in films directed by Gu Changwei* [Dissertation]. Northwest Normal University.

Lu, S. (2015). *A study on minor characters in films directed by Ning Hao* [Dissertation]. Guangxi Arts Institute.

Lu, Z. (2022). *A study on the narrative of Chinese dramatic cinema films on the subject of diseases* [Dissertation]. Macao University of Science and Technology.

Uhrig, S. C. N. (2005). *Cinema is good for you: The effects of cinema attendance on self-reported anxiety or depression and 'happiness'*. https://www.iser.essex.ac.uk/files/iser_working_papers/2005-14.pdf

Wald, P. (2008). *Contagious: Cultures, carriers and the outbreak narrative*. Duke University Press. https://doi.org/10.1215/9780822390572

Xie, X. (2010). Minor characters in recent urban films. *Movie Literature, 19*, 34–35.

Ziran Lu earned a master's degree in Film Studies and a bachelor's degree in Chinese-Portuguese Translation and Interpretation. Lu is currently a freelancer with an interest in film and language studies.

Li Li Ph.D. in Translation Studies, is currently a professor at Macao Polytechnic University. She has published widely in children's literature, Translation Studies and corpus linguistics. She is also a translator of many books such as *Introducing Translation Studies: Theories and Practice* and *The Last Battle*.

Conclusion

Selçuk Çolakoğlu and Kam Tong Chan

The pandemic is not over yet. There is an evolving process of the pandemic with many difficulties, changes, and uncertainties. No one knows or predicts what is next in terms of public health polices, economic uncertainties, sustainability of global supply chain, and social and political repercussions of the pandemic at the national and global levels. Writing a book with the aim of covering all these national & global social, cultural, economic, and political dynamics is not an easy task. Preparation of this book has taken more than one year with lots of changing dynamics, trends, and facts during the process. Researchers at the Beijing Normal University-Hong Kong Baptist University United International College made an enormous effort to present this volume catching up the latest research trends and themes of the pandemic. Despite all these difficulties, contributors to this book have successfully approached and reflected major ambiguities of the pandemic. The resultant book series help us to find out issues, problems, shortcomings in the process of pandemic studies and to think about possible solutions, if any.

The new era of the pandemic, ushered in by the pervasiveness of more infectious variants and the availability of COVID vaccines, has pushed the academia to respond with collaborative, comparative, and inclusive studies. This book is a clear example in the three dimensions. This book marked another effort by BNU-HKBU United International College researchers to address the COVID-19 pandemic after their first book series titled *COVID-19 Pandemic, Crisis Responses and the Changing World: Perspectives in Humanities and Social Sciences* in 2021. During the writing process, faculty members of BNU-HKBU UIC collaborated with student assistants. The book thus constitutes a model with a teacher-student nexus in the studies of fighting the pandemic. Secondly, as the title indicates, this book series offer a comparative perspective in addressing the pandemic control policies, most notably, the zero COVID strategy and "live with COVID" strategy. As analyzed in various chapters in the book, the dichotomy of control policies was the direct result of the emergence of the Delta and Omicron variants and the availability of COVID

© The Editor(s) (if applicable) and The Author(s), under exclusive license to Springer Nature Singapore Pte Ltd. 2023

S. X. B. Zhao et al. (eds.), *Comparative Studies on Pandemic Control Policies and the Resilience of Society*, https://doi.org/10.1007/978-981-19-9993-2

vaccines. This comparative perspective is also an indication of the timeliness of the book. Thirdly, this book is a major contribution to pandemic studies in humanities and social sciences area. The chapters of the book cover public health, social policy, social resilience, international relations, globalization, global governance, global leadership, domestic politics, civil societies, social and cultural changes, media, and social media as research subjects.

All these different subjects are grouped in three parts to reach a further consistency. Each part has addressed different perspectives of the pandemic and its outcomes. The book includes some interesting and insightful findings for the use of every kind of decision-makers, civil servants, scholars, journalists, businesspeople, and students. We highlight some of our interesting findings below.

Five chapters in Part I have particularly devoted to public health and social economic studies. They have collectively attested that any effective control policy must be rooted in concrete political, social, economic, and cultural settings of a country. No pandemic control policy should be dogmatic to the degree of doctrine. Any effective policy should also adjust to the evolution of the pandemic. In *Between Zero COVID and Live with COVID: Comparative Studies in Pandemic Control Policies of China, Singapore, and Hong Kong at the Stage of Omicron*, **Simon X. B. Zhao, Bo Yan,** and **Yutong Chen** select cases with the most representativeness in comparative studies of pandemic control: (Mainland) China, Singapore, and Hong Kong. China is the world's most prominent and the last holdout of a continuing zero COVID strategy. Singapore's quick switch from pursuing zero COVID to live with COVID provided insights and experiences to many nations wanting to bring economy and social lives back to normal. We could see that New Zealand's measures to return to normal in September 2022 are clearly modelled after Singapore's gradual relaxations in April and August 2022. Hong Kong is a telling example of why the zero COVID strategy, effective in China but unable to be tried elsewhere. The chapter firstly presents that China's unique effectiveness of zero COVID strategy lies in China's ability of carrying its mass testing, mobilization of population at grassroots level, and a vertical administrative structure. Then, the chapter analyzes that Singapore's swift actions to transition to live with COVID was attributable to its advanced preparedness. In the end, the chapter argues that Hong Kong's lack of both China's zero COVID social conditions and Singapore's scientific preparedness led to its fifth pandemic wave in February 2022.

The effectiveness of the zero COVID strategy was praised in China's pandemic control performance though, the constituent restriction measures were questioned for their negative impacts upon China's economy in 2022. **Simon X. B. Zhao, Bo Yan, Yu Liu, and Jiaqi Liu** analyze the economic cost of the "zero COVID" strategy and the "live with COVID" strategy in *Economic Rationale of Different Pandemic Control Policies and their Impact on Economy*. The authors documented the economic performances influenced by different control policies at different stage of the pandemic. Economic costs are important dimensions in measuring the effectiveness of a pandemic control policy in addition to human lives lost/saved. The economic rationales of different control policies have further denied the existence of a "universally" effective control policy.

Part II of the book "International Relations in the Post-Pandemic World: Globalization, Global Governance and Leadership" also has chapters with significant and interesting findings. In *How the Pandemic Affects Global Leadership: A Comparative Analysis of the G2 (United States and China)*, **Selçuk Çolakoğlu, Chaofan Chen,** and **Jiawei Chen** focus on chances in global leadership in the post-pandemic world. The chapter examines the United States and China in the framework of the G2 to analyze the global leadership and the pandemic's impacts. The chapter concludes that although China is on the rise, the United States is not in decline, and the pandemic has no or little affect to global leadership capacities of the G2 (United States and China) and their willingness to contribute to global public goods. Thus, the Kindleberger Trap is not sufficiently valid in the 2020s world. The authors anticipate that the United States and China can play a joint global leadership role under the framework of the G2 to handle the global challenges.

Chaofan Chen, Jiawei Chen, and **Bo Yan** trace the rationales behind countries' differing pandemic control policies, notably zero COVID strategy and "live with COVID" strategy from a political science perspective. In *Political Considerations and Rationales for Different Control Policies in COVID-19 Pandemic: An Analysis from the State Theory*, the authors have unearthed that the zero COVID strategy is embedded in the Neo-Weberian approach of the state theory. The liberalist approach of state theory, rooted in the idea of social contract, has resulted in a limited capacity of state, which is unable to ignore citizens' dissatisfaction of some restriction measures and to mobilize the masses to implement a zero COVID strategy like China. From the perspective of the state theory, this comparative chapter has again confirmed that each pandemic control is rooted in very concrete social, political, and cultural conditions.

Case studies has been a notable feature of the book series. Sweden, with its unique navigation through the pandemic almost opposite to China's control measures, came in **Liming Liu, Chally Chai Lee Lim, Chengxi Yao,** and **Zhiyi Liao**'s analysis in a comparative study for the understanding of how Chinese social media constructs the understanding of Sweden's pandemic strategy among Chinese people in their chapter. For this end, two specific groups of Chinese people living in Sweden and China were selected to compare their perceptions regarding Sweden's pandemic on Chinese social media within different residents' contexts. The authors argue that the understanding of Sweden's pandemic is largely comprehended by applying China as a benchmark to construct Sweden's lack of scientificity among Chinese in China. Contrary to this, Chinese in Sweden construct Sweden's priority as the understanding of the pandemic and justify Swedish strategy with large scientificity through explaining their real life. The authors also examine that Chinese in China construct Sweden's less favorable national image, while Chinese in Sweden consider Sweden's stable national image before and during the pandemic. The authors clarify that Chinese residing in China and Sweden both perceive Sweden's pandemic on Chinese social media has been largely mediatized; however, these two groups have interpreted it in different ways.

Part III "Cultural and Social Media Responses: Perceptions, Emotions, and Narratives" includes many valuable findings from the aspects of media, culture, and society regarding comparative studies of the pandemic. Not only scientific knowledge or policymaking, but also specific groups' interest demands, cultural types, and value orientation shape risk perception and responses. In *Body Politics and the Collective Well-being: A Comparative Study of the Cultural Motives of Mask Wearing During COVID-19*, **Qiaoyun Zhang** and **Yushan Wang** propose a sociocultural framework to analyze the risk processes. Specifically, the authors argue that cultural perceptions of risk can be analyzed relationally in the individual and collective perspectives. The view of individual including ideas of death, reincarnation, body boundary and pollution reveals the effects of identity politics on risk response. The collective view pertaining to the classification of social groups and blame of the "other" exposes how the construction of groupness and order influences the evaluation of risk. People's different decisions of mask wearing are extensively influenced by their perceptions of the risk of COVID-19, which in turn are shaped by the sociopolitical structure and tradition with which they identify themselves. The chapter provides advice on how public health policymaking can attend to people's particular sociocultural mapping of the pandemic to be culturally sensitive and socially effective.

Siu Han Chan, Ruijie Guo, and **Xinyu Huang** examine "Media discourses and China's Social Mobilization at the Early Crisis Stage of the COVID-19 Pandemic". The authors argue that media discourses help the Chinese government to effectively mobilize the society directly from the top, which is the key for the rapid control of the early pandemic outbreak in China in 2020. Successful crisis management then becomes the merit of the Chinese government and contributes to the reproduction of the national imagined community. Nevertheless, China's zero-COVID policy is becoming less effective in dealing with evolved variants with augmented virality. The social and economic cost of maintaining a zero-COVID policy makes social mobilization more difficult than it was in the earlier stage. The authors suggest that the ongoing negotiation on the pandemic control management between Chinese citizens and the Chinese government through social media is a highly interesting research topic for the future.

Qingyuan Zhao, Liang Nie, and **Xiaying Xu** analyze netizens' sentiments and demands in the context of normalized epidemic prevention and control in China in a comparative manner. The cities of Guangzhou, Xi'an, and Harbin were chosen to explore the demands and emotions of netizens as expressed by their Weibo posts during the first three days of local COVID-19 outbreaks. Their study confirms that the Chinese government should ensure the timely provision of mental health support services across all regions of China affected by the pandemic. In the meantime, their chapter shows that maintaining adequate general and medical supplies—as well as continuing to run transportation services—is essential for preventing and controlling the pandemic. The authors also identify the relationships between a city's level of economic development and the demands and sentiments of its Internet users. In those cities whose GDPs are relatively high, the government's administrative capacity is accordingly higher, with better healthcare resources and social security available. The authors indicate that netizens in wealthier cities were less disapproving of government prevention and control measures, demanded such measures less frequently, and expressed fewer basic living requirements, unlike their counterparts in lower-GDP urban areas.

Ziran Lu and Li Li's chapter introduces key elements of the Chinese epidemic narrative on film, with a reference to feature films that dramatize China's response to the COVID-19 pandemic. The authors survey recurrent aspects of the characterization of key agents in the epidemic narrative as it appears in feature films: the wise and caring government, the heroic medical and military professionals, and the vulnerable sufferers, as well as minor characters who embody virtues such as benevolence and self-restraint. They consider the necessary simplifications that the Chinese epidemic narrative demands and identifies some of the complexities absent in such narratives to date. The authors also argue that the Western outbreak narrative is characterized by a conventional xenophobia largely absent from its Chinese counterpart and underplays the value of systemic social responses to the causes of disease while overplaying lone individuals' heroic actions in the face of hostile institutions.

In conclusion, the pandemic of COVID-19 has continued to disrupt the world. The book with comparative studies on pandemic control policies and observations on social resilience, highlights that there is not a single, universal, "one-size-fits-all" solution for the pandemic prevention and control. Neither the zero COVID strategy nor the "live with COVID" strategy can offer an easy way out from pandemic control to economic recovery and social normalization. Each country, each region, and each city may have its own experience and solutions on how to deal with the pandemic from public health governance to economic openings. The book also reveals that cultural, social, and psychological effects of the pandemic on humankind are not clear yet. It is anticipated that a more concerted global governance is required to diminish social, economic, and medical suffers of the international community. The book is an in-depth research that goes far beyond mere description of responses to the pandemic. However, further and more nuanced studies are necessary to advance future research of the pandemic and expand people's knowledge. It is our sincere wish that this book is a stepping stone in the journey of ending the pandemic and fighting the future diseases.

SPRINGER NATURE

GPSR Compliance

The European Union's (EU) General Product Safety Regulation (GPSR) is a set of rules that requires consumer products to be safe and our obligations to ensure this.

If you have any concerns about our products, you can contact us on ProductSafety@springernature.com

In case Publisher is established outside the EU, the EU authorized representative is:

Springer Nature Customer Service Center GmbH
Europaplatz 3
69115 Heidelberg, Germany

The manufacturer's authorised representative in the EU is Springer Nature Customer Service Centre GmbH, Europaplatz 3, 69115 Heidelberg, Germany. If you have any concerns regarding our products, please contact ProductSafety@springernature.com

Printed and bound by CPI Group (UK) Ltd, Croydon, CR0 4YY
01/10/2026
02225851-0013